Advanced Theory and Principles of Materials

Advanced Theory and Principles of Materials

Edited by **Ricky Peyret**

NY RESEARCH PRESS

New York

Published by NY Research Press,
23 West, 55th Street, Suite 816,
New York, NY 10019, USA
www.nyresearchpress.com

Advanced Theory and Principles of Materials
Edited by Ricky Peyret

International Standard Book Number: 978-1-63238-022-7 (Hardback)

Printed in China.

Contents

Preface

Materials play a major role and have a great impact on our society and industry today. To understand the effect of materials science and the field of materials studies one must observe the great evolution of materials, which includes all the important technological developments, no matter what sector is considered.

All the work done by researchers back in time in the field of quantum mechanics in the early 19th century has helped us study the microscopic structure of materials, which can be organic or inorganic. Materials science plays a vital role in order to develop new solids or to transform those which are part of daily human life. Today, the complete understanding of all the different types of materials is dependent on the concept related to their molecular, atomic and micro structures.

It is necessary to have a unified teaching in the field of materials study, especially where the intense scientific innovations are used to study the case. This book can be useful in this global vision. Its contents' arrangement is well balanced between fundamental notions and applications.

I wish to thank all the contributing authors for their extensive research and hard work. I also wish to thank my publisher for finding me worthy of this unparalleled opportunity. Lastly, I wish to convey my regards to my friends and family who have supported me at every step.

Editor

Simulation of Water Vapor Ingress into PV-Modules under Different Climatic Conditions

Philip Hülsmann, Markus Heck, and Michael Köhl

Fraunhofer ISE, HeidenhofstraBe 2, 79110 Freiburg im Breisgau, Germany

Correspondence should be addressed to Philip Hülsmann; philip_huelsmann@web.de

Academic Editor: Guoping Chen

This work deals with the simulation of water vapor ingress into wafer-based PV-modules for long-term exposure under different climatic conditions. Measured material parameters together with climatic data sets from four test sites (tropic, moderate, alpine, and arid) were used to calculate the water concentration inside of the encapsulant between solar cell and glass for a lifetime of 20 years. Two back-sheet materials (PET-based and PA-based) combined with EVA as encapsulant were used in respect to their influence on water ingress. The results show faster water ingress for warmer regions, but the highest concentrations were found for the moderate test site. The water ingress was additionally influenced by the used encapsulant and back-sheet combination. In particular the temperature dependency of the mass transfer, which differs from material to material, was the focus of this investigation.

1. Introduction

PV-modules are globally installed and, thus, exposed to different climates with continuously changing environmental conditions as, for example, UV-irradiation, temperature cycles, or snow loads over a life time of 20 years and more.

The ambient atmosphere and its containing air humidity constitute one of these loads. Polymers interact with the water vapor in a manner of absorption and desorption. Therefore, water can enter a PV-module and causes hydrolysis of the polymers [1, 2] or corrosion of metallic components, but it can also be part of other chemical reactions, for example, acetic acid production [3]. In addition, also mechanical properties of the encapsulation and back-sheet materials (polymers) such as viscoelastic behavior can be changed by water uptake [4].

The absorption of air humidity is primarily controlled by the microclimate at the interface PV-module to ambient atmosphere and by specific temperature controlled material parameters such as water vapor transmission rate (WVTR) and diffusion coefficient (D). Water vapor therefore enters the PV-module via the whole surface of the back sheet and molecules diffuse through the encapsulation polymer until they reach the area between solar cell and front glass [5]. Depending on the water concentration inside of the polymers and the air humidity of the microclimate at the surface, the mass transfer of water molecules will be directed towards the inside or the outside of the PV-module.

The present paper deals with the investigation of water concentration behavior inside polymeric materials used in PV-modules, depending on different climates, and over a planned lifetime of 20 years. Due to the geometrical structure of wafer-based cSi PV modules and the long-lasting life time, measurements of water concentration inside the module are viable. Therefore, a combination of real measured data and 2-D FEM-simulation (finite element method) was used to solve this problem. The measurements include the determination of the temperature dependent permeability of water vapor through different encapsulation and back-sheet materials, but also the microclimate which was measured at four test sites representing different climates (moderate, alpine, tropic, and desert). Compared to prior publications addressing this issue [6], this paper focuses on the behavior of microclimates and on how different polymeric materials, respectively, their diffusion properties, (back sheet and encapsulant) influence the water uptake with respect to geometrical situation by using 2-D FEM simulations.

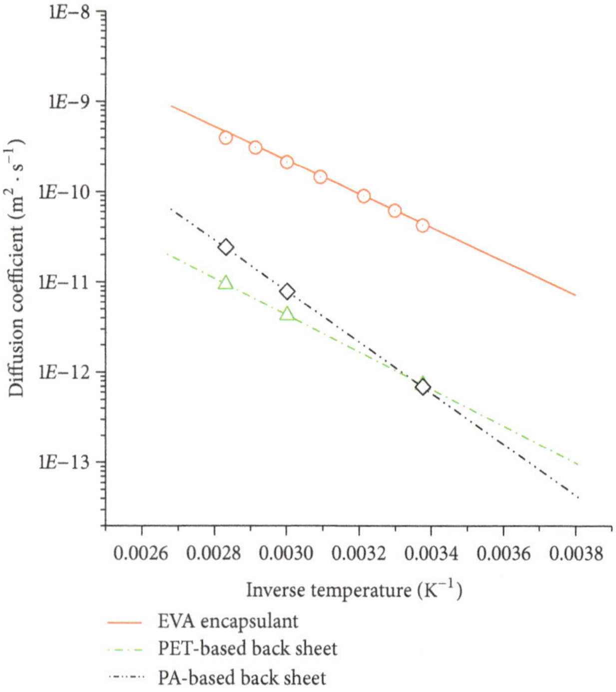

FIGURE 1: Measured effective diffusion coefficients of the materials used in the simulation [6-modified].

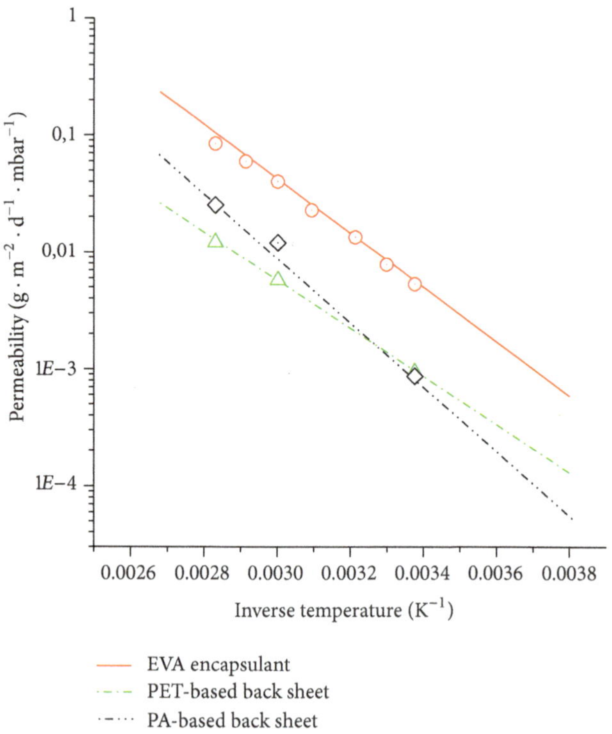

FIGURE 2: Measured permeability of materials used in the simulation of water ingress [6-modified].

2. Measurements

2.1. Materials. In this study, three commercial products—one encapsulant and two back sheets—were chosen for the analysis of their influence on water vapor ingress. The encapsulant was a standard EVA (fast cure) which is well established on the market.

Back sheets generally consist of three layers of polymeric films which are glued together by adhesives. The core material between the two outer layers is, because of its thickness, the dominant part regarding the permeability of gases. Results of a prior work [6] show that back sheets can be distinguished by their core material with respect to its temperature dependency of the permeability. Here, two polymeric material groups were analyzed. One represents polyethylene-terephthalate- (PET-) based back sheets, and the other one represents a polyamide- (PA-) based material group. The PET back sheet consists of a 250 μm thick PET core glued between two outer layers of PVF (polyvinylfluoride) all in all having a total thickness of 350 μm. The same thickness (350 μm) can be stated for the PA back sheet.

2.2. Measured Parameters. Two material parameters, permeability (P) and diffusion coefficient (D), were determined by a permeation test device that was developed by Fraunhofer ISE. This test setup is capable of measuring the permeation of water vapor and other gases through polymeric films or laminates over a wide range of temperature [7]. In this manner, the temperature dependency of mass transport was quantified and, therefore, the activation energy E_A for this process was calculated (see Table 1).

The results of P and D measurements for various polymeric materials were published in an earlier paper [6]. Here, the results for the materials used in the simulation are shown below in Figures 1 and 2. Both graphs show the measured values and an Arrhenius fitted extrapolation to a temperature range PV-modules can be exposed to. The back-sheet materials were measured as a laminate that consists of the core and both outer layers. Therefore, Figure 1 shows effective diffusion coefficients for the back sheets.

2.3. Climate Data. To simulate the water ingress under realistic climatic conditions, knowledge of the local climate the PV-module is exposed to is needed. Therefore, the climate data such as air temperature (T_{air}), relative air humidity (rH), and the PV-module temperature (T_{module}) were measured at four outdoor test sites which are located in different climate zones:

(i) Freiburg, Germany (moderate climate),

(ii) Negev desert, Israel (arid climate),

(iii) Zugspitze (altitude: 2650 m), Germany (alpine climate),

(iv) Serpong, Indonesia (tropic climate).

Figures 3, 4, 5, 6, 7, 8, 9, and 10 show the development of the measured climate parameters for the year 2008 at the four test sites in 5-minute increments. For each test site, the

TABLE 1: Calculated activation energies E_A of the diffusion process of water vapor for the materials used in the FEM simulations. The values of the back sheet-materials are effective E_A measured for the layered structure.

Material	Thickness (μm)	E_A of diffusion process [kJ·mol^{-1}·K^{-1}]
PET-based (back sheet)	350	39,2
PA-based (back sheet)	350	54,1
EVA (encapsulant)	1000	35,7

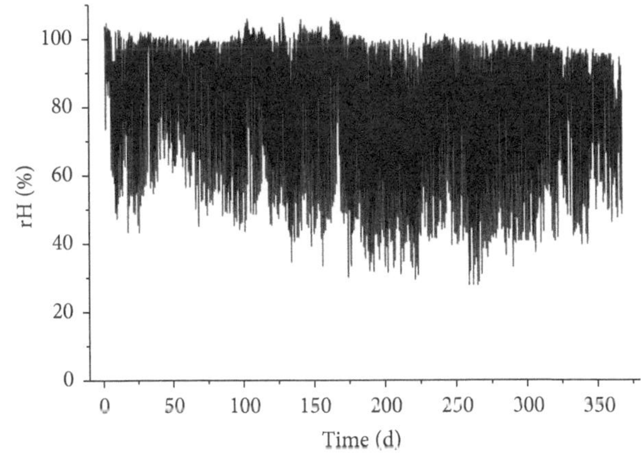

FIGURE 3: Relative air humidity: tropic.

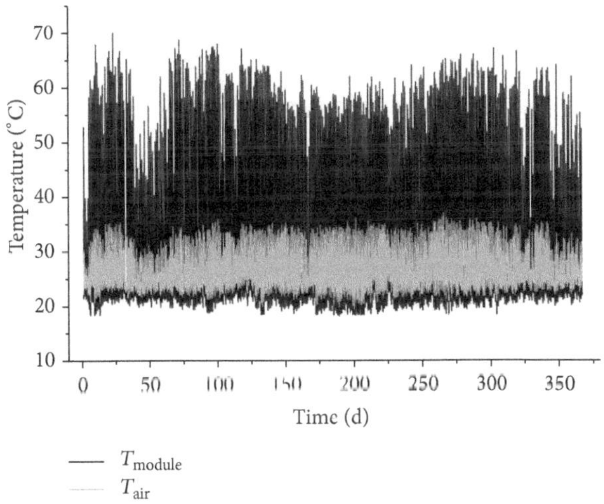

FIGURE 4: Module and air temperature: tropic.

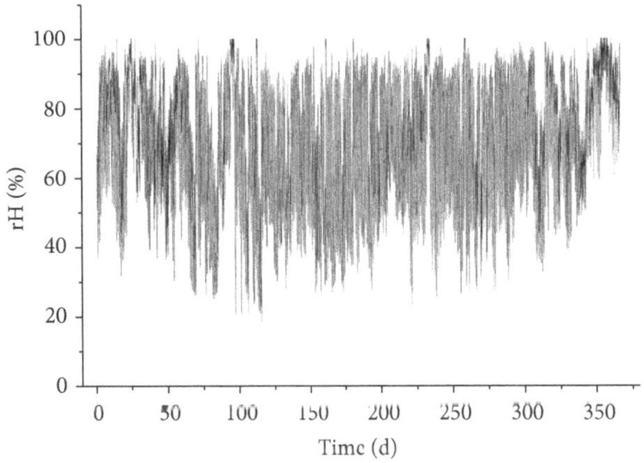

FIGURE 5: Relative air humidity: arid.

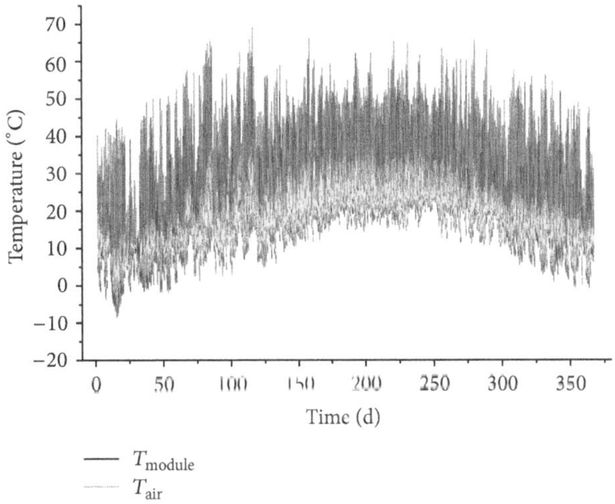

FIGURE 6: Module and air temperature: arid.

air humidity of the microclimate was calculated from these data sets via the so-called Magnus-formula [8].

Microclimate is defined in this study as the boundary layer between the atmospheric side of the back sheet and the ambient air. This air layer with a thickness of approximately 1 mm holds a temperature balance with the PV-module whose temperature is, compared to the ambient air temperature, elevated by solar irradiation during the day and lowered by thermal emission during the night. Thus, the humidity a PV-module is exposed to at its back sheet is determined by the humidity of the microclimate, which provides dryer conditions at day and wetter conditions at night because of the temperature deviation between PV-module and air.

Figures 11 and 12 show, by way of example, a closer view on the diurnal cycles of the climate data derived from the arid test site in the Negev desert (Israel). The strong deviation of module temperature and ambient temperature can cause very dry conditions during day and wet conditions at night (Figure 12). Here, also condensation of air humidity can be observed. To sum up, daytime presents faster diffusion conditions, while at night, the mass transport of water vapor and also other gases is slowed down.

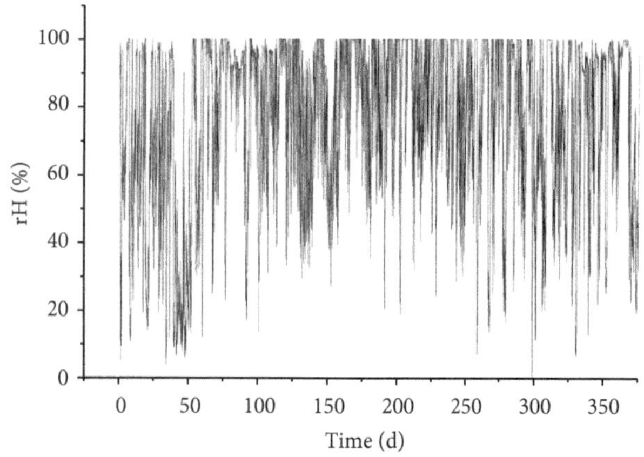

FIGURE 7: Relative air humidity: alpine.

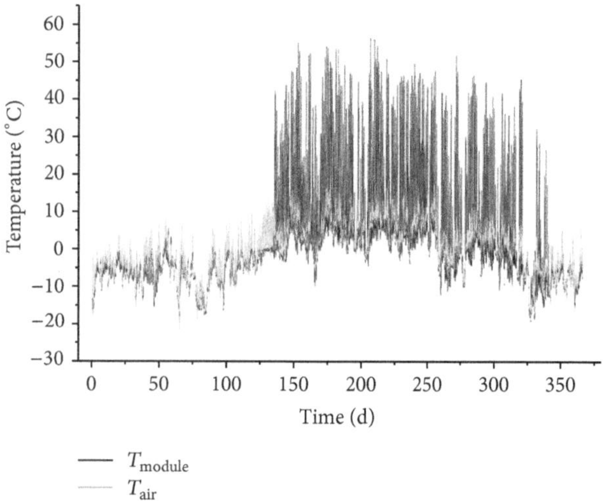

FIGURE 8: Module and air temperature: alpine.

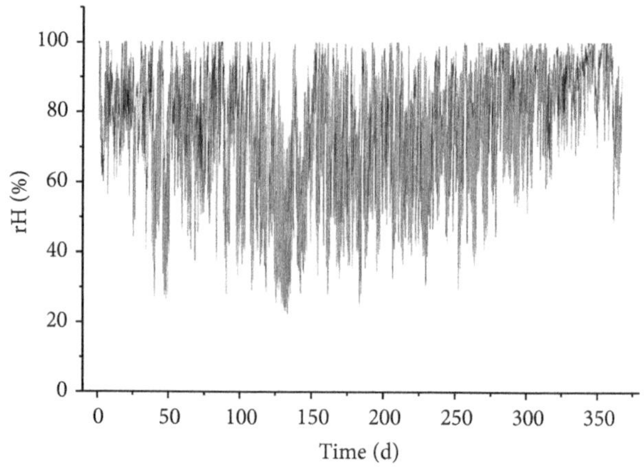

FIGURE 9: Relative air humidity: moderate.

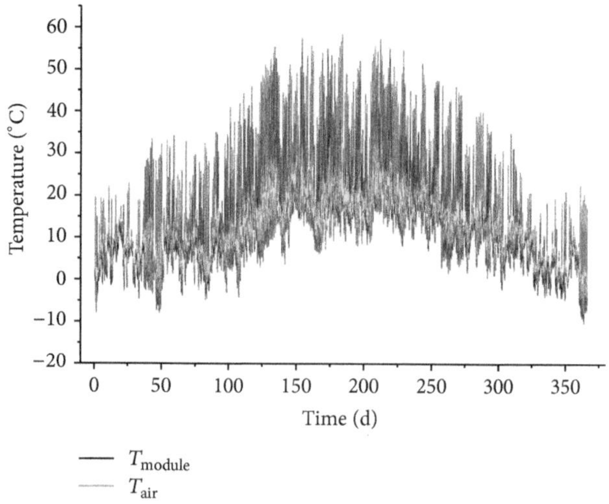

FIGURE 10: Module and air temperature: moderate.

3. Simulation

The FEM-based simulation was carried out with a commercial software tool. The following sections discuss geometry and layout of the used model including boundary conditions and discuss the results of the simulation.

3.1. Geometry and Boundary Conditions. The geometry used for the simulation presents a simple, symmetrical, and 2-dimensional constructed model (Figure 13). It is composed of a back-sheet layer, followed by the encapsulation integrated with two half-sized solar cells (6-inch cells) which show a gap of 3 mm in between them. The atmospheric side of the back sheet layer provides the feed-in zone for the water molecules. This layer is infinitely thin and, therefore, permanently in sorption equilibrium with the air humidity of the microclimate.

The possible pathway of a water molecule is illustrated by plotted arrows in Figure 13. A molecule will follow a concentration gradient and diffuse either deeper into the PV-module or back to the feed-in zone (atmospheric side of the

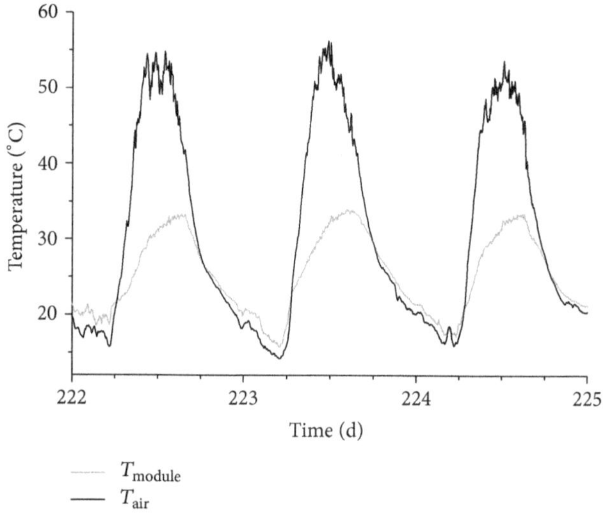

FIGURE 11: Diurnal cycles of measured air and module temperature: arid.

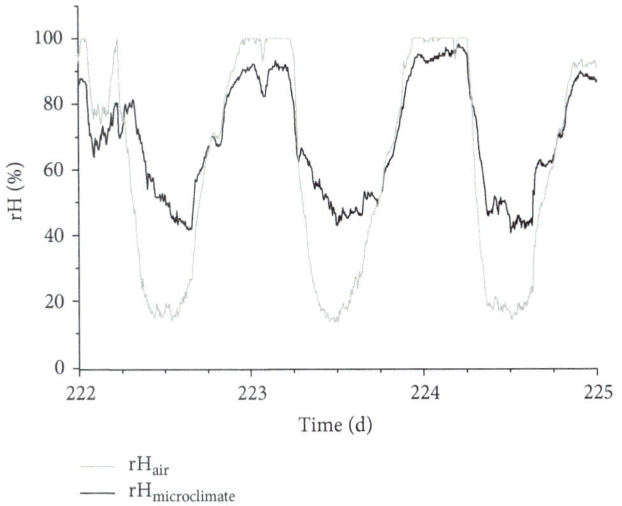

FIGURE 12: Diurnal cycles of measured air humidity and calculated microclimate humidity: arid.

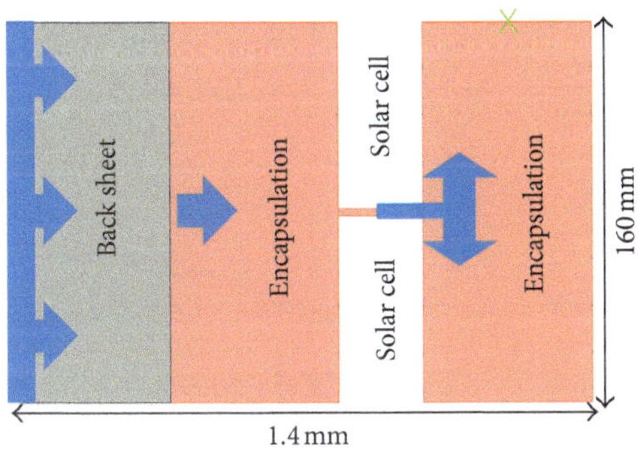

FIGURE 13: Assembling and geometry of the simulated model including pathway of water molecules.

back sheet), depending on the gradient. At the beginning, the molecule will diffuse perpendicular to the module level, passing the back sheet, followed by the first encapsulation layer, and has then to turn onto the module level past the small gap between the cells. Then, it continues until it reaches the center area between solar cell and front glass. All in all, the maximum possible way of diffusion adds up to approximately 80 mm in this geometry.

The main advantage of using a 2-D model is that also permeated cross-sectional areas are considered which is given, for instance, by the ratio of space between the cells (here, 3 mm) and the distance of cell to glass (here, ~0.4 mm). Please, mind the unequal dimension of the x- and the y-axis in Figure 13.

The input parameters used for this simulation were given by the measured temperature dependency of permeability of the encapsulation and back-sheet materials which were described in Section 2.2, including the already published data [6] and the climatic data sets from the outdoor test sites shown in Section 2.3. These one-year data sets were assembled to time series of 20 years by means of repetition.

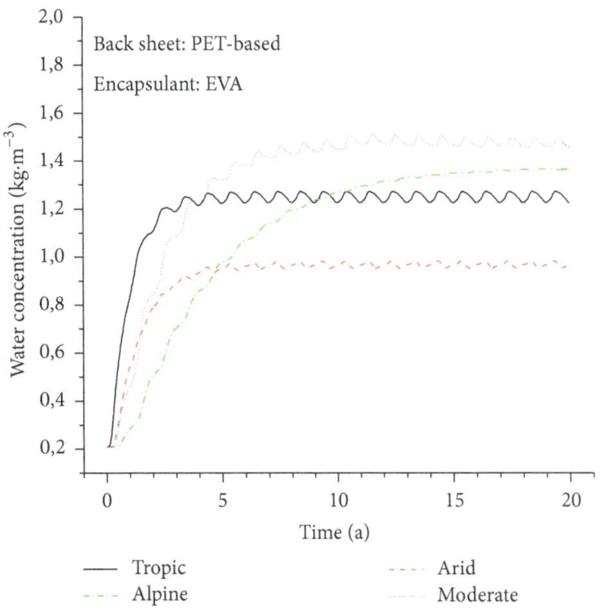

FIGURE 14: Simulated water ingress into a PV-module for different climates. Here, an EVA encapsulant and a PET-based back sheet were used.

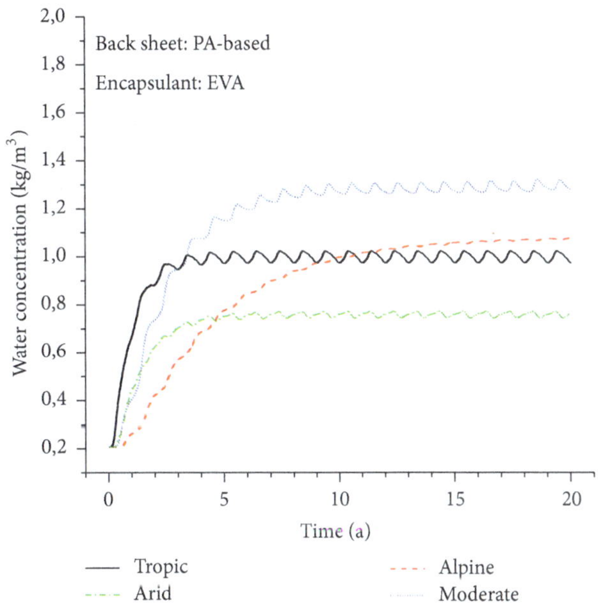

FIGURE 15: Simulated water ingress into a PV-module for different climates. The use of the PA-based back sheet causes significantly lower water concentrations.

This way, it was possible to simulate the slope of the water concentration inside a PV-module over a planned life time of 20 years and for different climate zones under continuously changing ambient temperature and humidity conditions. Additionally, a small initial concentration of water was to be considered in the encapsulant even after lamination process.

3.2. Results and Discussion. The water ingress was simulated for various back-sheet and encapsulant material combinations which are listed below:

(i) PET-based back sheet/EVA (fast cure) encapsulant,

(ii) PA-based back sheet/EVA (fast cure) encapsulant.

Figures 14 and 15 show the development of simulated water concentration inside the encapsulant over a time period of 20 years. The place inside the encapsulant represents the maximum diffusion length a water molecule can achieve in a PV-module. It is marked by an "x" in Figure 13 in the upper right corner.

Both graphs (Figures 14 and 15) show an increase of water concentration which runs faster in warmer climates, for example, at tropic sites, and much slower in colder regions (e.g., alpine). This effect is based on the acceleration of the mass transfer by temperature, the so-called temperature dependency [7, 9]. In detail, this effect causes differences in the time lag, which represents the time a molecule needs to diffuse from the outside of the back sheet towards the space between cell and glass, and is also visible in a sharper gradient of the concentration. The alpine site, for example, shows a time lag of more than ten years until equilibrium (Figure 14) whereas at the other test sites molecules will diffuse much faster because of higher temperatures (also Figure 14). Besides the local climate (temperature), also the encapsulant influences the time lag and the slope on concentration increase.

In addition to the speed of water ingress, also the maximum water concentration that is reached after several years depends on the climate (Figures 14 and 15). The arid test site shows the lowest concentration which also provides the driest climate. Interestingly, the tropic site features a lower water concentration after 20 years than the alpine and the moderate sites. This phenomenon is caused by the transient microclimate at each test site which provides drying and moisturizing conditions as a diurnal cycle. Depending on ratio between dry and wet conditions in cooperation with temperature and its effect on transport mechanism higher or lower water concentrations are reached inside the encapsulant before the center of the cell.

The influence of the back sheet can be determined by focusing on the concentration that is in equilibrium after several years but oscillates around a mean concentration. The mean equilibrium concentration differs from site to site (see above). But within the same site, the equilibrium concentration is significantly influenced by the used back sheet (Figure 16). It can be stated that this concentration is lower in samples with a PA-based back sheet (see Figures 15 and 16). This effect can be explained by a higher E_A of PA compared to PET (see Table 1), together with a transient microclimate. PA and PET have a comparable permeability for water vapor at 30°C, but PA has the higher temperature dependency which leads to a higher permeability at elevated temperatures and to a lower permeability at temperatures below 30°C. As presented in Section 2.3, the microclimate offers dry conditions and high temperatures during the day and wet and cold conditions (condensation) at night. The higher temperature dependency of the PA provides a more breathable back sheet in dry conditions (day) and a better barrier in wet conditions (night).

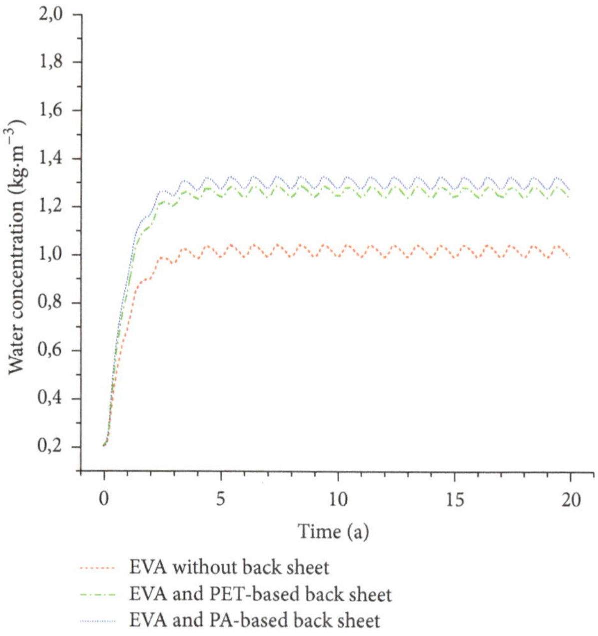

FIGURE 16: Simulated water ingress for the tropic test site and the effect of different back sheets.

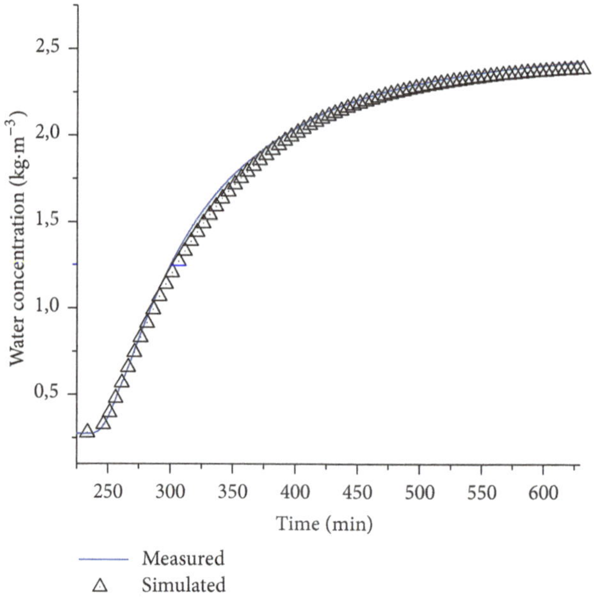

FIGURE 17: Simulated versus measured permeation through a 1 mm EVA film (at 38°C) which shows a very good agreement.

3.3. Validation. A first validation of the simulation of water ingress is shown in Figure 17. In this case, the permeation through a 1 mm thick EVA sample was measured with the test device mentioned in Section 2.2, followed by a simulation using the same geometry and boundary conditions. The simulation agrees very well with the measured values, also in regard to the time lag and the slope of concentration increase.

4. Summary and Conclusion

In this study, the water ingress into PV-modules was simulated for an outdoor exposure of 20 years in different climates. Real measured parameters such as permeability of encapsulant and back sheets, together with monitored data sets of four climatic sites were used as basis for this simulation. The results show the influence of different climates on water concentration inside the encapsulant between solar cell and glass. Interestingly, the tropic test site does not thereby provide the highest concentrations. But also the effect of using different encapsulants in combination with different back-sheet materials on water ingress was shown. In particular the temperature dependency of the mass transfer influences the water ingress, depending on the material combination that is used. Results show, for example, for PA-based back sheets, that a higher E_A in combination with a transient microclimate (humidity and temperature) causes lower water concentrations over long periods of outdoor exposure.

In summary, a method was presented, existing of measured parameters and simulation, which is well suited for material and design studies regarding water ingress—not only for PV-modules.

Acknowledgments

This work was partly funded by the German Federal Ministry for the Environment, Nature Conservation and Nuclear Safety (BMU FKz 0329978) and sponsored by the industrial partners Schott Solar, Solarfabrik, Solarwatt, SolarWorld, and Solon.

References

[1] G. Oreski and G. M. Wallner, "Aging mechanisms of polymeric films for PV encapsulation," *Solar Energy*, vol. 79, no. 6, pp. 612–617, 2005.

[2] F. J. Pern and A. W. Czanderna, "Characterization of ethylene vinyl acetate (EVA) encapsulant: effects of thermal processing and weathering degradation on its discoloration," *Solar Energy Materials and Solar Cells*, vol. 25, no. 1-2, pp. 3–23, 1992.

[3] M. D. Kempe, G. J. Jorgensen, K. M. Terwilliger, T. J. McMahon, C. E. Kennedy, and T. T. Borek, "Acetic acid production and glass transition concerns with ethylene-vinyl acetate used in photovoltaic devices," *Solar Energy Materials and Solar Cells*, vol. 91, no. 4, pp. 315–329, 2007.

[4] G. Ehrenstein, *Polymer-Werkstoffe: Struktur-Eigenschaften-Anwendung*, Carl Hanser, Berlin, Germany, 2003.

[5] M. D. Kempe, "Modeling of rates of moisture ingress into photovoltaic modules," *Solar Energy Materials and Solar Cells*, vol. 90, no. 16, pp. 2720–2738, 2006.

[6] P. Hülsmann, K.-A. Weiß, and M. Köhl, "Temperature-dependent water vapour and oxygen permeation through different polymeric materials used in photovoltaic-modules," *Progress in Photovoltaics: Research and Applications*, 2012.

[7] P. Hülsmann, D. Philipp, and M. Köhl, "Measuring temperature-dependent water vapor and gas permeation through high barrier films," *Review of Scientific Instruments*, vol. 80, no. 11, Article ID 113901, 2009.

[8] D. Sonntag, "Important new values of the physical constants of 1986, vapor pressure formulations based on ITS-90 and psychrometer formulae," *Zeitschrift für Meteorologie*, vol. 70, pp. 340–344, 1990.

[9] W. R. Vieth, *Diffusion in and through Polymers: Principles and Applications*, Hanser, New York, NY, USA, 1991.

2

On High-Temperature Materials: A Case on Creep and Oxidation of a Fully Austenitic Heat-Resistant Superalloy Stainless Steel Sheet

A. Kanni Raj

Faculty of Mechanical Engineering, The Indian Engineering College, Tirunelveli 627116, Tamil Nadu, India

Correspondence should be addressed to A. Kanni Raj; akanniraj@gmail.com

Academic Editor: Te-Hua Fang

The creep behavior of AISI 310S stainless steel taken from SAIL's Salem stainless steel plant has been investigated by constant load tensile creep test at the temperatures of 973, 1023, and 1073 K and loads of 66.6, 74.8, 86.6, and 94.8 MPa. It exhibits steady-state creep behavior in most test conditions. The double logarithm plot of rupture life and applied stress yielded straight lines at all the three test temperatures indicating that power-law creep due to dislocation climb is the operating mechanism of creep deformation. Linear relationship was obtained for plots of logarithm of rupture life against inverse temperature obeying Arrhenius type of temperature dependence with activation energy of 340 kJ/mol. The stress-rupture data yielded a master curve of Larson-Miller parameter. The plot of Monkman-Grant relationship is typical indicating that rupture is controlled by growth of grain boundary cavities. The metallographic examination of crept samples revealed formation of grain boundary voids and cracks leading to intergranular creep fracture. Deformation twins and carbide precipitates were also observed. Oxidation tests were also carried out isothermally at 973 K, 1023 K, and 1073 K in dry air. The plots of mass gain versus square root time were linear at all the three test temperatures obeying parabolic kinetics of oxidation. It was found that scales are well adherent to the substrate. The plot of parabolic rate constant and inverse temperature was linear giving an activation energy value of 210 kJ/mol. The metallographic examination of an oxidized sample reveals duplex types of scales. Finally, rupture properties are compared with that of AISI 600 iron-based superalloy and oxidation weight gain analysis with surface nanocrystalline AISI 310S stainless steel to analyze quantitatively its behavior.

1. Introduction

Materials operating at high temperature are subjected to creep, hot corrosion, erosion, and phase change which severely affect the temperature capability and load bearing capacity of materials [1]. AISI 310S stainless steel contains higher chromium and nickel content than common iron-based alloys [2–4]. It is a standardized high-temperature steel for use at temperatures of up to 1323 K in dry air. It has a unique combination of high strength and superior oxidation resistance for application up to 1323 K. It is also very ductile and has good weldability enabling its widespread usage in many applications. It finds wide application in all high-temperature environments where scaling and corrosion resistance, as well as high temperature strength and good creep resistance, are required. Because of its superalloy properties, this alloy is a likely candidate material for high-temperature applications. It is a nonmagnetic alloy that cannot be hardened by heat treatment. It has forming capabilities similar to that of AISI 304 stainless steel. It can be easily formed into most shapes [5]. Extended capabilities of this stainless steel, beyond that of conventional stainless steels, make it potentially viable to replace nickel-based super alloys. A quantum of information connected with processing, tensile properties, oxidation, and hot corrosion properties of this alloy is available in the literature. The current research project focuses on the creep and hot corrosion of sheet samples taken from production run of Indian public sector stainless steel giant SAIL [5]. AISI 310S stainless steel flat rolled coil is primarily used for high-temperature situations. So, the alloy in the form of 1 mm thick sheets was used in the tests. Further, a comparison of creep data of this alloy with those of an iron-based superalloy AISI 600 is presented, as it is also a candidate material for jet engines and steam turbines [6, 7]. Similarly, a comparison of oxidation data is made with surface nanocrystalline AISI 310S stainless steel [8, 9].

2. Experimental

The materials investigated in creep rupture have the nominal composition of AISI 310S stainless steel. The alloy was cut into 10 cm × 10 cm × 1 mm. The sheets were solutionised at 1323 K for 2 hours in the furnace of Mayes creep testing machine and quenched in water. Soaking time is sufficiently high to ensure uniformity of temperature. It is 12 times higher than that required for mild steel sample of same thickness. This treatment ensures that all carbides are in solution. The composition was obtained using spectrometer. The specimens were longitudinally midsectioned, polished by conventional polishing techniques, and etched. The etched specimens were examined by optical microscopy. Creep specimens 1 cm gauge width and 10 cm gauge length were machined from the heat-treated pieces that correspond to ASTM E8M subsize specifications. All creep specimens were produced out of the centre of plates transverse to rolling direction. Creep tests were carried out in air at temperatures 973–1073 K in stress range 60–100 MPa in a Mayes constant load creep testing machine. Test temperature (T) was controlled by means of three PtRh-Pt thermocouples and kept constant at $T \pm 1$ K by means of three PID control units. For creep measurement, the specimens were equipped with a double-coil extensometer, and the creep behavior as specimen elongation was registered continuously. The resolution was about 0.001 mm (1 : 1250). Loading took place in normal atmosphere (air) via lever arm (1 : 15) using weights. The sample was allowed to soak at test temperature for 1 hour before the test load was applied. A number of different tests have been performed with standard specimens in the usual stress range (60–100 MPa) at temperatures 973–1073 K.

For oxidation kinetics, specimens with dimensions of 10 mm × 10 mm × 1 mm were cut from the heat-treated sheets, ground to 600-grit surface finish, and finally cleaned in ethanol. Isothermal oxidation tests were carried out in dry air at 973, 1023, and 1073 K. The specimens were exposed for 200 h. All specimens were used in the same condition. During testing, the specimens were tied in stainless steel wire, hanged in Mayes creep test unit's furnace, and both top and bottom of the furnaces were closed using ceramic wool, oxidized at the mentioned temperatures, and removed from the furnace at regular time intervals for weight change measurements [10, 11]. Electronic balance was used to measure weights accurately for 5 decimal points. After oxidation, microstructure of oxidized surfaces was investigated.

3. Results and Discussion

The chemical composition was obtained accurately using vacuum evaporation spectrometer of the alloys considered in the analysis (Table 1). Usual optical metallographic procedure was used to get microstructure which consists mainly grain

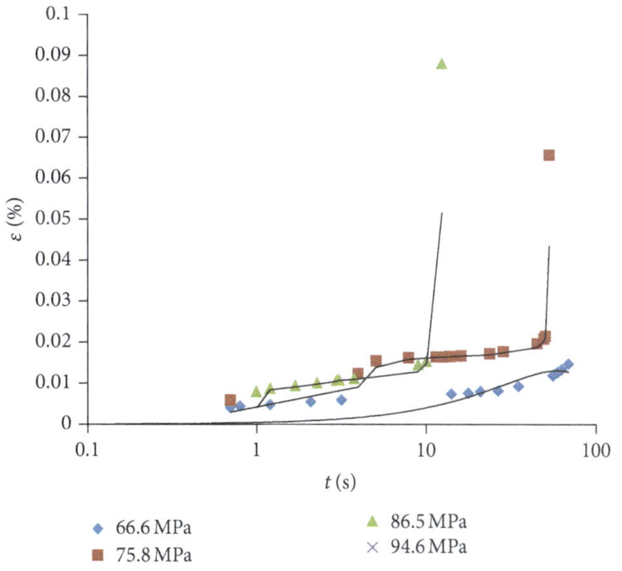

FIGURE 1: Creep curves of strain measured as a function of time for test temperature 1023 K.

with a grain size of 100 μm. The microstructure consists of two phases, namely, gamma and carbide phases. Twins are also seen. It is very hard to see distinctly anything at carbides at low magnifications. But optical micrographs clearly reveal the presence of the carbides inside grains.

Present test temperatures 973–1073 K and stresses 60–100 MPa fall in power-law creep (dislocation creep) region in deformation map. The region is governed by diffusion-controlled climb-plus-glide processes such as (a) based on lattice diffusion-controlled climb (high-temperature creep), (b) based on core diffusion-controlled dislocation climb (low-temperature creep), and (c) transition from climb-plus-glide to glide alone (power law breakdown). Typical creep curves at 973 K for loads of 66.6, 74.8, 86.6, and 94.8 MPa are shown in Figure 1. For the first three loads, all three regions of creep curves such as primary, secondary, and tertiary are available. Well-defined and longer secondary steady-state creep is noted well. But for the highest load, full strains are accumulated only during tertiary stage. Figure 2 shows typical creep curves for 86.5 MPa for temperatures 973, 1023, and 1073 K. For the first two temperatures, all three regions of creep curves such as primary, secondary, and tertiary are available. Well-defined and longer secondary steady-state creep is noted well. But for the highest temperature, full strains are accumulated only during tertiary stage. Tests were conducted for all the three test temperatures and all the four loads. But for want of avoiding unnecessary discussion, other plots are not shown, and graphs available with full clarity are taken for presentation. The main factor which determines the shape of creep curve besides development of twins and dislocations is investigated here. The factor is creation, coalescence, and growth of creep cavities [6]. Figure 3 shows the stress dependence of steady-state creep rate for all the three temperatures. The data can be represented by a power law with a stress exponent of −6.5. An important

TABLE 1: Chemical composition of AISI 310S stainless steel used in creep and oxidation studies along with surface-nanostructured AISI 310S stainless steel and heat-resistant AISI 600 iron-based superalloy taken for comparison.

The alloys	C	Mn	P	S	Si	Cr	Ni	Mo	Ti	Al	V
Ordinary AISI 310S stainless steel	0.08	1.25	0.045	0.030	0.96	24.8	18.5	—	—	—	—
Surface nanocrystalline-structured AISI 310S	0.04	0.97	0.016	0.001	0.61	25.0	19.5	—	—	—	—
AISI 600 iron-base-superalloy	0.08	1.0	0.025	0.025	1.0	15.0	26.0	1.0	2.0	0.2	0.1

significance of this observation is that the steady-state creep rate is controlled by the growth of the cavities nucleated at grain boundaries. Fractographic studies were not carried out, but optical metallographic images showed extensive voids. It is interesting to note that the dependence of steady-state creep rate on applied stress can be represented by a power law at all the three temperatures. Steady-state creep rate ($\acute{\varepsilon}_s$) could be represented by the formula $\acute{\varepsilon}_s \alpha \sigma^{-n} e^{-Q_c/RT}$. Plots were also made to get activation energy. Logarithm of steady-state creep rate against inverse of temperature was used to get activation energy. Was obtained from the plot is 340 kJ/mol and is a little higher than the self-diffusion value for gamma iron, that is, 300 kJ/mol. Figure 4 is the plot showing the effect of temperature on steady-state creep rate for stresses 75.8 and 86.5 MPa. Logarithm of steady state creep rate against inverse of temperature was used. Intercept of the plot is obtained, that is, the value of the constant C in Larson-Miller parameter equation. One of the most popularly used techniques in representing creep-rupture data is Larson-Miller parameter. This parameter can be derived from the stress and temperature dependence of the creep rate or time to rupture. 'Ihe rate equation generally can be written in the form of the Arrhenius equation and expressed as $P_{LM} = T[C + \log(t_r)]$. The Larson-Miller model is used for experimental tests so that results at certain temperatures and stresses can predict rupture lives of time spans that would be impractical to reproduce in the laboratory (the plot of ln σ versus P_{LM} is a line (Figure 5)). To further compare the heat-resistant fully austenitic AISI 310S stainless steel with heat-resistant AISI 600 iron-based superalloy, the time to fracture is plotted against applied stress (Figure 6). Superalloy took the longest time to fail more than 3 times to that of the stainless steel. This indicates that solid solution strengthening, dispersion strengthening, and precipitation hardening are not significant compared to superalloy. The microstructures of crept samples compare well with typical of the literature data, and a typical micrograph is shown in Figure 7 [5].

The oxidation curves of weight gain per unit area versus time for all alloys after exposure at temperatures 973, 1023, and 1073 K are depicted in Figure 8. After 720 ks exposure at all test temperatures, the alloy formed thicker oxide scales. Alloy showed the highest weight gain of 0.003 kg/m^2 at 1073 K, followed by 1023 K with 0.002 kg/m^2, while for 973 K, the weight gain was rather small which is below 0.001 kg/m^2. From the kinetics of oxidation at 973 K, the corrosion behavior is similar to that of alloys with high chromium showing a protective oxidation, while the oxidation rate becomes

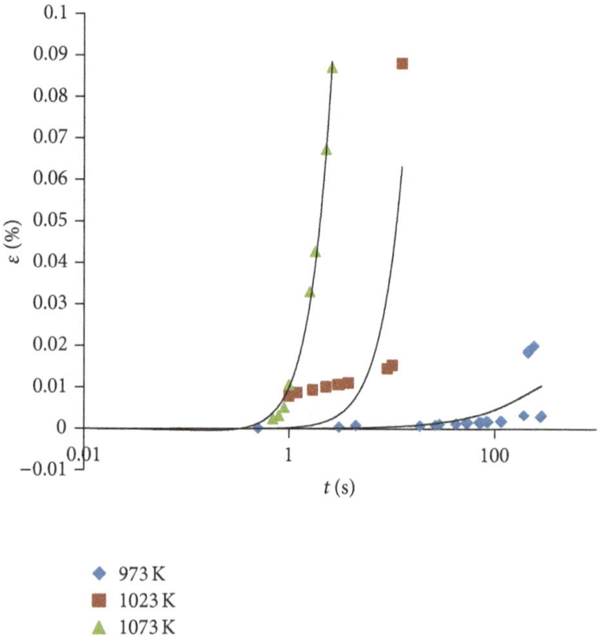

FIGURE 2: Creep curves of strain measured as a function of time for applied stress 86.5 MPa.

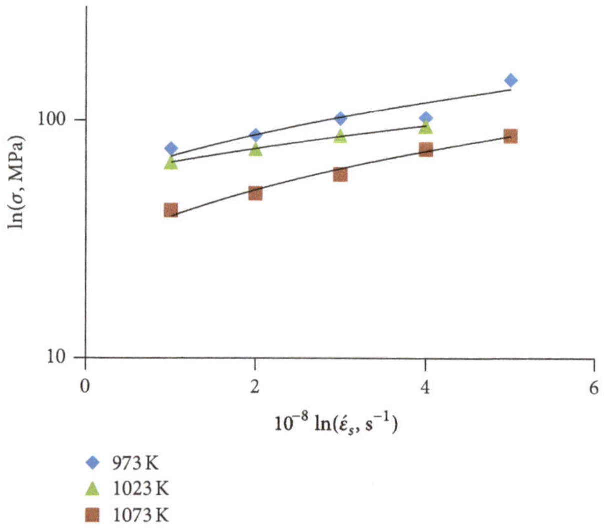

FIGURE 3: Effect of applied stress on steady-state creep rate at all test temperatures 973, 1023 and 1073 K.

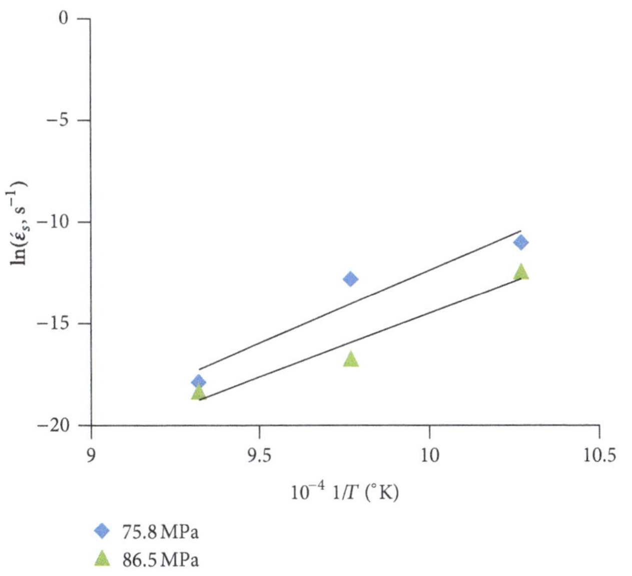

FIGURE 4: Effect of temperature on rupture life for stresses 75.8 and 86.5 MPa.

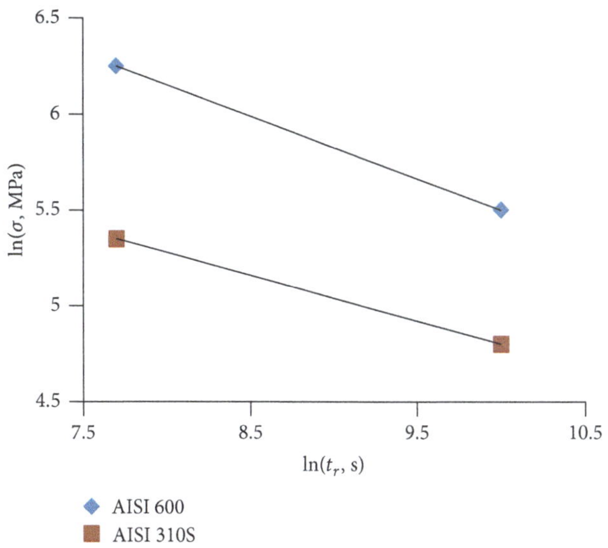

FIGURE 6: Effect of applied stress on rupture life at test temperature 923 K (AISI 600 compared).

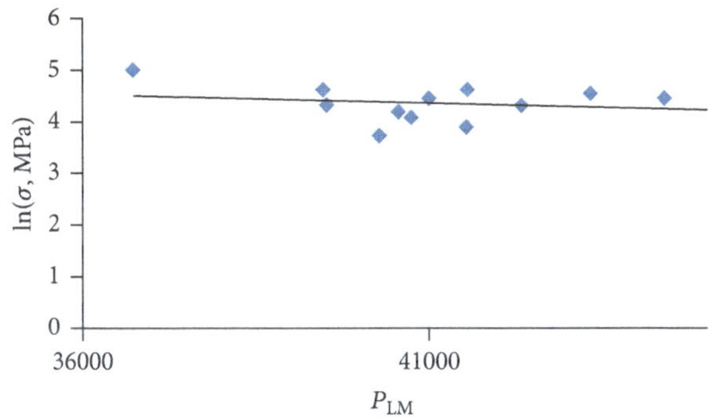

FIGURE 5: Master curve for temperatures 973–1073 K and stresses 66.6–94.6 MPa.

significantly larger at 1023 and 1073 K. The average final weight gain of oxidation at 1073 K is higher than oxidation at 1023 K which in turn is higher than that at 973 K without any sudden phase change. The plot of square of weight gain per unit area versus time for all alloys after exposure at temperatures 973, 1023, and 1073 K is linear, and so it is inferred that the oxidation kinetics under this test condition follows parabolic growth law ($\Delta w^2 \propto te^{-Q_c/RT}$). The slope provides oxidation rate constant. The rate constant is plotted versus inverse temperature to get activation energy. also Q_o = 210 kJ/mol (Figure 9). The cross-sectional analysis of the samples provides supporting evidence to this phenomenon. In terms of weight gain, similar trend of oxidation of that at all temperatures was observed. The slowest oxidation rate is in the alloy at 973 K showing protective behavior. Inspection of the surface of the oxide scales covering the specimens after the isothermal exposure for 720 ks revealed that the part of the oxide grown spalled off (Figure 10). The finding is that at all temperatures (973 K, 1023 K and 1073 K), oxide layer showed considerable influence against further oxidation. Contrary to oxidation at lower temperature, exposure at 1073 K shows remarkable findings for the alloys. Even though the final weight gain for these alloys was relatively higher compared to exposure at the lower temperatures, they exhibit a parabolic behavior only even after 720 ks exposure. The parabolic nature of the oxidation kinetics demonstrates that the initial oxidation is due to diffusion of the metal cation to the surface forming oxide films. Consequently, the oxidation proceeds by diffusion of oxygen through the initial layers towards the metal. From the experimental results, it is found that the main oxidation products are Fe_2O_3 and a mixture of Fe_2O_3 and spinel, whereby Fe_3O_4 is detected as the outermost oxide layer. This is because the Fe_3O_4 kinetics is typically greater than Fe_2O_3. The presence of Cr as a ternary element in the bulk alloy at 1073 K might reduce the outward diffusion of Fe and thus slow down the oxidation process. In addition, the addition of Cr may be incorporated with the oxidation reaction of the base alloy and result in modification of the structure of the oxides scale. These oxidation data are proved by microscopic examination of oxidized samples. Also, the oxidation behavior of it in normal condition and surface-modified condition provides interesting outcome. In oxidation studies, it is oxidized isothermally at 973 K, 1023 K, and 1073 K in dry air. Previous researchers have indicated that addition of surface nanostructure has a beneficial effect on the oxidation resistance. Therefore, attention is also paid to review the effect of nanostructured surface as a comparison from the literature data in this study [8]. It is clear that the nanostructured surface is 30% better in oxidation weight change and 100% better in spalling resistance.

4. Conclusions

Creep rupture followed power law with stress exponent of 6.5. The activation energy for creep (340 kJ/mol) was found

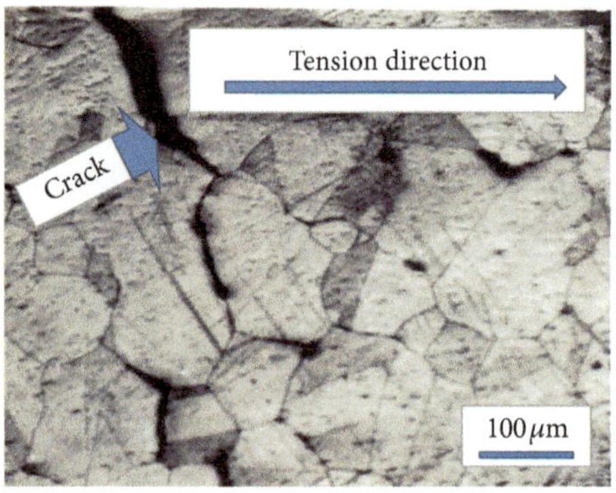

FIGURE 7: Optical micrograph of creep-ruptured sample at test temperature 1073 K and applied stress 66.6 MPa.

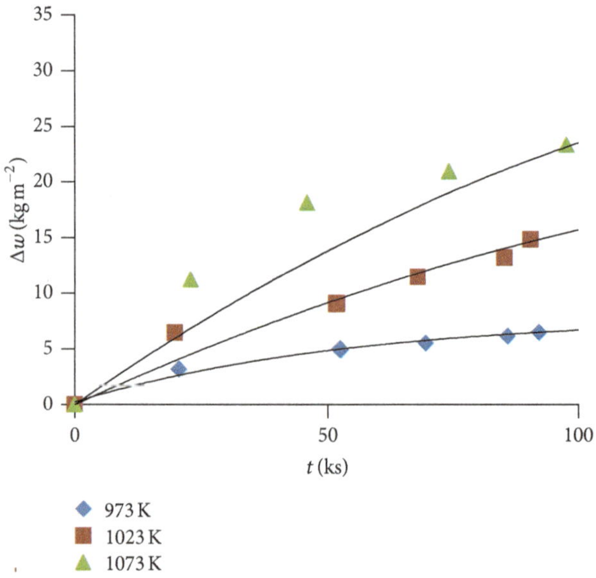

FIGURE 8: Plot of weight gain per unit area versus time for test temperatures 973, 1023, and 1073 K.

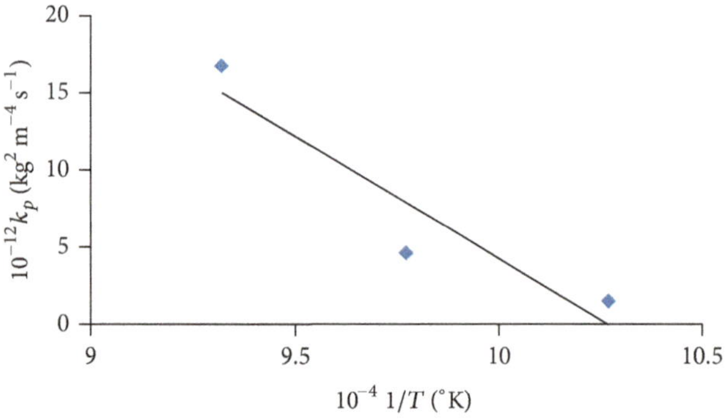

FIGURE 9: Plot of oxidation rate constant versus inverse of temperature. The activation energy from the plot is 210 kJ/mol.

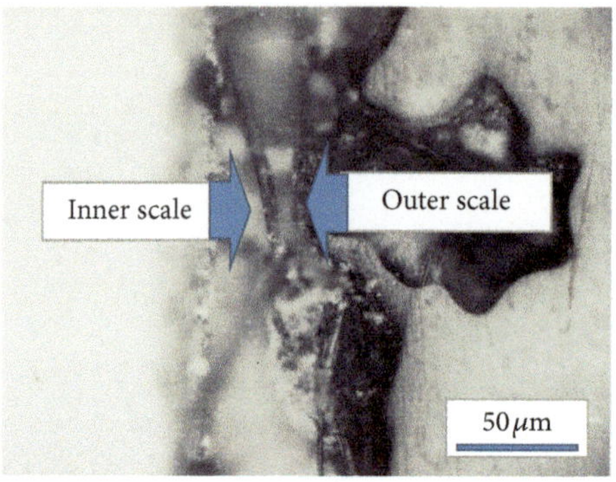

FIGURE 10: Optical micrograph of oxidized sample, taken in cross-section at 1023 K after 720 ks.

to be 10% higher than for lattice self-diffusion of γ-iron (300 kJ/mol). The stress dependence of steady-state creep rate ($\dot{\varepsilon}_s$) could be represented by a power law of form $\dot{\varepsilon}_s \alpha \sigma^{-n} e^{-Q_c/RT}$ (with $n = 6.5$ and $Q_c = 340$ kJ/mol), for all the threes test temperatures. On comparison, rupture life of AISI 600 superalloy was found as 3 times better. Micrograph showed clearly the intergranular creep fracture. The crack runs perpendicular to load axis. Oxidation follows a parabolic kinetics, $\Delta w^2 \alpha t e^{-Q_o/RT}$ (with $Q_o = 210$ kJ/mol), revealing the formation of protective oxide scale. Oxidized samples reveal an adherent inner layer and a spalling outer layer.

Acknowledgment

The project work was supported by GATE scholarship contingency grant of the National Institute of Technology, Tiruchirappalli, Tamil Nadu, India.

References

[1] M. E. Kassner and M. T. Perez Prado, "Five-power-law creep in single phase metals and alloys," *Progress in Materials Science*, vol. 45, no. 1, pp. 1–102, 2000.

[2] A. Kanni Raj, *Creep: Basic Theory and Dissertation*, AV, Saarbrücken, Germany, 2011.

[3] K. Maruyama, H. Ghassemi Armaki, and K. Yoshimi, "Multiregion analysis of creep rupture data of 316 stainless steel," *International Journal of Pressure Vessels and Piping*, vol. 84, no. 3, pp. 171–176, 2007.

[4] R. Voicu, J. Lacaze, E. Andrieu, D. Poquillon, and J. Furtado, "Creep and tensile behaviour of austenitic Fe-Cr-Ni stainless steels," *Materials Science and Engineering A*, vol. 510-511, pp. 185–189, 2009.

[5] S. S. M. Tavares, V. Moura, V. C. da Costa, M. L. R. Ferreira, and J. M. Pardal, "Microstructural changes and corrosion resistance of AISI 310S steel exposed to 600–800 °C," *Materials Characterization*, vol. 60, no. 6, pp. 573–578, 2009.

[6] S. Srinivas, D. V. V. Satyanarayana, and M. C. Pandey, "Creep testing," *Transactions of the Indian Institute of Metals*, vol. 49, pp. 625–640, 1996.

[7] H. De Cicco, M. I. Luppo, H. Raffaeli, J. Di Gaetano, L. M. Gribaudo, and J. Ovejero-García, "Creep behavior of an A286

type stainless steel," *Materials Characterization*, vol. 55, no. 2, pp. 97–105, 2005.

[8] Z. Liu, Y. He, and W. Gao, "Surface nanocrystallization of 310s stainless steel and its effect on oxidation behavior," *Journal of Materials Engineering and Performance*, vol. 7, no. 1, pp. 88–92, 1998.

[9] M. L. Lau and E. J. Lavernia, "Microstructural evolution and oxidation behavior of nanocrystalline 316-stainless steel coatings produced by high-velocity oxygen fuel spraying," *Materials Science and Engineering A*, vol. 272, no. 1, pp. 222–229, 1999.

[10] A. M. Huntz, A. Reckmann, C. Haut et al., "Oxidation of AISI 304 and AISI 439 stainless steels," *Materials Science and Engineering A*, vol. 447, pp. 266–276, 2007.

[11] A. Kanni Raj, "Creep and hot corrosion behavior of fully austenitic heat resistant SAIL's AISI 310S stainless steel," in *Proceedings of the 6th International Conference on Creep, Fatigue and Creep-Fatigue Interaction*, Kalpakkam, India, 2012.

3

Reactive Pulsed Laser Deposition of Titanium Nitride Thin Films: Effect of Reactive Gas Pressure on the Structure, Composition, and Properties

R. Krishnan,[1] C. David,[1] P. K. Ajikumar,[1] R. Nithya,[1] S. Tripura Sundari,[1] S. Dash,[1] B. K. Panigrahi,[1] M. Kamruddin,[1] A. K. Tyagi,[1] Vikram Jayaram,[2] and Baldev Raj[3]

[1] *Materials Science Group, Indira Gandhi Centre for Atomic Research, Kalpakkam, Tamil Nadu 603102, India*
[2] *Department of Materials Engineering, Indian Institute of Science, Bangalore, Karnataka 560012, India*
[3] *President, PSG Institutions, Coimbatore, Tamil Nadu 641004, India*

Correspondence should be addressed to R. Krishnan; krish@igcar.gov.in

Academic Editor: Ram Gupta

Titanium nitride (TiN) thin films were deposited by reactive pulsed laser deposition (RPLD) technique. For the first time, the composition evaluated from proton elastic backscattering spectrometry, in a quantitative manner, revealed a dependence on the partial pressure of nitrogen from 1 to 10 Pa. Grazing incidence-XRD (GI-XRD) confirmed the formation of predominantly nanocrystalline TiN phase with a crystallite size of around 30 nm. The hardness showed maximum value of ~30 GPa when the composition is near stoichiometric and the friction coefficient was found to be as low as 0.3. In addition, a systematic optical response was observed as a function of deposition pressure from the surface of the TiN films using spectroscopic ellipsometry.

1. Introduction

Titanium nitride (TiN), by virtue of its high hardness, high melting point, low density, chemical stability, corrosion resistance, low coefficient of friction, lower wear rate, and so forth, is the most extensively used protective coating material [1–4]. All these properties depend on the coating techniques and vary as a function of deposition conditions. Amongst the gamut of available techniques, reactive pulsed laser deposition (RPLD) has distinct advantages to deposit ceramic films using metallic targets.

Despite large volume of published work on TiN synthesis using RPLD, none brings out a clear recipe for synthesizing high-quality TiN films with desired stoichiometry, structure and morphology. For many applications, quantitative evaluation of the composition is very essential as the properties of TiN are closely related to its stoichiometry [1]. Till date there is only one research article that attempted to study the systematic evolution of structure in RPLD grown TiN films and a review article which summarized those results was published in 2002 [5, 6]. Even that article did not report quantitative compositional variation with nitrogen pressure and proposed the deposition of nanocrystalline TiN films over a range of deposition pressures (0.7 to 7 Pa) without accounting for the finer variations in the given range. In fact, the ellipsometry was used in-situ by Heil et al. to monitor, in-situ, the thickness of the films rather than to characterize the optical properties [7]. Though qualitative, in a recent work, we presented a novel secondary-ion-mass-spectrometry (SIMS) based methodology for optimization of three important process parameters, namely, deposition pressure, substrate temperature and pulse energy [8]. As relative sensitivity of elemental factors arising from the matrix effects are little known, obtaining quantitative estimates of the elemental composition from SIMS data is difficult.

In this study, we present the systematic evolution of structure through grazing incident X-ray diffraction (GIXRD) analysis and quantitative composition variation of RPLD grown TiN films as a function of deposition pressure in the range of 1.0 to 10.0 Pa using proton elastic back scattering

(PEBS) spectrometry. PEBS is superior to the conventional Rutherford backscattering spectrometry (RBS) using alpha particles, especially for analyzing low Z elements embedded or present as films on higher Z substrates [9]. In the particular case of obtaining compositional information of TiN using conventional RBS it is imperative to deposit these films on a lower Z substrate (usually a carbon/silicon substrate). But the application of PEBS facilitates determination of composition of the TiN films coated over the structural material itself [AISI 304 SS]. In the present study, the effect of deposition pressure in influencing the structure, composition, and morphology of TiN films is addressed.

2. Experimental

A complete description of the experimental facility is given in our earlier publication [10]. A Q-switched Nd:YAG laser, operating at 1064 nm with a fluence of 4 J/cm^2 and 10 Hz repetition rate, was introduced into an ultra-high vacuum (UHV) compatible reaction chamber through a quartz window and was focused to approximately 2 mm diameter circular spot at the target surface at an oblique incidence of 45°. The laser pulse duration was 5 ns. A titanium disc of 24.5 mm diameter, 10 mm thickness, and having purity of 99.9% was used as a target. The target was rotated at 3 rpm to prevent crater formation and to expose fresh material for laser ablation. Prior to deposition, the AISI 304 SS substrates (20 mm diameter × 3 mm thick) were degreased, cleaned, and dried. Part of the ablating plume was made to deposit on the substrates positioned on a substrate heater, located just in front of the target. A target-substrate distance of 40 mm was maintained during the deposition, and 1,00,000 laser pulses were fired to get approximately 1 μm thick coatings. The chamber was evacuated to a base pressure of 5.0×10^{-5} Pa using a turbomolecular pumping system. Ultra-high pure N_2 gas was bled into the chamber using a mass flow controller. The coatings were characterized for structural, compositional, and morphological information by GIXRD, PEBS, and AFM, respectively. Hardness of the coatings was evaluated by nanoindentation. The optical studies were carried out using a SOPRA ESVG model rotating polarizer type ellipsometer in the energy range 1.38 to 5 eV at an angle of incidence of 75 degree. Low-load (25–100 mN range) friction and wear measurement were carried out using a nanotribometer (M/s CSM Instruments, Switzerland).

3. Results and Discussion

3.1. Structural Determination. The GIXRD patterns of the TiN_x films, deposited on SS substrates with a constant pulse energy and substrate temperature, are shown in Figure 1. The nitrogen pressure inside the chamber was varied from 1 to 10 Pa in 5 steps. But for the small fraction of Ti_2N, TiON, and TiO_2 phases, all the patterns indicated the formation of single-phase cubic TiN (space group: Fm$\bar{3}$m) with a lattice parameter of 4.234 Å. This is in agreement with the JCPDS data file: 38–1420. While the films deposited at lower pressures (≤3 Pa) showed the presence of Ti_2N, deposition at higher pressures (above 5 Pa) indicates the incorporation of oxygen into TiN leading to the formation of TiON or even TiO_2. At higher pressures, even a small increase of oxygen in the background gas leads to a preferential formation of oxides and oxynitrides as reported in literature [5, 11]. Formation of nearly stoichiometric and polycrystalline TiN is observed for films deposited at around ~5 Pa. Thus the XRD patterns confirm the formation of stoichiometric TiN by RPLD process and the presence of minor amount of Ti_2N and TiON phases. The crystallite size was evaluated from (200) peak using the Scherer formula and found to be around 28 nm for all the films.

Table 1: Composition of TiN thin films as measured using PEBS.

Deposition pressure P_{N_2} (Pa)	N_N/N_{T_i}	N_0/N_{T_i}
1.0	0.44	0.56
3.0	0.82	0.18
5.0	0.90	0.10
7.0	0.76	0.24
10.0	0.59	0.41

3.2. Compositional Evaluation. The proton backscattering studies $^{14}N(p, p)^{14}N$ were carried out at an incident proton energy of 2.0 MeV so as to utilize a energy regime with enhanced scattering cross section [9]. The proton backscattering studies are standardized by obtaining the composition of a bulk TiN compound. The experimental spectrum obtained for the bulk TiN was simulated using SIMNRA program [12], and it is found that the estimated composition is in agreement with the certified value. The PEBS spectra and the fits obtained using SIMNRA for the TiN_x films deposited at 7 Pa are shown in Figure 2(a). The various edges (Ti, substrate, O and N) are indicated. To obtain the quantitative estimates, the experimental spectrum is simulated using the SIMNRA code [12].

Cumulative composition of the films as a function of deposition pressure was computed, and the results are tabulated in Table 1. The tabulation clearly indicates the trend in the N/Ti ratio. For the film deposited at 5 Pa, the N/Ti ratio approaches stoichiometry and then drops. Oxygen incorporation in titanium nitride films could be reasoned as follows: titanium rich films deposited at lower nitrogen pressures pickup more oxygen when exposed to air than stoichiometric TiN films. But at higher deposition pressures the quantity of impurity oxygen increases, and hence its incorporation is more.

3.3. Morphological Investigations. The surface morphology was analyzed using AFM over an area of 5 × 5 μm and the results are shown in Table 2. The films grown at ~1 Pa revealed RMS roughness of 1.9 nm. Large number of particles with average diameter of 100–200 nm were found to be distributed. At low pressures, the incoming plume will have high energy with high mean free path leading to high flux and nonuniformity. With increase in background pressure, the incoming flux gets thermalized due to decrease in mean free

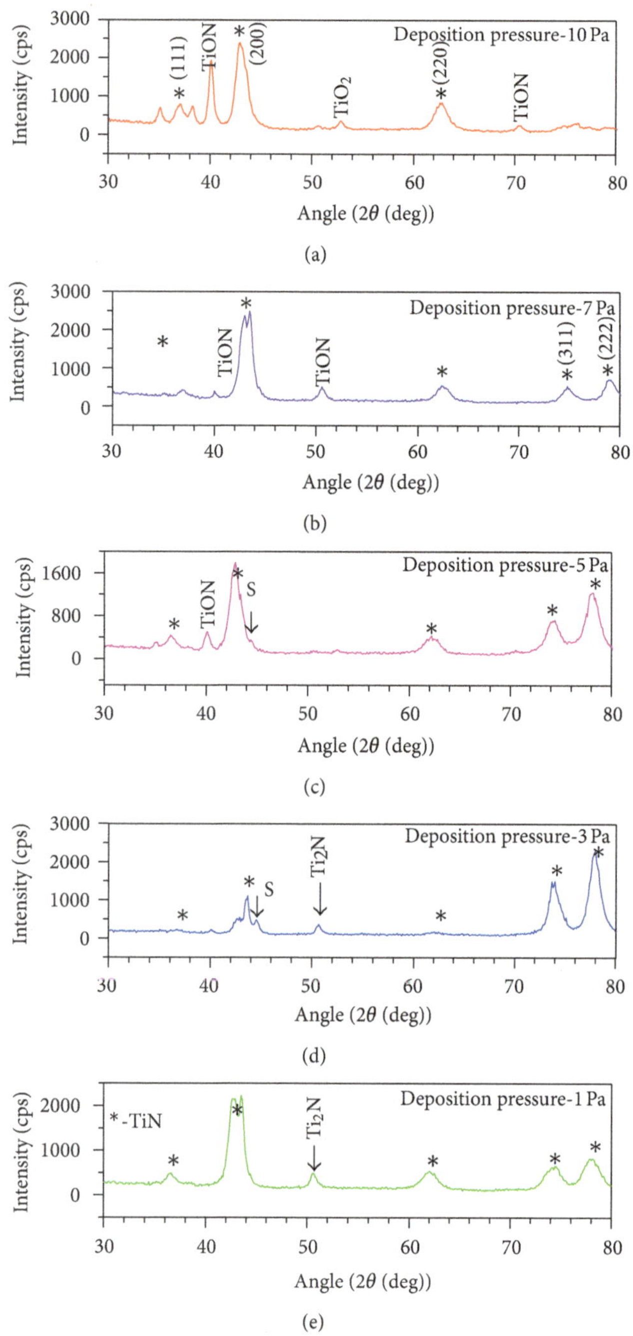

Figure 1: X-ray diffraction patterns of TiN films deposited at different nitrogen pressures. (∗) and (S) denote peaks due to TiN and SS substrate, respectively.

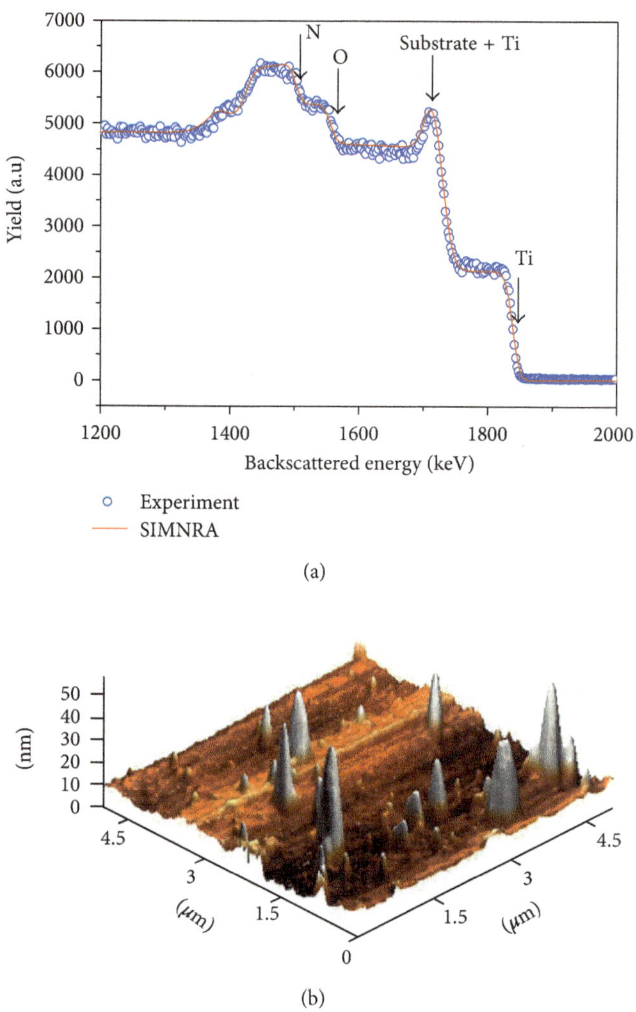

Figure 2: (a) PEBS spectra and (b) AFM image of TiN films grown at 7 Pa.

Table 2: Roughness of the films as measured from AFM.

Deposition pressure (Pa)	RMS roughness (nm)
1.0	1.9
3.0	2.2
5.0	6.6
7.0	4.5
10.0	4.3

path. This leads to coalescence of the particles leading to fewer and bigger particles. Hence the RMS roughness value at 3 Pa increases to 2.2 nm and then further rises to 6.6 nm at 5 Pa. However, at 7 Pa, due to the increased availability of nitrogen and the collision events, the number of particles reaching the surface reduces leading to the reduction in both number density and their average height leading to reduction in RMS roughness to 4.3 nm. Figure 2(b) shows the representative AFM image of TiN films grown at 7 Pa. This trend continues for the deposition pressure of 10 Pa.

3.4. Optical Properties. A qualitative behavior of the optical response of the films grown at different partial pressures is presented. The real part of the dielectric functions for specimens deposited between 1 Pa to 7 Pa is shown in Figure 3(a). From the figure, it is seen that the magnitude of the real part of the dielectric function $\langle \varepsilon_1 \rangle$ increases with increase in partial pressure of nitrogen, varying from ~−5 @ 1.5 eV for 1 Pa to ~0.2 @ 1.5 eV for 7 Pa. The plasma frequency of the films is 2.37 and 2.09 (energy at which $\varepsilon_1 = 0$) for films deposited at 1 Pa and 3 Pa, respectively. The typical value of $\langle\varepsilon_1\rangle$ @ 1.5 eV and plasma frequency for a reference TiN is typically −15 and 2.7, respectively. The increase in $\langle\varepsilon_1\rangle$ and plasma frequency is attributed to the formation of TiO_2 which is semiconducting and has a band gap of ~3.2 eV. For the film deposited at 5 Pa, the material is predominantly TiO_2 (more than 50%). It is to be pointed out that, in ellipsometry, the thickness that is probed depends on the material and is typically 100 nm for the TiN owing to its metallic nature. The percentages obtained from ellipsometry cannot be compared one to one with that obtained from PEBS as PEBS probes deeper into the film. The refractive indices shown in Figure 3(b) differ from that reported by Jeyachandran et al. [13] for very low nitrogen concentration. The film shows an evolution towards TiO_2-rich phase with increase in partial pressure of nitrogen. A clear systematic evolution is seen in the optical properties and indicates the potential for being used in-situ to characterize the relation between stoichiometry and thickness [14].

3.5. Hardness. In order to evaluate the hardness of these TiN coatings, samples were subjected to nanoindentation. Figure 4 shows the results of nanohardness measurements of a series of TiN coatings deposited at room temperature on a AISI-SS-304 substrates with a total thickness of 1 μm. The nanohardness of TiN samples reaches its maximum at the deposition pressure of 3–5 Pa and then decreases at higher deposition pressures. It is interesting to note that average hardness values of films deposited at around 3–5 Pa are greater than the hardness values of other deposition pressure in concomitant with the PEBS results showing higher nitrogen content with N_N/N_{Ti} values greater than 0.8, whereas the hardness values are lower for the higher oxygen values.

3.6. Tribological Properties. Figure 5 depicts the typical curves showing the variation of coefficient of friction (COF) with two different speeds and two different loads when the 100Cr6 steel ball was made to slide on the RPLD grown TiN films deposited at 3 Pa. At low speed (0.1 m/sec) and low load (100 mN), the surface regenerates itself after initial break down as revealed by the rapid rise in COF followed by a steady decline. But at higher load (700 mN), for the same speed of (0.1 m/sec), the COF falls initially due to smoothening of surface asperities followed by steady increase due to compound wear that consequently enhances COF. With the passage of laps, the surface gets smoothened, and hence there is a fall in COF which again rises much more steeply due to the faster wear rates coupled with smoothening. However, at higher speeds (0.5 cm/sec), a cyclic pattern in the variation of COF is observed for both the loads (100 and 700 mN), though with different periodicity.

4. Conclusions

Studies have been carried out on the synthesis of nanocrystalline TiN thin films using RPLD through ablation of high purity titanium targets in the presence of low-pressure nitrogen gas, with constant temperature and laser power using Nd:YAG laser operating at 1064 nm. Films were characterized using GIXRD, PEBS, and AFM for their structure, composition, and morphology, respectively. Ellipsometric and hardness measurements indicated a systematic evolution in both optical and mechanical properties. Results reveal that deposition pressure is crucial for obtaining desired stoichiometry. Nanocrystalline TiN close to stoichiometry was obtained for the film deposited with nitrogen pressure of 3–5 Pa with maximum hardness values as measured using nanoindentation. Also, the friction coefficient of the TiN

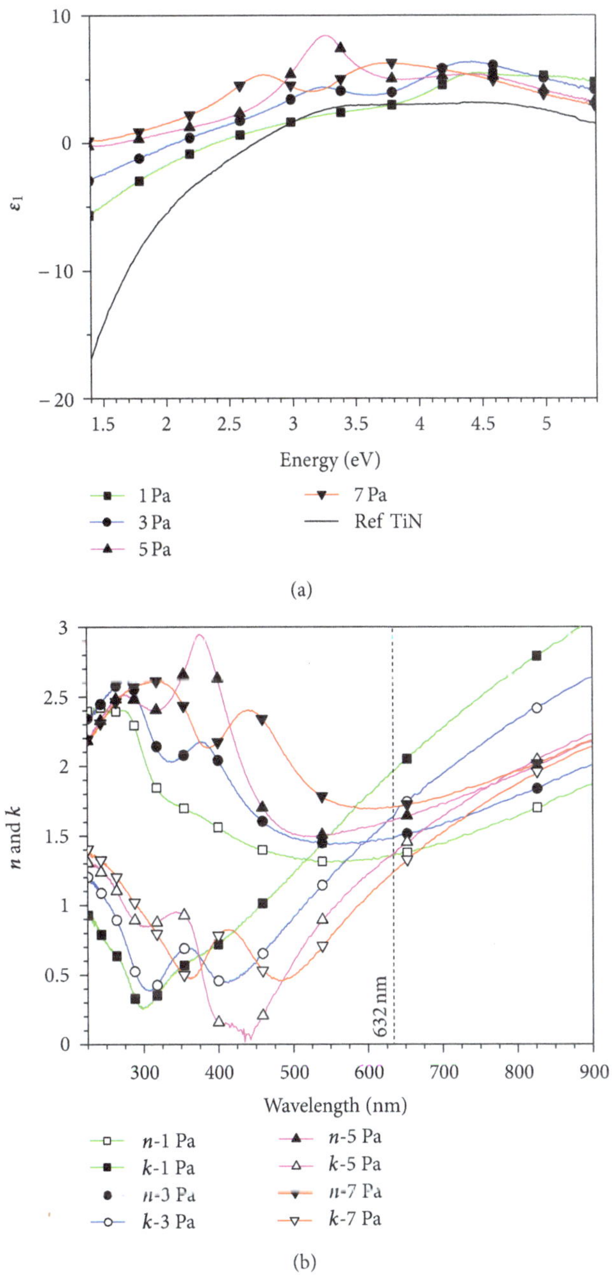

FIGURE 3: (a) Real part (ε_1) and (b) Refractive index (n and k) for specimen deposited at different nitrogen pressures.

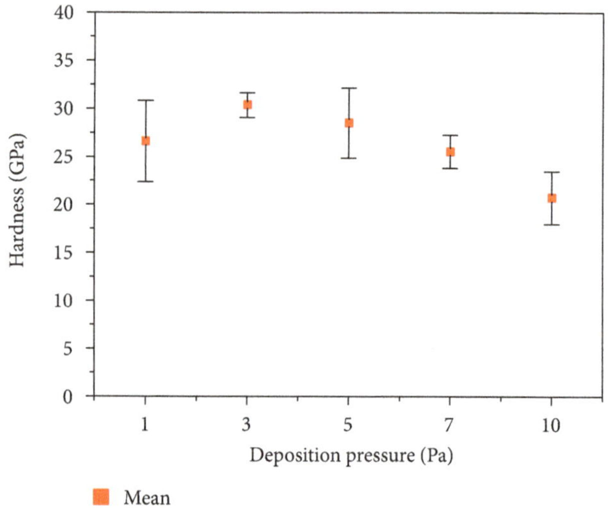

FIGURE 4: Hardness of the TiN films deposited at different nitrogen pressures.

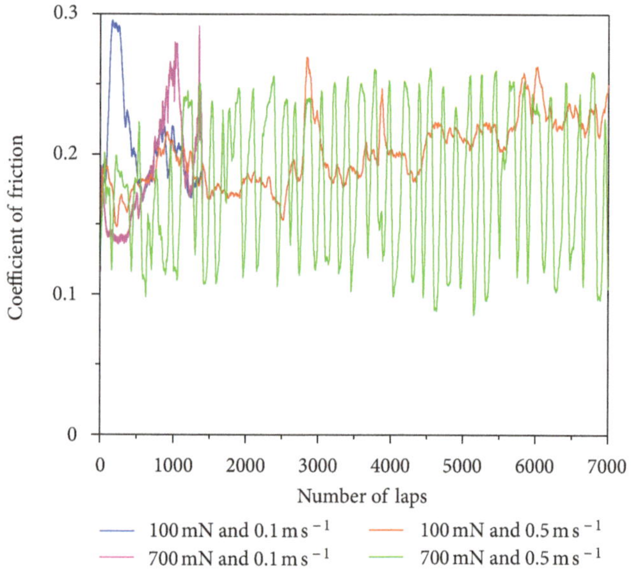

FIGURE 5: Variation of COF of monolithic TiN thin film deposited at 3 Pa under variable loads and speeds.

films, deposited by RPLD, was found to be low as compared to that of their microcrystalline counterparts.

References

[1] J. E. Sundgren, "Structure and properties of TiN coatings," *Thin Solid Films*, vol. 128, no. 1-2, pp. 21–44, 1985.

[2] P. Ettmayer and W. Lenguer, "Transition metal nitrides: solid state chemistry," in *Encyclopedia of Inorganic Chemistry*, R. B. King, Ed., John Wiley & Sons, 2nd edition, 1994.

[3] H. Randhawa, "Hard coatings for decorative applications," *Surface and Coatings Technology*, vol. 36, no. 3-4, pp. 829–836, 1988.

[4] C. Subramanian and K. N. Strafford, "Review of multicomponent and multilayer coatings for tribological applications," *Wear*, vol. 165, no. 1, pp. 85–95, 1993.

[5] I. N. Mihailescu, E. Gyorgy, N. Chitica et al., "A parametric study of the deposition of the TiN thin films by laser reactive ablation of titanium targets in nitrogen: the roles of the total gas pressure and the contaminations with oxides," *Journal of Materials Science*, vol. 31, no. 11, pp. 2909–2915, 1996.

[6] A. Perrone, "State-of-the-art reactive pulsed laser deposition of nitrides," *Japanese Journal of Applied Physics A*, vol. 41, no. 4, part 1, pp. 2163–2170, 2002.

[7] S. B. S. Heil, E. Langereis, F. Roozeboom, M. C. M. Van De Sanden, and W. M. M. Kessels, "Low-temperature deposition of TiN by plasma-assisted atomic layer deposition," *Journal of the Electrochemical Society*, vol. 153, no. 11, Article ID 045611JES, pp. G956–G965, 2006.

[8] R. Krishnan, T. Mathews, A. K. Balamurugan et al., "Reactive pulsed laser deposition of titanium nitride thin film: optimization of process parameters using secondary ion mass spectrometry," *Applied Surface Science*, vol. 256, no. 10, pp. 3077–3080, 2010.

[9] E. Rauhala, "Proton elastic scattering cross sections of carbon, nitrogen and silicon for backscattering analysis in the energy range 0.7–2.5 MeV," *Nuclear Instruments and Methods in Physics Research B*, vol. 12, no. 4, pp. 447–452, 1985.

[10] R. Krishnan, R. Ramaseshan, T. Mathews et al., "Synthesis of nanostructured titanium nitride films by PLD through reactive processing," *Surface Engineering*, vol. 25, no. 3, pp. 218–222, 2009.

[11] E. D'Anna, M. L. De Giorgi, G. Leggieri et al., "Oxidation interference in direct laser nitridation of titanium: relative merits of various ambient gases," *Thin Solid Films*, vol. 213, no. 2, pp. 197–204, 1992.

[12] M. Mayer, SIMNRA User's Guide, Report IPP 9/113, Max-Planck-Institut fur Plasmaphysik, Garching, Germany, 1997.

[13] Y. L. Jeyachandran, S. K. Narayandass, D. Mangalaraj, S. Areva, and J. A. Mielczarski, "Properties of titanium nitride films prepared by direct current magnetron sputtering," *Materials Science and Engineering A*, vol. 445-446, pp. 223–236, 2007.

[14] S. Adachi and M. Takahashi, "Optical properties of TiN films deposited by direct current reactive sputtering," *Journal of Applied Physics*, vol. 87, no. 3, pp. 1264–1269, 2000.

4

Porosity Analysis of Plasma Sprayed Coating by Application of Soft Computing

Ajit Behera,[1] S. C. Mishra,[2] Asit Behera,[3] and J. P. Dhal[4]

[1] *Department of Metallurgical & Materials Engineering, Indian Institute of Technology, Kharagpur, West Bengal 721302, India*
[2] *Department of Metallurgical & Materials Engineering, National Institute of Technology, Rourkela, Odisha 769008, India*
[3] *Department of Mechanical Engineering, National Institute of Technology, Rourkela, Odisha 769008, India*
[4] *Department of Chemistry, National Institute of Technology, Rourkela, Odisha 769008, India*

Correspondence should be addressed to J. P. Dhal, jyoti84.chem@gmail.com

Academic Editor: Marie-Louise Saboungi

The present piece of work describes the industrial wastes and low grade ores (fly ash + quartz + ilmenite, as the coating material), deposited on mild steel substrates. In many cases it is found that porosity is an important factor on the coating surface. Knowledge about the extent of these porosity imperfections is critical since they influence a wide range of spray coated properties and behaviors. To decrease the porosity by optimizing necessary operating parameters, artificial neural network (ANN) technique is used. The aim of this investigation is to find out appropriate input vectors in ANN model. ANN experimental results indicate that the projection network has good generalization capability to optimize the porosity.

1. Introduction

Plasma surfacing is an excellent technology in high-performance coating applications ranging from aerospace industry to biomedical industry [1–3]. Thermal spraying process implements a wide variety of materials (metal, ceramic, alloy and its composite) and processes (atmospheric plasma spraying, vacuum plasma spraying, etc.) for improving surface properties [4–6]. The conventional plasma-spraying process is commonly referred to as air or atmospheric plasma spraying (APS). To generate plasma, an inert gas typically argon/argon + hydrogen mixture is superheated by a DC arc [7]. Plasma temperatures range from approximately 6000°C to 15000°C in the power heating region, which are significantly above the melting point of any known material [8]. The powder mixture of fly ash + quartz + ilmenite (size from 40 μm to 100 μm) injected into a high-temperature plasma flame which is then rapidly heated and accelerated to a very high velocity by the plasma flame impacts the surface of the substrate material in the form of molten or semimolten state and very quickly cools to form a high-quality coating [7, 9, 10].

The quality control of plasma spray technique generally considers the monitoring of the molten feedstock particle characteristics deposited at the substrate surface, that is, to increase the coating efficiency [11]. Coating efficiency represents the effectiveness of the deposition process as well as the coatability of the powder under study [12]. The coating efficiency directly or indirectly depends on many other parameters during spraying, in which each one is interrelated with each other. Porosity is a microstructural feature and affects a lot of coating properties such as thermal conductivity, coefficient of thermal expansion, elastic modulus, and dielectric behavior [13]. Various methods are employed for quantitative measurement of porosity and form a necessary part of microstructural characterization of thermal spray coatings. The most used technique is image analysis which is a reliable, straightforward, versatile, and inexpensive technique for the characterization of porosity. ANN study and design of systems are capable of perceiving their environment and taking actions maximizing their chance of success (i.e., to decrease the coating porosity), by considering all of the previously interrelationships.

Table 1: Experimental operating parameters used during plasma spraying of fly ash + quartz + ilmenite.

Operating parameters	values
Plasma arc current (Amp)	270, 300, 400 and 420
Arc voltage (volt)	40, 45 and 50
Torch input power (KW)	11, 15, 18 and 21
Plasma gas (argon) flow rate (IPM)	28
Secondary gas (N_2) flow rate (IPM)	3
Carrier gas (Ar) flow rate (IPM)	12
Powder feed rate (gm/min)	12, 15 and 18
Powder size (μm)	40, 60, 80 and 100
Torch-to-base distance (TBD) (mm)	100

Table 2: Input parameters selected for ANN training.

Input parameters for training	Values
Error tolerance	0.003
Learning parameter (β)	0.002
Momentum parameter (α)	0.002
Noise factor (NF)	0.001
Maximum cycles for simulations	10,000,000
Slope parameter (£)	0.6
Number of hidden layer neurons	8
Number of input layer neurons (I)	5
Number of output layer neurons (O)	1

2. Experimental Procedure

Fly ash + quartz + illuminate mixture was taken with weight percentage ratio of 60 : 20 : 20 and mechanically milled in planetary ball mill for 3 hour for homogenization. This mixture used as feed stock for plasma spraying was first sieved. Four different sizes (i.e., 40 μm, 60 μm, 80 μm, and 100 μm) are separated out. Mild steel is used as substrate material, having dimensions of 1 inch diameter and 3 mm thickness. The substrates were grit-blasted at a pressure of 3 kg/cm^2 using alumina grit to make the surface roughness around 5 Ra. Then substrates surface was cleaned by acetone and plasma spraying was immediately carried out. In plasma torch input power level was varied from 11 KW to 21 KW. This is a typically DC power supplied nontransferred arc mode atmospheric plasma spray process. The injection of the powder was external from the torch nozzle and directed towards the plasma flow. Argon and hydrogen plasma mixture gas is used as carrier gas. The major subsystems of the set-up include the power supply, plasma spraying torch, powder feeder, and carrier gas supply, stand of distance of torch, control console, cooling water, and spray booth. A four-stage closed loop centrifugal pump water cooling system was used for cooling the system and regulated at a pressure of 10 kg/cm^2 supply. Plasma spray operating parameters specifications are given in Table 1. Flow rates of plasma gas (argon) and secondary gas (N_2 gas) are kept constant. With increasing power level, different parameters such as powder feed rate, powder size, and stand of distance of torch are varied. Experimental value of porosity was measured by image analyzer technique. In image analyzer technique the porosity of coatings was measured by putting polished cross-sections of the sample under a microscope (Neomate made) equipped with a CCD camera (JVC, TK 870E). This system is used to obtain a digitized image of the object. The digitized image is transmitted to VOIS image analysis software. The total area captured by the objective of the microscope or a fraction can be accurately measured by the software. Hence the total area and the area covered by the pores are separately measured.

3. Artificial Neural Network

A neural network is a soft computing model processing of a system, based on interrelationship between different input and output parameters and learning from data set through iteration, without requiring a prior knowledge on the relationships between the process variables [13]. This model is able to approximate various nonlinearities in the data series among other models [14–16] and can give an appropriate optimized data output (here porosity), more quickly [17]. ANN of simple processing elements (neurons) typically organized in layers (input layers, hidden layers, and output layers) is shown in Figure 1. A software package NEURALNET for neural computing is developed, by using back propagation algorithm as the prediction tool for output (coating coating porosity) [18, 19].

For developing an ANN there are no fixed rules, but in general framework it can be followed based on previous successful applications in engineering. The aim of an ANN is to normalize an input-output relationship of the form

$$y^m = f(x^n), \tag{1}$$

where x^n is an n-dimensional input vector represents variables $x_1, \ldots, x_i, \ldots, x_n$ and y^m is an m-dimensional output vector that represents the resulting variables $y_1, \ldots, y_i, \ldots, y_m$. In plasma spray modeling, values of x may be current, voltage, stand of distance, powder federate, and powder size.

4. Results and Discussion

In NEURALNET Software, based on least error criterion as shown in Table 2, is selected for training of the input-output data by taking 12 sets of parameters. The network training and testing process are conducted for 10,000,000 cycles for which stabilization of the error is obtained. Neuron numbers in the hidden layer are varied and in the optimized structure of the network, this number is 6 for Mild Steel. The three-layer network involved is shown in Figure 1.

4.1. Predicted Porosity Compared with Experimental Results Based on Different Feed Rate. Comparative study of ANN prediction value of coating porosity with that of experimental value for mild steel substrate is given in Figure 2. Here the

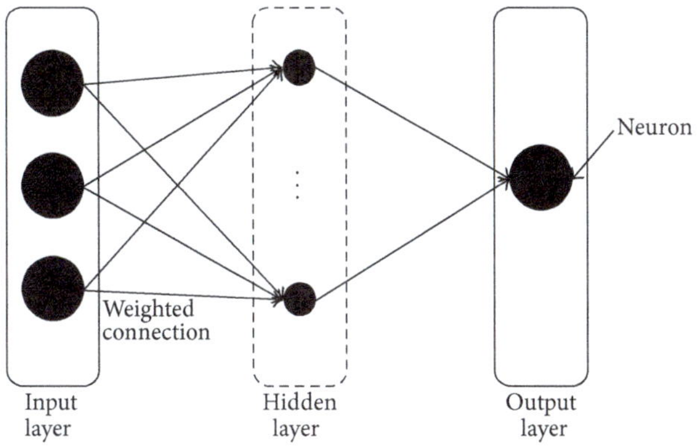

FIGURE 1: Architecture of artificial neural network.

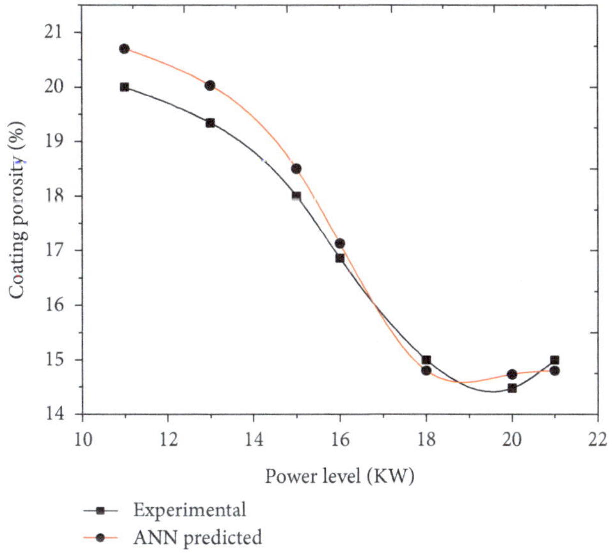

FIGURE 2: Comparison plot between ANN prediction porosity and experimental value at 12 gm/min feed rate (fly ash + quartz + ilmenite powder mixture is deposited on mild steel substrates).

fly ash + quartz + illmenite composite powder is deposited on mild steel at 12 gm/min feed rate, 100 mm torch stand of distance with varying power level. Two sigmoidal plots were found by varying input power level. From this figure, it can be seen that the porosity decreases in between 18 and 20 KW power level. In spraying the exact power level cannot be specified, because thermally sprayed coatings are very complex and incorporate process dependent defects such as splat gaps/interlamellar pores, globular pores, and cracks (for ceramics). It is clear that there is a close agreement of porosity measurement by the neural network and the experimental study, which indicates that the ANN model can be used for predicting the amount coating porosity. At 12 gm/min feed rate and 100 mm torch stand of distance; the lowest (14.48%) coating porosity is obtained at 20 kW.

From the figures it is found that by increasing power level the porosity value decreases up to a certain point and increases with further increase in power level. This increase

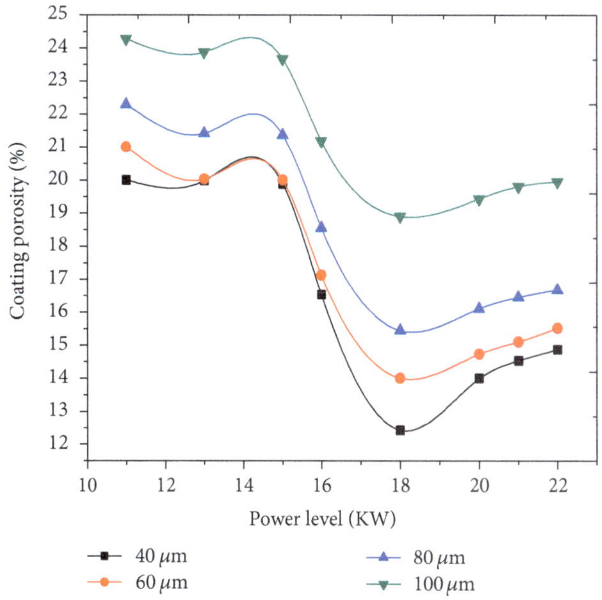

FIGURE 3: ANN results of coating porosity with respect to power level for different powder size at 12 gm/min feed rate.

is due to formation of vaporized molten powder particle. It is also known that the lesser the coating thickness, the lesser will be the coating porosity [20]. Porosity formation by intersplat position also depends on the mode of heat transfer through the substrate or prior deposited splats. This is also similar to splat quenching by the Duwez gun technique [21].

4.2. Prediction of Porosity Based on Variation of Powder Size. By keeping feed rate (12 gm/min) constant, all other parameters are changed in ANN plot of Figure 3. It is found that for lower powder particle size, porosity decreases rapidly with increase in power level. At 13 to 15 kW power level, shown in Figure 3, the porosity value increases. This may be due to inadequate amount of heat flow which results in that molten-semimolten particles gather in a way to form more pores. In Figure 4, here initially, porosity value remains the same with increase in power level and then suddenly decreases up to 19 kW and then becomes constant. In Figures 3 and 4, for both feed rates, 402009μm, 60 μm, and 80 μm powder particle sizes should be better for plasma spraying. In Figure 5, the constant feed rate is 18 gm/min. Here similar tendency is observed as previous plots, that is, Figures 3 and 4. Here it is found that only 40 μm and 60 μm power particles are better to spray.

Hence, if the feed rate will increase, the powder size should be lower for better plasma spray result.

4.3. Prediction Results of Porosity Based on Variation of Feed Rate. In plots of Figure 6, there are nearly similar results by increase in feed rate from 12 gm/min to 15 gm/min at 40 μm powder size. Figure 7 indicates that porosity is lower in the case of lower powder particle size, with lower feed rate at nearly 18 kW power level. But in case of 15 gm/min feed rate,

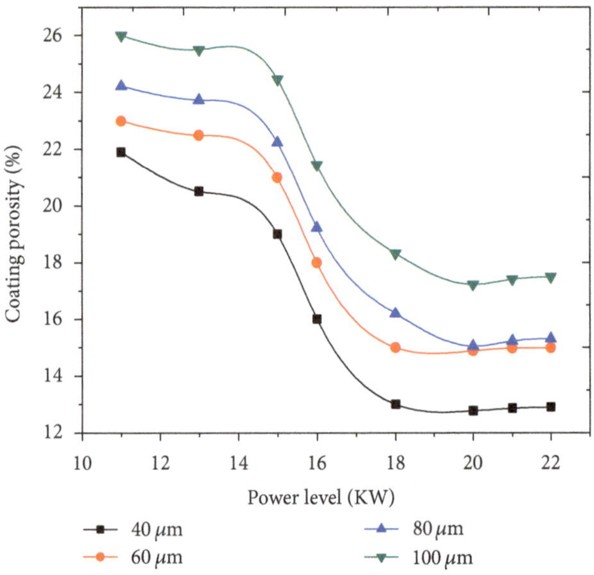

FIGURE 4: ANN results of coating porosity with respect to power level for different powder size at 15 gm/min feed rate.

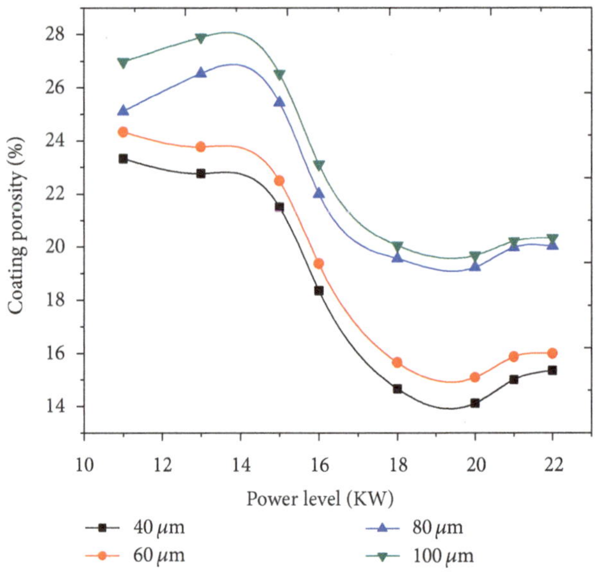

FIGURE 5: ANN results of coating porosity with respect to power level for different powder size at 18 gm/min feed rate.

porosity remains the same after 18 kW. For 18 gm/min, the plot is as the same as that of 12 gm/min, but porosity value is higher. In Figure 8, ANN result of porosity for 15 gm/min is lowest at higher power level. In Figure 9, it is observed that the better porosity results at a feed rate of 15 gm/min for 100 μm powder size. In these plots, after 19 kW power level, the peaks are increases due to the start of vaporization of powder particle.

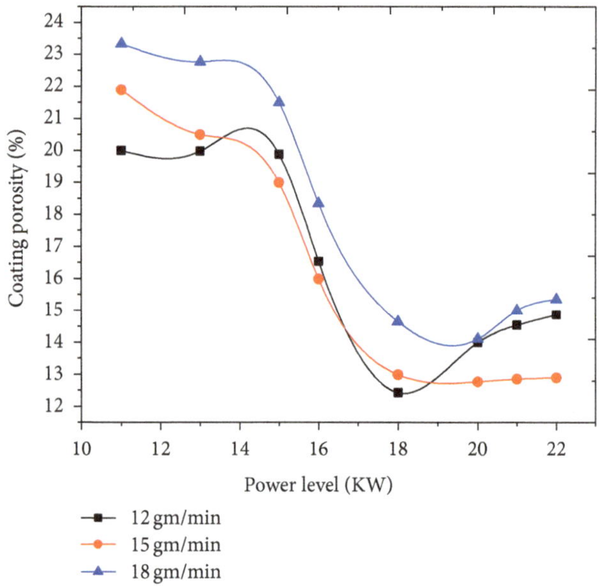

FIGURE 6: ANN results of coating porosity with respect to power level for different feed rate by keeping powder size 40 μm.

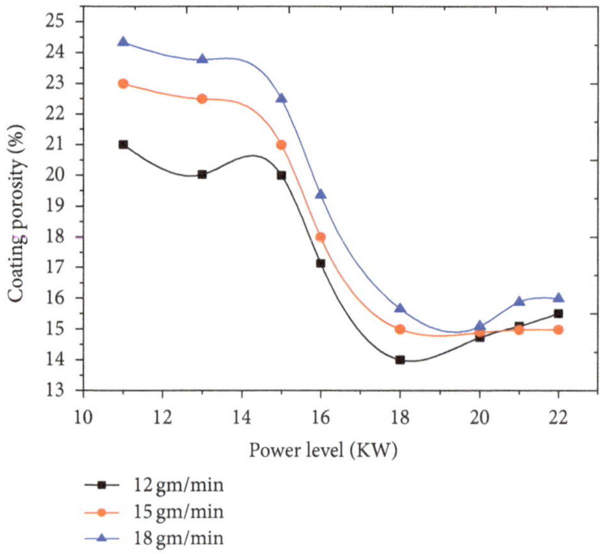

FIGURE 7: ANN results of coating porosity with respect to power level for different feed rate by keeping powder size 60 μm.

5. Conclusions

ANN technique is found to be a convenient and reliable means for optimizing the parameters of thermal spray coatings. The results obtained from ANN have been complemented with those obtained experimentally. From porosity (percentage) value, it is indicated that neural networks can yield fairly accurate results and can be used as a practical tool in plasma spraying manufacturing process. ANN provides the ability to solve the mapping problems much faster than conventional methods.

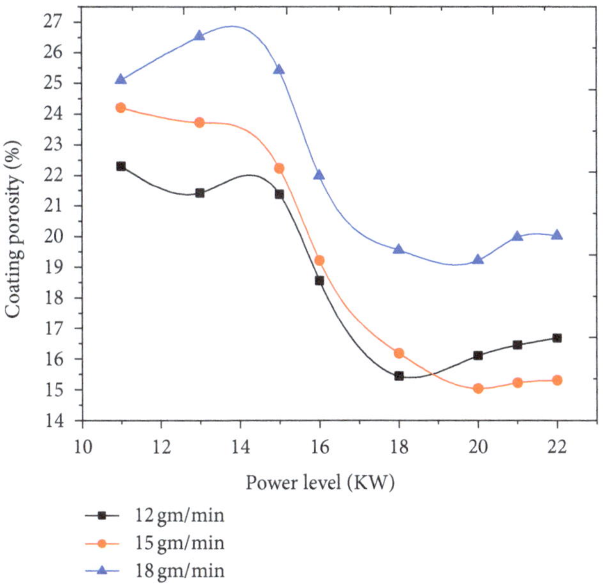

FIGURE 8: ANN results of coating porosity with respect to power level for different feed rate by keeping powder size 80 μm.

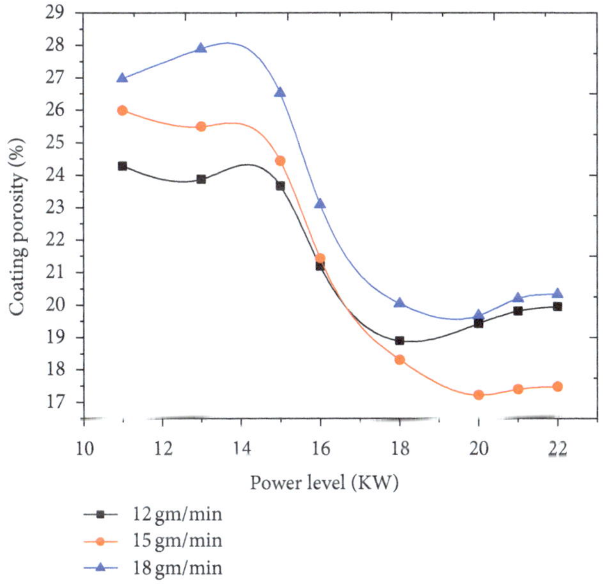

FIGURE 9: ANN results of coating porosity with respect to power level for different feed rate by keeping powder size 100 μm.

Conflict of Interest

It is to inform the reader that the "VOIS image analysis software" used in this analysis is being licensed by the authors. Their coauthor Professor S. C. Mishra is having the image analyzer system bearing this software. So there is no conflict of interest of the software.

References

[1] B. Friedrich-Wilhelm, A. Laarmann, and T. Wenz, *Modern Surface Technology*, pp. 254–259, John Wiley & Sons, New York, NY, USA, 2nd edition, 2007.

[2] D. Wixson, "Thermal sprayed-deposits shield structures from corrosion," *Indian Welding Journal*, vol. 88, no. 7, pp. 67–69, 2009.

[3] H. Shahriar, *Design of experiment analysis of high velocity oxy-fuel coating of hydroxyapatite [master degree Thesis]*, School of Mechanical and Manufacturing Engineering, Dublin City University, Dublin, Ireland, 2009.

[4] E. Pfender, in *Thermal Spray: Advances in Coatings Technology*, D. L. Houck, Ed., vol. 14, pp. 14–17, ASM International, Materials Park, Ohio, USA, 1987.

[5] W. X. Guo, "A method to calculate the evaporation rate of thermal cathode," *Journal of Electronics*, vol. 6, p. 166, 1984.

[6] J. R. Davis, *Handbook of Thermal Spray Technology*, Thermal Spray Society Training Committee, ASM International, Materials Park, Ohio, USA, 2004.

[7] M. Zhang, X. Wang, J. Luo, Q. Zhao, and X. Liao, "Influence of plasma spraying on the performance of oxide cathodes," *IEEE Transactions on Electron Devices*, vol. 58, no. 7, pp. 2143–2148, 2011.

[8] R. B. Heimann, *Plasma Spray Coating: Principles and Applications*, WILEY-VCH, New York, NY, USA, 2008.

[9] P. Fauchais, "Understanding plasma spraying," *Journal of Physics D*, vol. 37, pp. 86–108, 2004.

[10] C. Moreau, P. Gougeon, M. Lamontagne et al., *Thermal Spray Industrial Applications*, ASM International, Materials Park, Ohio, USA, 1994.

[11] S. C. Mishra and A. Satapathy, "Neural network analysis for deposition of nickel-aluminide coatings on steel by plasma spraying," in *Proceedings of the 21st National Symposium of Plasma Science Society of India*, MNIT, December 2006.

[12] S. S. Sablani, O. D. Baik, and M. Marcotte, "Neural networks for predicting thermal conductivity of bakery products," *Journal of Food Engineering*, vol. 52, no. 3, pp. 299–304, 2002.

[13] S. Deshpande, A. Kulkarni, S. Sampath, and H. Herman, "Application of image analysis for characterization of porosity in thermal spray coatings and correlation with small angle neutron scattering," *Surface and Coatings Technology*, vol. 187, no. 1, pp. 6–16, 2004.

[14] M. Friis and C. Persson, "Process window for plasma spray processes," in *Thermal Spray 2001: New Surfaces for a New Millenium*, C. C. Berndt, K. A. Kohr, and E. F. Lugscheider, Eds., pp. 1313–1319, ASM International, Materials Park, Ohio, USA, 2001.

[15] S. Kundas, "Computer simulation of plasma spraying processes," in *Thermal Spray 2001: New Surfaces for a New Millenium*, pp. 925–932, ASM International, Materials Park, Ohio, USA, 2001.

[16] P. Fauchais, G. Montavon, M. Vardelle, and J. Cedelle, "Developments in direct current plasma spraying," *Surface & Coatings Technology*, vol. 201, no. 5, pp. 1908–1921, 2006.

[17] H. Çetinel, M. Toparli, and L. Özsoyeller, "Finite element based prediction of the microstructural evolution of steels subjected to the Tempcore process," *Mechanics of Materials*, vol. 32, no. 6, pp. 339–347, 2000.

[18] V. Rao and H. Rao, *C++ Neural Networks and Fuzzy Systems*, BPB, New Delhi, India, 2000.

[19] A. Verikas and M. Bacauskiene, "Using artificial neural networks for process and system modelling," *Chemometrics and Intelligent Laboratory Systems*, vol. 67, no. 2, pp. 187–191, 2003.

[20] A. Behera and S .C. Mishra, "Dependence of adhesion strength of plasma spray on coating surface properties," *Journal of Materials & Metallurgical Engineering*, vol. 2, no. 1, pp. 23–30, 2012.

[21] P. Duwez, "Structure and properties of alloys rapidly quenched from the liquid state," *ASM Transactions*, vol. 60, pp. 607–633, 1967.

5

Dielectric Relaxation and Hopping Conduction in $La_2NiO_{4+\delta}$

Woo-Hwan Jung

Division of Electrical Engineering, Howon University, 64, Howondae 3 gil, Impi, Gunsan 573-930, Republic of Korea

Correspondence should be addressed to Woo-Hwan Jung; jungwoohwan@gmail.com

Academic Editor: Iwan Kityk

An ac conductivity as well as dielectric relaxation property of $La_2NiO_{4.1}$ is reported in the temperature range of 77 K–130 K and in the frequency range of 20 Hz–1 MHz. Complex impedance plane plots show that the relaxation (conduction) mechanism in this material is purely a bulk effect arising from the semiconductive grain. The relaxation mechanism has been discussed in the frame of electric modulus spectra. The scaling behavior of the modulus suggests that the relaxation mechanism describes the same mechanism at various temperatures. The logarithmic angular frequency dependence of the loss peak is found to obey the Arrhenius law with the activation energy of ~0.09 eV. The frequency-dependent electrical data are also analyzed in the frame of ac conductivity formalism. The ac conductivity has been found to follow a power-law behavior at a limited temperature and frequency region where Anderson localization plays a significant role in the transport mechanism for $La_2NiO_{4.1}$.

1. Introduction

Perovskite oxides with high dielectric constant play an important role in microelectronics [1] and have been used as memory devices and capacitors. In the literature, the dielectric relaxation with extremely high permittivity has been reported in many materials, in which the contribution of conduction carriers to dielectric polarization may play an important role. However, the explanation of this phenomenon is controversial based on different models [2, 3]. The electron localized on an ion can cause displacements of neighboring ions from their positions in the crystal lattice. The quasi-particle formed by an electron and corresponding lattice displacements is called polaron [4, 5].

Crystal chemical considerations indicate that among the ternary transition metal oxides of the type A_2BO_4, only La_2NiO_4 should be stable in the K_2NiF_4 structure [6–8]. The natures and properties of La_2NiO_4 are highly sensitive to the content of excess oxygen. In fact, $La_2NiO_{4+\delta}$ exhibits the metal-insulator transition [8]. Although $La_2NiO_{4+\delta}$ remains semiconducting nature up to $\delta = 0.25$, hole doping due to excess oxygen induces increasing structural distortion and changes electronic structures which modify the band gap width [9]. $La_2NiO_{4+\delta}$ exhibits semiconducting behavior at low temperatures with thermally activated or variable range hopping-type conductivity [10, 11]. The previous experimental results suggest a high possibility that the major carrier at low temperature in $La_2NiO_{4+\delta}$ is a polaron [10, 11]. However, the details of the transport kinetics are still unknown. Although there are many ways of addressing polaron dynamics, dielectric relaxation measurements are useful metrics because the hopping process of polaron has a high probability of involving a dielectric relaxation process [12–16]. Furthermore, many perovskite materials have shown high dielectric or ferroelectric properties and the dipole unit in these systems plays an important role in the electronic transport phenomena [17, 18].

In this paper, we investigate the electrical relaxation properties of K_2NiF_4 type $La_2NiO_{4+\delta}$ ceramic in the temperature range 77 K–300 K and frequency rang 20 Hz–1 MHz by means of dielectric spectroscopy. These studies may give valuable information on the electrical properties, which are very important for practical applications of perovskite oxides.

2. Experimental

$La_2NiO_{4+\delta}$ polycrystalline ceramic was prepared by conventional solid state synthesis technique. La_2O_3 (>99.99% purity) and NiO (>99.9% purity) powders were used. First, the mixed powder was heated in air at 1273 K for 12 h. After being ground and mixed well, the powder was heated again in air at 1373 K for 12 h. Then the powder was pressed into the pellets and

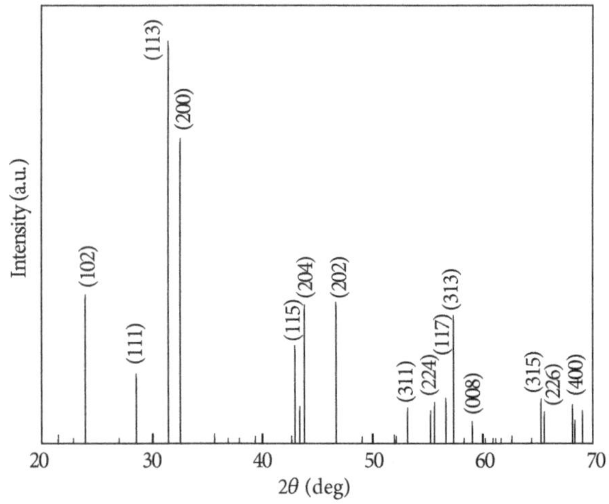

FIGURE 1: X-ray diffraction pattern of $La_2NiO_{4.1}$.

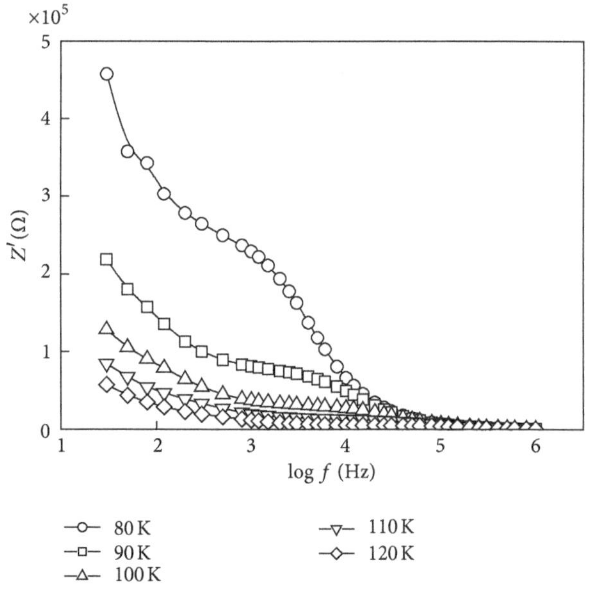

FIGURE 2: Frequency dependence of the real part (Z') of impedance at various temperatures for $La_2NiO_{4.1}$.

sintered finally in pure flowing oxygen at 1423 K for 12 h. The powder diffraction using a Cu Kα X-ray indicates the tetragonal structure with the lattice constants $a = 3.8615$, $c = 12.6484$ Å. Figure 1 shows the XRD spectrum of the sintered La_2NiO_4 sample recorded at room temperature. The oxygen content of the sample was determined by iodometric titration. The experimental error of the oxygen content is within ±3%. δ is found to be 0.1. The chemical composition was confirmed by inductively coupled plasma (ICP). The composition was found to be nearly identical to the nominal one within the accuracy of ICP.

Capacitance and impedance were obtained as functions of temperature by four-terminal pair ac impedance measurement method using a HP 4284A precision LCR meter. The measured values of capacitance and impedance were correlated by calibrating capacitance and resistance of leads to zero. Flat surfaces of the specimens were coated with an In-Ga alloy in the ration 7 : 3 by a rubbing technique for an electrode.

3. Results and Discussion

Figure 2 shows the frequency dependence of real part (Z') of impedance of $La_2NiO_{4.1}$ as a function of temperature. In Figure 2, Z' decreases with an increase in frequency for all temperatures implies relaxation process in the specimens. The magnitude of Z' decreases at lower frequencies with an increasing temperature which indicates increase in the ac conductivity and this can be also attributed to the fact that Z' values decrease with increase of temperature indicating reduction of grains, grain boundaries, and electrode interface resistance. The merge of Z' values at higher frequencies is observed, which may be due to the release of space charge [19, 20].

Figures 3(a) and 3(b) show the frequency dependence of imaginary (Z'') part of impedance of $La_2NiO_{4.1}$ as a function of temperature. The pattern of variation is characterized by the occurrence of peaks with asymmetric broadening which indicates relaxation in the materials. Double peaks are observed at 130 K. The peaks in higher and lower frequency regions are found due to effect of bulk and grain boundaries, respectively. Thus, the observed lower frequency region peaks above 140 K are due to the effect of grain boundaries. The peak values of Z'' are observed to decrease with increasing temperature. It indicates an increase of loss in the resistive property of the materials. Shifting of peaks toward the higher frequency sides with increasing temperatures indicates the temperature dependence of the relaxation time. Furthermore, the broadening of the peaks with rise in temperature indicates nonunique relaxation timescale, whose variance/distribution widens at higher frequency side [19, 21].

Figure 4 represents the variation of Z''/Z''_{max} with $\log(f/f_{max})$ at different temperatures. It shows the scaling behavior of the materials. The overlap of the curves for all the temperatures into a single master curve indicates that relaxation described the same mechanism at different temperatures in this material.

We have adopted the modulus formalism to study the relaxation mechanism. The usefulness of the modulus representation in the analysis of the relaxation properties has been demonstrated for poly crystalline ceramics. The advantage of adopting complex electric modulus formalism is that it can discriminate against electrode polarization and grain boundary conduction mechanism [22]. It is also suitable in detecting bulk phenomena properties as apparent conductivity relaxation time [23]. The other advantage of the electric modulus is that the electrode effect can be suppressed [23]. Electric modulus corresponds to the relaxation of electric field in the material when the electric displacement remains constant. Thus the electric modulus represents the real dielectric relaxation process. The complex electric modulus $M^* = M' + jM''$ can be expressed as [24]

$$M^*(\omega) = M_\infty \left[1 - \int_0^\infty \left(\frac{-d\phi(t)}{dt} \right) \exp(-j\omega t)\, dt \right], \quad (1)$$

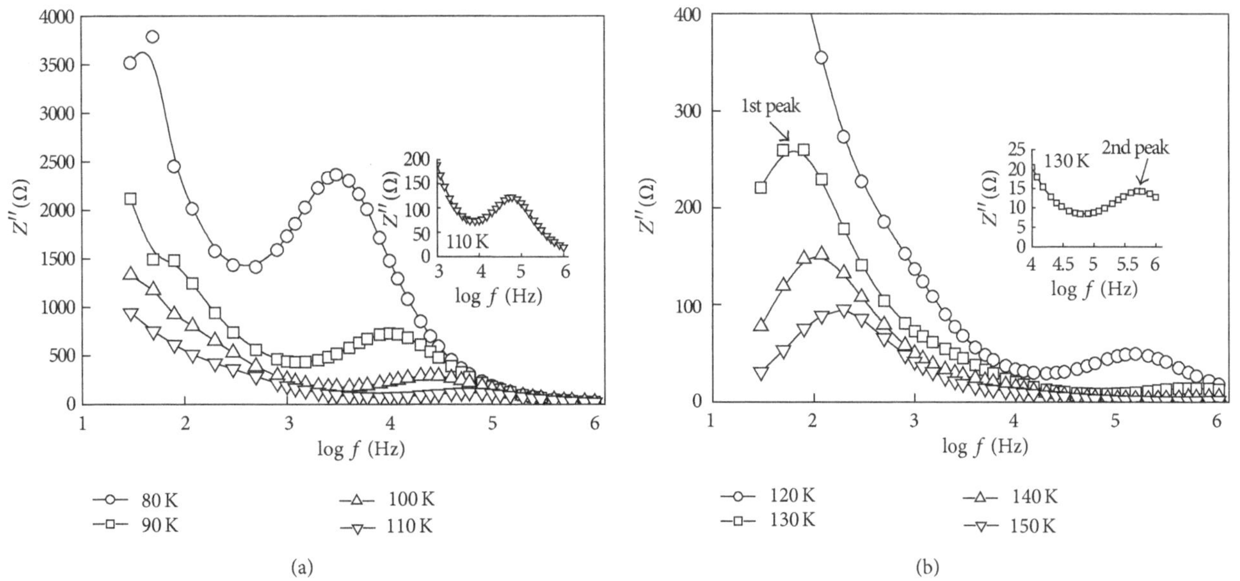

FIGURE 3: Frequency dependence of the imaginary part (Z'') of impedance at various temperatures for $La_2NiO_{4.1}$; (a) $T < 120$ K (b) $T > 120$ K.

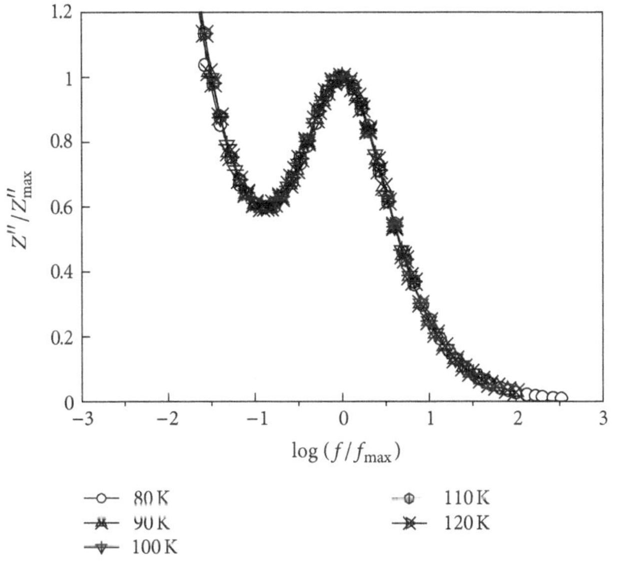

FIGURE 4: Scaling behavior of Z'' at various temperatures for $La_2NiO_{4.1}$.

where M' and M'' are real and imaginary part of M^*, $\omega = 2\pi f$ is the angular frequency, $M_\infty = 1/\varepsilon_\infty$ is the asymptotic value of $M'(\omega)$, and $\phi(t) = \exp[-(t/\tau_M)\beta]$ represents time evolution of the electric field within the material where β $(0 < \beta < 1)$ is the stretched exponent and τ_M is the conductivity relaxation times. The M^* formalism is widely used to study conductivities of materials. The variation of imaginary component M'' as a function of frequency at different temperatures provides useful information of the charge transport mechanism such as electrical transport and conductivity relaxation process. Peaks are observed in the frequency spectra of M'' which indicate the existence of conductivity relaxation process. The frequencies at which the peaks in M'' spectroscopic plots are observed follow the relation $\omega_{max}\tau_m = 1$, where ω_{max} is the angular frequency corresponding to the peak maximum. It is typically observed that τ_m follows the Arrhenius law given by [22, 23]

$$\tau_M = \tau_{M0} \exp\left(E/k_B T\right), \tag{2}$$

where τ_{M0} is the preexponential factor and E is the activation energy.

The frequency dependence of $M'(\omega)$ is shown in Figure 5(a). At low frequency, the value of $M'(\omega)$ decreases as the temperature increases. Also at a given temperature, $M'(\omega)$ increases with increase in the frequency and takes almost a constant value at higher frequency. At low frequency and in the high temperature region, $M'(\omega)$ approaches zero, confirming an appreciable electrode and/or ionic polarization [25]. Figure 5(b) shows the variation in $M''(\omega)$ with frequency at different temperatures where the peaks are shifted toward higher frequencies with increasing temperature. We have fitted the relaxation times (τ_m) obtained from the peak frequencies with inverse of temperature according to the Arrhenius law given by (2) where the fit is shown by solid line in inset of Figure 6. It is observed that the value of relaxation time is found to be decreasing with increasing temperature which indicates semiconductor behavior of the compound. For a typical lattice dipole relaxation, τ_{M0} is on the order of $10^{-10} \sim 10^{-13}$ s [18]. The value of E and τ_{M0} obtained from the Arrhenius fit is $E \approx 0.09$ eV and $\tau_{M0} \approx 2 \times 10^{-10}$ s. This experimental result indicates that the dielectric loss anomaly (polaron-like relaxation) observed in this compound is arising from the interaction between free charges and lattice, which induces the local dipole. The lattices distortions introduced by the excess oxygen and associated vacancies are sufficient to produce more than one off-center equilibrium position for Ni^{3+} ions. The observed relaxation is associated with thermal activation

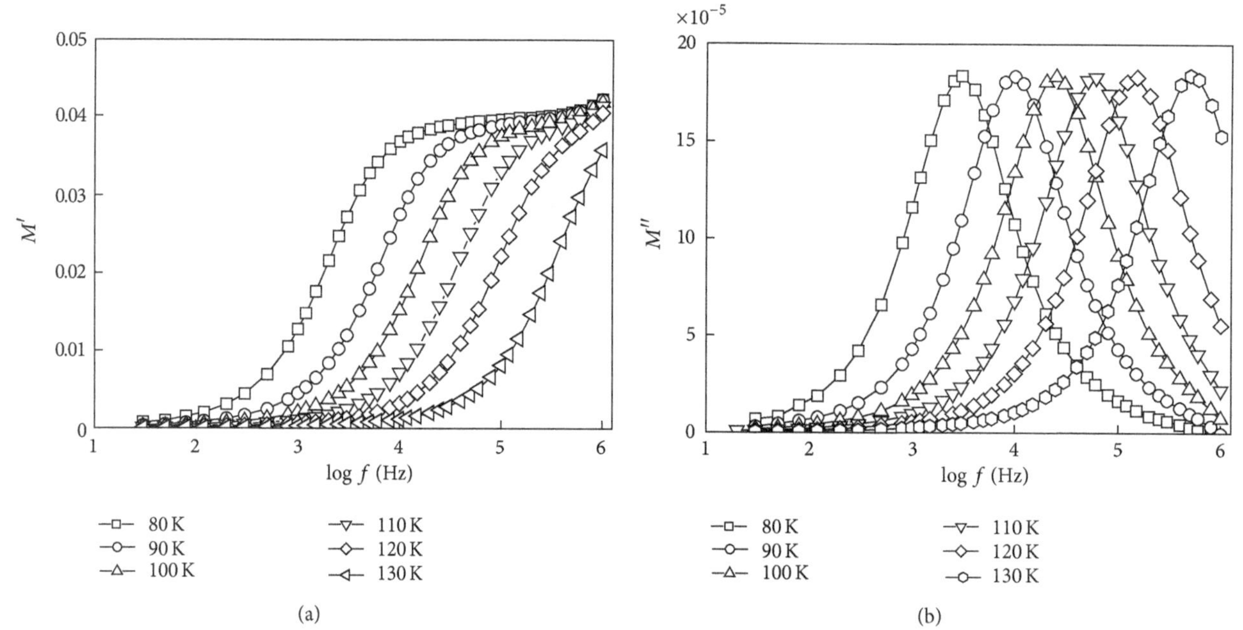

FIGURE 5: (a) Frequency dependence of the real part (M') of electric modulus at various temperatures for $La_2NiO_{4.1}$. (b) Frequency dependence of imaginary part (M'') of electric modulus at various temperatures for $La_2NiO_{4.1}$.

motion between these equivalent potential minima [26]. Under the action of an ac field, the charge may hop between equivalent available lattice sites depending on the crystal symmetry. This hopping is equivalent to the reorientation of an electric dipole, which gives frequency-dependent, complex, dielectric properties exhibiting Debye-type behavior. Thus the experimental results suggest that spectral intensity of the electric modulus observed in $La_2NiO_{4.1}$ is thermally activated like dielectric relaxation occurring in the other various materials in which the process of hopping of small polaron dominates the electrical transport [18, 20, 26–28].

Scaling of the electric modulus can give further information about the dependence of the relaxation dynamics on the temperature, structure, and also on the concentration of the charge carrier [29]. Figure 6 shows the scaling results at different temperatures where $M''_{\max}$ and $(f/f_{\max})$ are used as the scaling parameters for M'' and f, respectively. Clearly all modulus spectra can be seen to completely overlap and are scaled to a single master curve indicating that relaxation dynamics does not change with temperature.

In Figure 7, the variation of normalized parameter $(M''/M''_{\max})$ and $(\tan\delta/\tan\delta_{\max})$ as a function of logarithmic angular frequency measured at 100 K is shown. The overlapping peak position of $(M''/M''_{\max})$ and $(\tan\delta/\tan\delta_{\max})$ curves is an evidence of delocalized or long-range relaxation. However, for the present system the $(M''/M''_{\max})$ and $(\tan\delta/\tan\delta_{\max})$ peaks do not overlap but are very close suggesting the components from both long-range and localized relaxation [29].

Figure 8 shows the variation of ac conductivity (σ_{ac}) as a function of frequency at different temperatures. It is obvious that the conductivity increases with increasing frequency and increasing temperature. The observed behavior is typical of the existence of polarons in the system [13, 16–19, 30]. The very basic fact about ac conductivity in complex perovskite oxides is that conductivity is an increasing function of frequency (any hopping model has this feature). Anderson-localized charge carriers contribute to conductivity by hopping processes which lead to an increase in complex conductivity with frequency (ω). This behavior can be described by using a power law with exponent $s < 1$ which is usually represented as "universal dielectric response" (UDR) [31–33]. By considering a dc conductivity term, the real part of the ac conductivity is described as

$$\sigma(\omega) = \sigma(0) + A\omega^s, \tag{3}$$

where $\sigma(\omega)$ is the total conductivity, $\sigma(0)$ the frequency independent conductivity, that is, dc conductivity, A depends on the temperature, and s is the dimensionless frequency exponent. In case of measurement at sufficiently high frequency, a crossover from power law, $\sigma \propto \omega^s$, to a linear increase of ω, $\sigma \propto \omega$, is frequently observed in the variety of materials which is defined as the "second universality" (SU) [31–33]. The microscopic picture of the origin of SU region at high frequency is still not established.

It is seen from Figure 8 that at low frequency a step-like feature is observed possibly due to the grain boundary effect where a corresponding relaxation peak is observed in the Z'' spectrum. With further increasing frequency, a crossover to the linear $\sigma(\omega) \propto \omega$ dependence above ~10^5 Hz is observed. Linearity of the frequency dependence is shown by broken line in the figure. The linear SU region is shifted toward higher frequencies with increasing temperature and consequently the SU region is shifted out of the frequency window above 130 K. The UDR and SU contributions to the intrinsic ac conductivity in the present investigation are consistent with results for single crystal or polycrystalline of $LaMnO_3$ [13, 18].

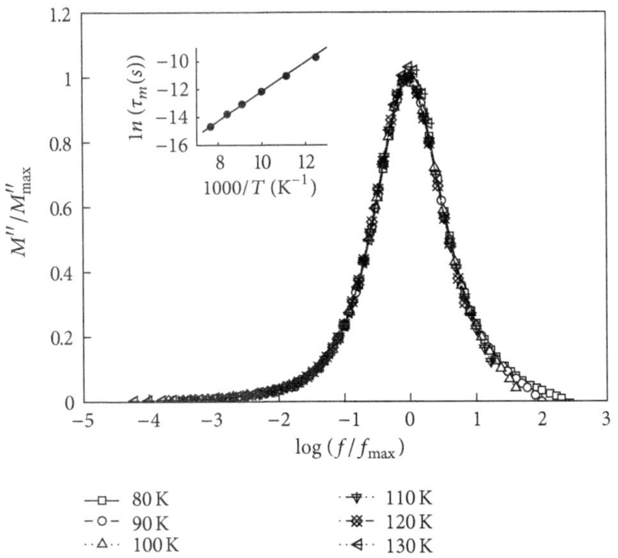

FIGURE 6: Scaling behavior of M'' at various temperatures for $La_2NiO_{4.1}$. The temperature dependence of the most probable relaxation time (τ_m) obtained from the frequency-dependent plot of M'' is shown in the inset where the symbols are the experimental data and the solid line is the least-squares straight line fit.

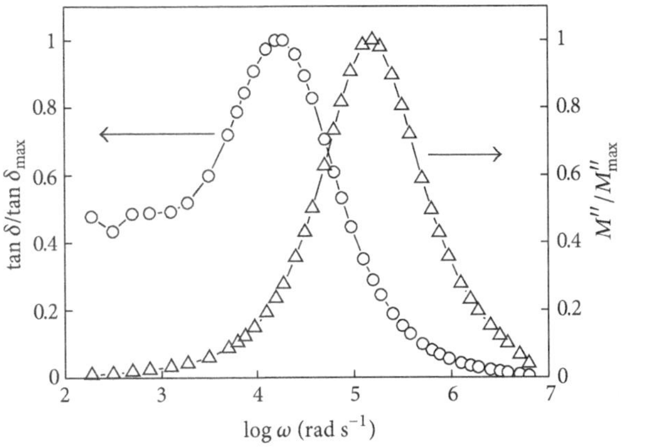

FIGURE 7. Logarithmic angular frequency dependence of normalized peaks. $\tan\delta/\tan\delta_{max}$ and M''/M''_{max} for $La_2NiO_{4.1}$ at 100 K.

For $La_2NiO_{4.1}$, $\sigma(\omega)$ increases with increasing frequency and therefore the observed ac conductivity must be related to the bound carriers trapped in the specimen. The frequency dependence of ac conductivity at various temperatures resembles that of hopping type conduction between these trap centers. Ni^{3+} is one kind of charge trap. Hopping of charge carriers among the trap leaves situated in the band gap of the material may give rise to a frequency dependence ac conductivity with a frequency dependence ~ ω^s ($s < 1$). Since UDR contribution to the ac conductivity is found in a very small region of the frequency, a large error is noted while analyzing the ac conductivity by power law for $La_2NiO_{4.1}$. The temperature dependence of the frequency exponent s is shown in the inset of Figure 8. For $La_2NiO_{4.1}$, s varies between 1.0 and 0.8. There are vast numbers of theoretical approaches to deduce this behavior form the microscopic transport properties of various classes of materials [34]. The small polaron tunneling (SPT) model, which corresponds to the variable range hopping model for nonzero frequencies, predicts a temperature-independent frequency exponent s near 0.8 [31–33]. The results of a previous experiment showed that the dc conduction mechanism of $La_2NiO_{4+\delta}$ is well described by variable range hopping conduction [10, 11]. This suggests that the SPT model is an appropriate mechanism for the low temperature ac conductivity of the present materials. Unfortunately, due to the small frequency range where the ω^s power law is seen, an unambiguous determination of ac conduction mechanism is difficult but the overall response is very similar to that found for Sr-doped $LaMnO_3$ [13, 18] which is a well-described SPH model.

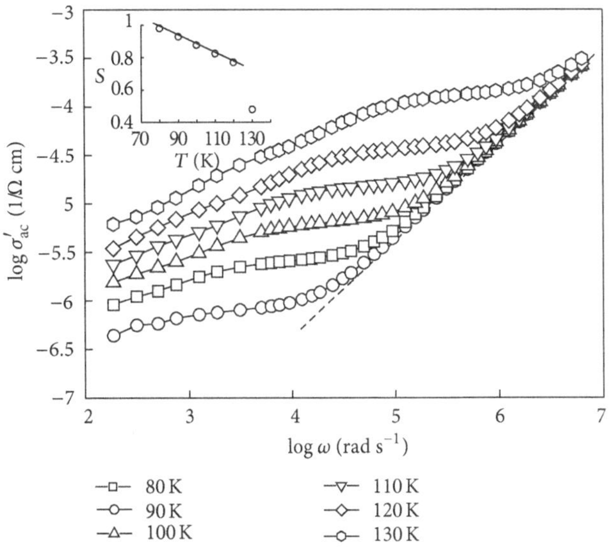

FIGURE 8: Frequency dependence of ac conductivity at some selected temperature. Inset: the temperature dependence of the exponent parameter, s.

4. Conclusions

The frequency dependence of electric modulus, impedance, and ac conductivity of $La_2NiO_{4.1}$ ceramic synthesized by solid-state reaction method is investigated in the temperature range from 80 K to 130 K and frequency range 20 Hz to 1 MHz. The relaxation mechanism has been discussed in the framework of electric modulus. The shifting of the peak position to the higher frequency side with increasing temperature indicates the polaron transport in this material. The logarithmic angular frequency dependence of loss peak is found to obey the Arrhenius law with the activation energy of 0.09 eV. This value of activation energy suggests that the bulk conduction maybe due to polaron hopping based on electron carriers. The scaling behavior of the electric modulus and impedance suggest that relaxation mechanism describes the same mechanism at various temperatures. The frequency-dependent ac conductivity data are also analyzed in the framework of conductivity formalism.

Acknowledgment

This work was supported by Howon University.

References

[1] R. J. Cava, "Dielectric materials for applications in microwaves communications," *Journal of Materials Chemistry*, vol. 11, no. 1, pp. 54-62, 2001.

[2] S. Komine, "Dielectric relaxation study on $Ce_{0.9}$ $Gd_{0.1}$ $O_{1.95}$ ceramics," *Physica B*, vol. 392, no. 1-2, pp. 348–352, 2007.

[3] E. Iguchi and K. J. Lee, "Dielectric relaxations in $SrTiO_3$ doped with La_2O_3 and MnO_2 at low temperatures," *Journal of Materials Science*, vol. 28, no. 21, pp. 5809–5813, 1993.

[4] E. Iguchi, H. Satoh, H. Nakatsugawa, and F. Munakata, "Correlation between hopping conduction and transferred exchange interaction in $La_2NiO_{4+\delta}$ below 300 K," *Physica B*, vol. 270, no. 3-4, pp. 332–340, 1999.

[5] C. N. R. Rao, D. J. Buttrey, N. Otsuka et al., "Crystal structure and semiconductor-metal transition of the quasi-two-dimensional transition metal oxide, La_2NiO_4," *Journal of Solid State Chemistry*, vol. 51, no. 2, pp. 266–269, 1984.

[6] K. K. Singh, P. Ganguly, and J. B. Goodenough, "Unusual effects of anisotropic bonding in Cu(II) and Ni(II) oxides with K_2NiF_4 structure," *Journal of Solid State Chemistry*, vol. 52, no. 3, pp. 254–273, 1984.

[7] M. Sayer and P. Odier, "Electrical properties and stoichiometry in La_2NiO_4," *Journal of Solid State Chemistry*, vol. 67, no. 1, pp. 26–36, 1987.

[8] D. C. Johnston, J. P. Stokes, D. P. Goshorn, and J. T. Lewandowski, "Influence of oxygen defects on the physical properties of La_2CuO_{4-y}," *Physical Review B*, vol. 36, no. 7, pp. 4007–4010, 1987.

[9] Th. Strangfeld, K. Westerholt, and H. Bach, "Electronic charge transport and magnetism of the system $(La_{1-x}Sr_x)_2NiO_{4+\delta}$," *Physica C*, vol. 183, no. 1–3, pp. 1–10, 1991.

[10] T. Katsufuji, T. Tanabe, T. Ishikawa, Y. Fukuda, T. Arima, and Y. Tokura, "Optical spectroscopy of the charge-ordering transition in $La_{1.6}Sr_{0.33}NiO_4$," *Physical Review B*, vol. 54, no. 20, pp. R14230–R14233, 1996.

[11] J. M. Bassat, J. P. Loup, and P. Odier, "Progressive change with T from hopping to random phase propagation in $La_{2-x}NiO_{4-\delta}$," *Journal of Physics Condensed Matter*, vol. 6, no. 40, pp. 8285–8293, 1994.

[12] A. Mansingh, J. M. Reyes, and M. Sayer, "Dielectric behaviour in a vanadium phosphate glass," *Journal of Non-Crystalline Solids*, vol. 7, no. 1, pp. 12–22, 1972.

[13] A. Seeger, P. Lunkenheimer, J. Hemberger et al., "Charge carrier localization in $La_{1-x}Sr_xMnO_3$ investigated by ac conductivity measurements," *Journal of Physics Condensed Matter*, vol. 11, no. 16, pp. 3273–3290, 1999.

[14] H. Jhans, D. Kim, R. J. Rasmussen, and J. M. Honig, "ac-conductivity measurements on $La_2NiO_{4+\delta}$," *Physical Review B*, vol. 54, no. 16, pp. 11224–11229, 1996.

[15] W. H. Jung, J. H. Sohn, J. H. Lee, J. H. Sohn, M. S. Park, and S. H. Cho, "Alternating-current electrical properties of $CaMnO_3$ below the Neel temperature," *Journal of the American Ceramic Society*, vol. 83, no. 4, pp. 797–801, 2000.

[16] C. Ang, Z. Yu, Z. Jing, P. Lunkenheimer, and A. Loidl, "Dielectric spectra and electrical conduction in Fe-doped $SrTiO_3$," *Physical Review B*, vol. 61, no. 6, pp. 3922–3926, 2000.

[17] G. Chern, W. K. Hsieh, M. F. Tai, and K. S. Hsung, "High dielectric permittivity and hole-doping effect in $La_{1-x}Sr_xFeO_3$," *Physical Review B*, vol. 58, no. 3, pp. 1252–1260, 1998.

[18] A. Karmakar, S. Majumdar, and S. Giri, "Polaron relaxation and hopping conductivity in $LaMn_{1-x}Fe_xO_3$," *Physical Review B*, vol. 79, no. 9, pp. 094406–094413, 2009.

[19] M. Ram, "A.c. Conductivity and dielectric properties of $LiNi_{3/5}Cu_{2/5}VO_4$ ceramics," *Physica B*, vol. 405, no. 5, pp. 1359–1361, 2010.

[20] Y. Q. Lin, Y. J. Wu, X. M. Chen, S. P. Gu, J. Tong, and S. Guan, "Dielectric relaxation mechanisms of $BiMn_2O_5$ ceramics," *Journal of Applied Physics*, vol. 105, no. 5, Article ID 054109, 2009.

[21] R. S. Kumar and K. Hariharan, "AC conductivity and electrical relaxation studies on 10CuI-60AgI-30V_2O_5 glasses," *Materials Chemistry and Physics*, vol. 60, no. 1, pp. 28–38, 1999.

[22] P. K. Bajpai and K. N. Singh, "Dielectric relaxation and ac conductivity study of $Ba(Sr_{1/3}Nb_{2/3})O_3$," *Physica B*, vol. 406, no. 6-7, pp. 1226–1232, 2011.

[23] P. Kumar, B. P. Singh, T. P. Sinha, and N. K. Singh, "AC conductivity and dielectric relaxation in $Ba(Sm_{1/2}Nb_{1/2})O_3$ ceramic," *Physica B*, vol. 406, no. 2, pp. 139–143, 2011.

[24] H. Mahamoud, B. Louati, F. Hlel, and K. Guidara, "Impedance and modulus analysis of the $(Na_{0.6}Ag_{0.4})_2PbP_2O_7$ compound," *Journal of Alloys and Compounds*, vol. 509, no. 20, pp. 6083–6089, 2011.

[25] D. K. Mahato, A. Dutta, and T. P. Sinha, "Impedance spectroscopy analysis of double perovskite Ho_2NiTiO_6," *Journal of Materials Science*, vol. 45, no. 24, pp. 6757–6762, 2010.

[26] W. H. Jung, "Investigations of the charge transport properties in $(LaMn)_{0.956}O_3$ at low temperatures," *Journal of Physics Condensed Matter*, vol. 18, no. 29, pp. 6691–6698, 2006.

[27] M. Idrees, M. Nadeem, and M. M. Hassan, "Investigation of conduction and relaxation phenomena in $LaFe_{0.9}Ni_{0.1}O_3$ by impedance spectroscopy," *Journal of Physics D*, vol. 43, no. 15, Article ID 155401, 2010.

[28] A. Dutta and T. P. Sinha, "Impedance spectroscopy study of $BaMb_{1/3}Nb_{2/3}O_3$: frequency and time domain analyses," *Physica B*, vol. 405, no. 6, pp. 1475–1479, 2010.

[29] S. Saha and T. P. Sinha, "Low-temperature scaling behavior of $BaFe_{0.5}Nb_{0.5}O_3$," *Physical Review B*, vol. 65, no. 13, Article ID 134103, pp. 1341031–1341037, 2002.

[30] A. Levstik, C. Filipič, V. Bobnar, S. Drnovšek, J. Holc, and M. Kosec, "Polaron conductivity mechanism in $0.5Pb(Zr_{0.5}75Ti_{0.425})O_{30.5}Pb(Fe_{2/3}W_{1/3})O_3$," *Physica B*, vol. 405, no. 20, pp. 4271–4273, 2010.

[31] S. R. Elliott, "A.c. conduction in amorphous chalcogenide and pnictide semiconductors," *Advances in Physics*, vol. 36, no. 2, pp. 135–218, 1987.

[32] W. K. Lee, J. F. Liu, and A. S. Nowick, "Limiting behavior of ac conductivity in ionically conducting crystals and glasses: a new universality," *Physical Review Letters*, vol. 67, no. 12, pp. 1559–1561, 1991.

[33] A. S. Nowick, A. V. Vaysleyb, and B. S. Lim, "Further evidence for a "second universality" in alternating-current conductivity relaxation," *Journal of Applied Physics*, vol. 76, no. 7, pp. 4429–4431, 1994.

[34] A. R. Long, "Frequency—dependent loss in amorphous semiconductor," *Advances in Physics*, vol. V 31, no. 5, pp. 553–637, 1982.

6

Growth and Neutron Diffraction Investigation of $Ca_3NbGa_3Si_2O_{14}$ and $La_3Ga_{5.5}Nb_{0.5}O_{14}$ Crystals

I. A. Kaurova,[1] G. M. Kuz'micheva,[2] V. B. Rybakov,[3] A. Cousson,[4] O. Zaharko,[5] and E. N. Domoroshchina[2]

[1] *Moscow State Open University, 3 Entuziastov Street, Aleksandrov 601655, Russia*
[2] *Lomonosov Moscow University of Fine Chemical Technology, 86 Vernadskogo Prospekt, Moscow 119571, Russia*
[3] *Lomonosov State University, Vorobyovy Gory, Moscow 119992, Russia*
[4] *Laboratoire Leon Brillouin, Cea/Saclay, 91191 Gif-sur-Yvette Cedex, France*
[5] *Laboratory for Neutron Scattering, ETZ Zurich & Paul Scherrer Institute (PSI), CH 5232 Villigen, Switzerland*

Correspondence should be addressed to I. A. Kaurova; kaurchik@yandex.ru

Academic Editor: Necmettin Maraşlı

Langanite ($La_3Ga_{5.5}Nb_{0.5}O_{14}$, growth atmosphere: 99% Ar + 1% O_2) and kanigasite ($Ca_3NbGa_3Si_2O_{14}$, growth atmosphere: 100% Ar) crystals grown by the Czochralski technique in Ir crucibles along <0001> direction have been firstly investigated by neutron diffraction. The difference between the compositions of upper $(La_{2.935(2)}\square_{0.065})(Ga_{0.450}Nb_{0.550(3)})Ga_3(Ga_{1.965(4)}\square_{0.035})(O_{13.90(1)}\square_{0.10})$ and lower $(La_{2.940(1)}\square_{0.060})(Ga_{0.590}Nb_{0.410(2)})Ga_5(O_{13.82(1)}\square_{0.18})$ parts of orange langanite crystal was found. It was established that the colorless $Ca_3NbGa_3Si_2O_{14}$ crystal grown by using the single-crystal charge has the composition $(Ca_{2.95}\square_{0.05(1)})NbGa_3Si_2O_{14}$ and is less defective in comparison to the yellow one grown by using the charge prepared by conventional solid-state reaction. For $Ca_3NbGa_3Si_2O_{14}$ and $La_3Ga_{5.5}Nb_{0.5}O_{14}$ crystals the possibility of microtwin formation (two unit cells connected by the translation: 1/2 z) was revealed for the first time. It was found that the difference between the color of crystals is attributed to the qualitative differentiation of oxygen vacancies.

1. Introduction

Kanigasite ($Ca_3NbGa_3Si_2O_{14}$, **CNGS**) and langanite ($La_3Ga_{5.5}Nb_{0.5}O_{14}$, $La_3(Ga_{0.5}Nb_{0.5})Ga_5O_{14}$, **LGN**) crystals have langasite-type structure (sp.gr. $P321$, $Z = 1$) similar to langasite ($La_3Ga_5SiO_{14}$, $La_3Ga_4(GaSi)O_{14}$, **LGS**) and langatate ($La_3Ga_{5.5}Ta_{0.5}O_{14}$, $La_3(Ga_{0.5}Ta_{0.5})Ga_5O_{14}$, **LGT**) ones. Langasite family crystals possess a number of properties that ensure the successful application in piezo-, opto-, and acoustoelectronics. These properties include the absence of structural phase transitions up to the melting temperature, high thermal stability, high values of electromechanical coupling coefficients and acoustic Q, and low loss propagation of the elastic waves in the crystal. Langanite is characterized by the largest values of piezoelectric modules ($d_{11} = -7.41 \pm 0.2$ C/N; $d_{14} = 6.16 \pm 0.5$ C/N) [1] among the langasite family crystals, whereas the **CNGS** has high values of the acoustic quality factor ($Q = 56000$) due to the fact that every atom in its structure is located in an individual position. The crystallographic orientation with a zero TCF (temperature coefficient of frequency) at room temperature was found for this crystal. The traditional Czochralski method is the primary method of obtaining large-size **LGN** [2–7] and **CNGS** [6–16] single crystals up to 200 mm in diameter.

However, the use of these compounds depends on the composition (in other words, on the type and concentration of point defects), which does not coincide with the composition of the initial charge. There are quite a few works devoted to the structural investigation of **CNGS** [6, 8, 11] and **LGN** [6, 17, 18] crystals. All of them were performed using X-ray, and the occupancies were refined only in some cases. In the literature there are no results of neutron investigation of **LGN** and **CNGS** crystals, as the most correct to identify defects in the crystallographic positions, in particular, oxygen ones.

Neutron diffraction investigations of langasite [19] and langatate [20–22] single crystals have revealed that the point defects are present in all positions of the structure. In addition, the relationship between the crystal color and the oxygen vacancies contents has been established. Probably the other crystals with langasite-type structure, in particular **LGN** and **CNGS**, will have the same relationship.

The aim of this work is to determine the compositions of **CNGS** and **LGN** crystals grown by the Czochralski technique using neutron diffraction.

2. Experimental Technique

$Ca_3NbGa_3Si_2O_{14}$ and $La_3Ga_{5.5}Nb_{0.5}O_{14}$ crystals were grown by the Czochralski technique along <0001> direction in a Kristal-3M growth system in Ir crucible (h = 120 mm, d = 120 mm, the thickness of wall and the bottom s = 2 mm) using a pure Ar (**CNGS**) or 99-98% Ar + 1-2% O_2 (**LGN**) growth atmospheres. The pulling rate was 1–3 mm/h, and the crystal rotation rate was 1-10 rpm. The **CNGS** and **LGS** seeds were used for $Ca_3NbGa_3Si_2O_{14}$ and $La_3(Ga_{0.5}Nb_{0.5})Ga_5O_{14}$ crystals growth, respectively. Either crushed **CNGS** single crystal (crystal 1) or polycrystalline material prepared by conventional solid-state reaction (Al_2O_3 crucible, T ~ 1300°C, 6 hours) (crystal 2) was used for growth of two **CNGS** crystals. Starting materials were prepared by mixing of 99.99% pure La_2O_3, Ga_2O_3, Nb_2O_5, SiO_2, and $CaCO_3$ powders at the stoichiometric ratio. A single-crystal charge was used for $La_3(Ga_{0.5}Nb_{0.5})Ga_5O_{14}$ crystal growth.

The colorless (crystal 1) and yellow (crystal2) $Ca_3Nb Ga_3Si_2O_{14}$ crystals 50 mm in diameter with 80 mm long cylindrical portion [23], as well as the orange $La_3Ga_{5.5} Nb_{0.5}O_{14}$ crystal 30 mm in diameter with 100 mm long cylindrical portion [24], were grown by the Czochralski technique. All the crystals were transparent. The faceting of the crystals was represented by poorly pronounced faces of the hexagonal prism.

The **CNGS** samples under investigation represent the upper parts of colorless (*CNGS-1*) and yellow (*CNGS-2*) single crystals (Figures 1(a) and 1(b)). Orange **LGN** samples were cut from the top (*LGN-1, LGN-1′*) and bottom (*LGN-2*) of the crystal (Figures 2(a) and 2(b)). It should be noted that the upper parts (*LGN-1, LGN-1′*) had a more intense orange color.

The neutron diffraction analysis (NDA) of *LGN-1* (size: ~4.3 × 4 × 5 mm) and *CNGS-2* (size: ~4 × 4 × 4 mm) samples was carried out at room temperature on the four-circle diffractometer installed at the hot source (5C2) of the Orphee reactor (LLB, France; λ = 0.83 Å, ω-scan mode), and investigation of *CNGS-1* (size: ~3 × 4 × 4 mm), *LGN-1′* (size: ~4 × 4 × 4 mm), and *LGN-2* (size: ~4 × 4 × 4.5 mm) samples was performed at room temperature on the four-circle single-crystal TriCS diffractometer (Paul Scherrer Institute, Switzerland; λ = 1.18 Å, ω-scan mode).

The X-ray diffraction analysis (XRD) of *CNGS-2* microregion ~0.3 × 0.3 × 0.3 mm was carried out on a CAD-4 four-circle diffractometer at room temperature (MoK_α radiation, graphite monochromator, ω-scan mode).

The crystal structure (atomic coordinates, occupancies of all positions) was refined using the full-matrix least squares procedure in the anisotropic approximation for all atoms with SHELXL-97 program package [25].

3. Results and Discussions

In the $Ca_3NbGa_3Si_2O_{14}$ structure each atom is located in the individual position: Ca in the dodecahedral (position 3*e*), Nb in the octahedral (position 1*a*), Ga in the tetrahedral (position 3*f*), and Si in a trigonal-pyramidal (position 2*d*) [8]. While in the $La_3(Ga_{0.5}Nb_{0.5})(1)Ga_3(2)Ga_2(3)O_{14}$ crystals lanthanum atoms are in the dodecahedral position of the structure, and gallium atoms occupy octahedral (Ga, Nb)(1) position, half replaced by the niobium atoms, tetrahedral Ga(2) and trigonal-pyramidal Ga(3) ones [17].

3.1. $Ca_3NbGa_3Si_2O_{14}$ Crystals. According to the neutron diffraction data obtained (Table 1), *CNGS-1* cut from the colorless crystal has a composition $(Ca_{2.95}\square_{0.05(1)})NbGa_3Si_2O_{14}$ with vacancies (□) located in the Ca position-V_{ca}''. The deficiency of other positions, as well as their occupation by the other atoms (antistructural defects), was not detected. Due to the fact that the electroneutrality condition for the refined composition is not satisfied, we assumed that a certain number of Ca^{2+} ions in the structure are interstitial ones, as for the $(La_{2.964}\square_{0.036})La_{i(0.036)}Zr_{0.5}Ga_5Si_{0.5}O_{14}$ [26]. The analysis of the residual nuclear density does not exclude such possibility. The peak with the coordinates x = 0.464, y = 0.069, and z = 0.136 close to the interstitial La_i atom coordinates (x = 0.418, y = 0.009, and z = 0.139) [26] was found. Interstitial Ca_i atoms (like La_i atoms) are displaced from the dodecahedral positions approaching one of four oxygen atoms, thus forming two reduced Ca–O (~ 2.06 Å) and two increased distances. According to Bokiy [27] "the apparent size of the interstitial atoms depends on their relative quantity: the smaller relative quantity of dissolved non-metal, the smaller the apparent size of its atoms."

Thereby, it allow us to rewrite the *CNGS-1* crystal composition as $(Ca_{2.95}\square_{0.05(1)})Ca_{i(0.05)}NbGa_3Si_2O_{14}$.

Refining the structure of *CNGS-2* cut from the yellow crystal by direct methods allows revealing both atomic coordinates similar to *CNGS-1* ones (the unit cell 1) and atomic coordinates associated with the previous ones by the transition matrix /100/010/00z+0.5/ (the unit cell 2). Both these unit cells are present in structure of *CNGS-2* crystal in the ratio of unit cell 1 : unit cell 2~4 : 1 (Table 2). For the structure determination the atomic coordinates of *CNGS-1* were taken as the initial ones. However, as a result of the calculation, fairly strong peaks of the residual nuclear density were found. The coordinates of peaks correspond to a crystal model with a shift 1/2 z. The refinement of the structure using the second model (the initial coordinates are shifted by 1/2 along the z-axis) based on electron density maps led to the appearance of strong peaks of the residual nuclear density with the coordinates of the initial model (without shift). These experimental results suggest the existence of the translation twin (like racemic twins) at the level of elementary cells. In

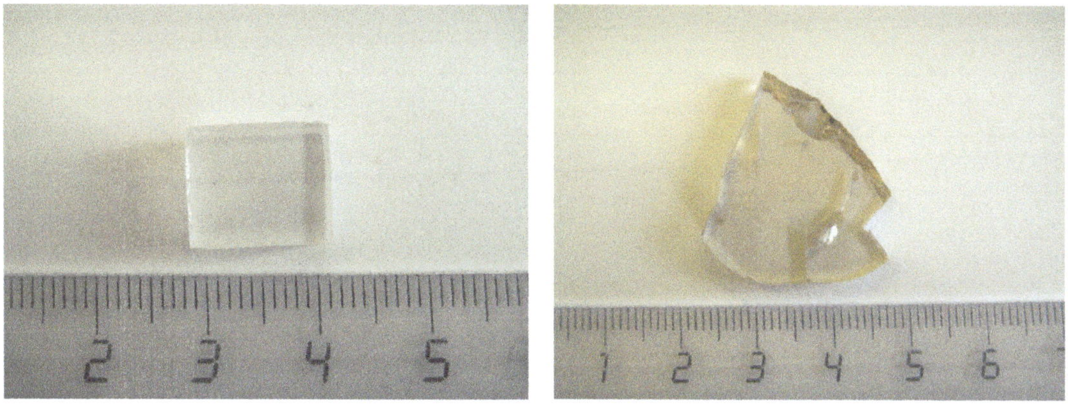

(a) (b)

FIGURE 1: Photo of the samples *CNGS1* (a) and *CNGS2* (b).

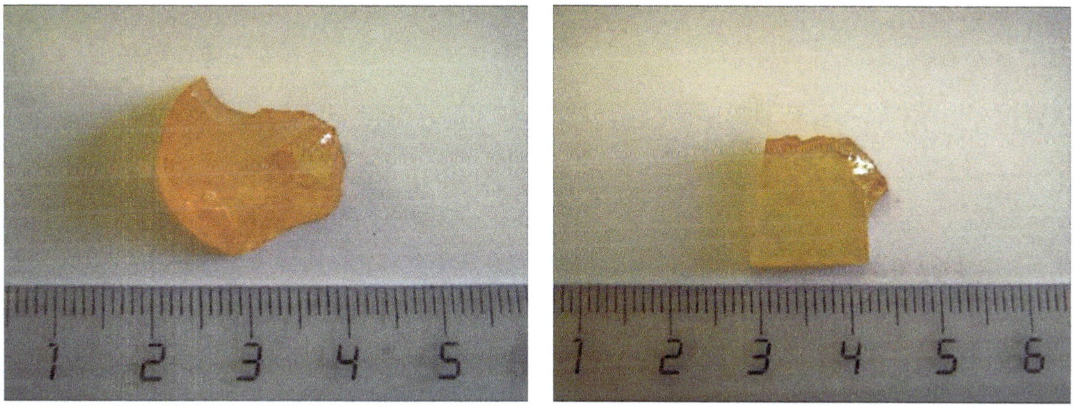

(a) (b)

FIGURE 2: Photo of the samples *LGN1* (a) and *LGN2* (b).

other words, in *CNGS-2* structure there are two unit cells which can be connected with the transition: 1/2 *z*. Thereby, the volume defect of the crystal structure (translation twin) is present in some regions of the *CNGS-2* sample. Such a defect has not been previously met for langasite family crystals.

According to our results of X-ray diffraction analysis (XRD) which were used for the refinement of the crystal structure of *CNGS-2* sample a microregion has a composition $Ca_3NbGa_3(Si_{1.985}\square_{0.015(10)})(O_{13.97(3)}\square_{0.03})$ (R_1 = 5.23%) with vacancies in trigonal pyramidal (V_{Si}'''') and oxygen ($V_O^{\bullet\bullet}$) positions.

Takeda et al. [5] and Kimura et al. [28] have suggested that the color of langasite family crystals depends on material of the crucible, in particular, Ir. However, the orange **LGN** and colorless **CNGS** have been grown using both Pt and Ir crucibles under the same growth conditions. It should be noted that, according to [29], the crystal color cannot be associated with the presence of iridium in the composition of the crystal, because, according to the inductively coupled plasma atomic emission spectroscopy data, the iridium content in the colored crystals is lower than 1 ppm. It can be assumed that the Ir content less than 1 ppm can cause the yellow color of **CNGS** crystal. However the crystal growth in Pt crucibles at the same growth conditions as growth in the Ir crucible leads to the orange color of the crystals too. Thus, the influence of crucible material on color of **LGN** and **CNGS** is quite controversial.

Kuz'micheva et. al. [20, 22], Kaurova et. al. [21], and Domoroshchina et. al. [29] have reported using X-ray or neutron diffraction analysis that the **LGS** and **LGT** crystals color depends on the presence of oxygen vacancies in the structure of crystals; that is, the absence of color for this crystals is due either to the absence of oxygen vacancies or to their large content. Thus, the light yellow and yellow crystals, as well as the colorless ones, are characterized by different compositions of oxygen positions [21, 22].

According to our data [21, 22] and the literature data [30], the annealing in vacuum of the orange crystal causes its discoloring. It is accompanied by oxygen vacancies content increasing in the crystal structure [21, 22]. Annealing in air of the same crystal leads to more saturated color tones [21, 22, 30]. It is accompanied by decreasing the content of oxygen vacancies [21, 22]. In turn, colorless crystals grown in argon and treated in vacuum do not change their color [30].

Table 1: Refined compositions of $Ca_3NbGa_3Si_2O_{14}$ and $La_3Ga_{5.5}Nb_{0.5}O_{14}$ crystals according to the results of the neutron diffraction analysis (NDA).

Crystal	Unit cell parameters, Å a, c	Refined composition (NDA)
CNGS-1	8.0911 (2) 4.9811 (1)	$(Ca_{2.95}\square_{0.05(1)})NbGa_3Si_2O_{14}$ $R = 4.15\%$
LGN-1	8.2283 (4) 5.1268 (3)	$(La_{2.935(2)}\square_{0.065})(Ga_{0.450}Nb_{0.550(3)})Ga_3(Ga_{1.965(4)}\square_{0.035})(O_{13.90(1)}\square_{0.10})$ $R = 8.10\%$
LGN-2	8.2293 (3) 5.1272 (2)	$(La_{2.940(1)}\square_{0.060})(Ga_{0.590}Nb_{0.410(2)})Ga_5(O_{13.82(1)}\square_{0.18})$ $R = 7.24\%$

Table 2: The atom coordinates in two-twinned unit cell of *CNGS-2* crystal.

Parameter		*CNGS-2* Unit cell 1	*CNGS-2* Unit cell 2
Ca	x	0	0
	y	0.5724	0.5936
	z	0	0.5
Nb	x	0	0
	y	0	0
	z	0	0.5
Ga	x	0	0
	y	0.2334	0.2501
	z	0.5	0
Si	x	1/3	1/3
	y	2/3	2/3
	z	0.4710	0.9700
O(1)	x	1/3	1/3
	y	2/3	2/3
	z	0.1853	0.7245
O(2)	x	0.1352	0.1553
	y	0.4529	0.4489
	z	0.3069	0.7638
O(3)	x	0.2255	0.2553
	y	0.1566	0.1511
	z	0.2404	0.7638

The results of NDA and XRD investigations of the **CNGS** crystals confirm these conclusions. The color of $Ca_3NbGa_3Si_2O_{14}$ is also caused by the oxygen vacancies, which are absent in the structure of the colorless crystal 1. It is not excluded that the annealing of crystals in the air will weaken their color, will reduce the quantity of point defects, and will improve their optical properties, as it was observed for langasite crystals [31]. According to our results, only oxygen vacancies which depend on growth and treatment conditions take part in color centers formation.

Thus, the colorless $Ca_3NbGa_3Si_2O_{14}$ crystal (crystal 1) grown by using the single-crystal charge is structurally more perfect in comparison with the yellow one (crystal 2) grown by the charge prepared by conventional solid-state reaction. In this way, colorless $Ca_3NbGa_3Si_2O_{14}$ crystal is more preferred for the practical application in piezoelectric and acoustic-electronic devices. The results of the dielectric properties investigations of **CNGS** crystals are the confirmation of this conclusion. For the colorless crystals the values of relative dielectric constants are higher compared with the ones for the yellow crystals [16, 32].

The same (upper) parts of crystals 1 and 2 were investigated by the diffraction methods. The compositions of the upper part of the crystal growing from polycrystalline charge differ from those of the lower part [19]. If this crystal will be crushed and regrown the composition will probably be averaged and oxygen content will be increased.

3.2. $La_3Ga_{5.5}Nb_{0.5}O_{14}$ Crystals. According to the neutron diffraction data obtained (Table 1), the composition of the upper part *LGN-1* of $La_3Ga_{5.5}Nb_{0.5}O_{14}$ crystal can be written as $(La_{2.935(2)}\square_{0.065})(Ga_{0.450}Nb_{0.550(3)})$ Ga_3 $(Ga_{1.965(4)}\square_{0.035})(O_{13.90(1)}\square_{0.10})$, that is, with Nb > Ga in the octahedral position (Ga,Nb)(1) and with vacancies in the dodecahedral (V_{La}''') and trigonal-pyramidal (V_{Ga}''') positions and in the oxygen ($V_O^{\bullet\bullet}$) one. The atomic coordinates which were used for the *LGN-1* calculation was also used for the refinement of the crystal structure and composition of the *LGN-1* sample cut from the top of the crystal. The refined composition of *LGN-1′* can be written as $(La_{2.940(2)}\square_{0.060})(Ga_{0.485}Nb_{0.515(6)})(Ga_{2.940(2)}\square_{0.060})Ga_2$ $(O_{13.84(1)}\square_{0.16})$ (R = 10.02%). The compositions of the *LGN-1* and *LGN-1′* samples are in good agreement with each other. However, for the *LGN-1′* sample as well as for the *CNGS-2* sample sufficiently strong peaks of the residual electron density with coordinates associated with the found ones by the transition matrix /100/010/00z+0.5/ were revealed. It also indicates the defect structure of the upper part of the $La_3Ga_{5.5}Nb_{0.5}O_{14}$ crystal what can be connected with the using of **LGS** as a seed.

The lower part of the crystal (*LGN-2*) has a composition $(La_{2.940(1)}\square_{0.060})(Ga_{0.590}Nb_{0.410(2)})Ga_5(O_{13.82(1)}\square_{0.18})$ with vacancies located in the dodecahedral (V_{La}''') and the oxygen ($V_O^{\bullet\bullet}$) positions and with Ga > Nb in the octahedral position (Ga, Nb)(1).

Thus, according to the neutron diffraction data obtained, there are the same types of point defects (the vacancies in the lanthanum and oxygen positions) in the compositions of the top and the bottom of the $La_3Ga_{5.5}Nb_{0.5}O_{14}$ crystal.

Different ratios of Ga/Nb and of oxygen vacancies concentrations (Table 1) indicate the composition and properties inhomogeneity over the volume of the crystal. It is not excluded that the **LGN** crystal color is also attributed to the qualitative differentiation of oxygen vacancies present in the structures. In orange crystals (*LGN-1, LGN-1′*) a higher oxygen content was found. In our opinion, the relationship between color and oxygen content for **LGN** must be the same as for **LGT** [21, 22] and **LGS** [29]. For example, the color of the crystals of composition $La_3(Ga,Ta^{5+})Ga_5O_y$ varies depending on oxygen content (the neutron diffraction results): $y = 13.872(6)$ is colorless crystal, $y = 13.918(7)$ yellow crystal, $y = 13.969(8)$ orange crystal, $y = 13.975(6)$ light yellow crystal, and $y = 14.0$ colorless crystal; that is, the colorless crystals may be characterized either by the absence of oxygen vacancies or by their large content [21, 22]. The reason of crystal color is believed to be the color center formation: $(V_O^{n\bullet}, ne')^{\times}$: $V_O^{n\bullet} > (V_O^{n\bullet}, ne')^{\times}$ colorless crystals and $V_O^{n\bullet} < (V_O^{n\bullet}, ne')^{\times}$ colored crystals.

4. Conclusions

The colorless and yellow $Ca_3NbGa_3Si_2O_{14}$ crystals (growth atmosphere: 100% Ar) and orange $La_3(Ga_{0.5}Nb_{0.5})Ga_5O_{14}$ crystal (growth atmosphere: 99%Ar + 1% O_2) were grown by the Czochralski technique in the Ir crucible along <0001> growth direction. The results of the neutron diffraction investigations of this crystals are reported.

It was established that the using of single-crystal charge is necessary for the growth of structural quality crystals.

The difference between the compositions of top and bottom of $La_3Ga_{5.5}Nb_{0.5}O_{14}$ crystal was found. The poor structural quality of $La_3Ga_{5.5}Nb_{0.5}O_{14}$ crystal upper part is due to the using of $La_3Ga_5SiO_{14}$ crystal as a seed.

The possibility of microtwins formation (two unit cells connected by the translation: 1/2 *z*) was revealed for $Ca_3NbGa_3Si_2O_{14}$ and $La_3(Ga_{0.5}Nb_{0.5})Ga_5O_{14}$ crystals.

It was found that the color of the $Ca_3NbGa_3Si_2O_{14}$ and $La_3(Ga_{0.5}Nb_{0.5})Ga_5O_{14}$ crystals is caused by oxygen vacancies present in the structures like the $La_3(Ga_{0.5}Ta_{0.5})Ga_5O_{14}$ and $La_3Ga_4(GaSi)O_{14}$ crystals.

References

[1] J. Bohm, E. Chilla, C. Flannery et al., "Czochralski growth and characterization of piezoelectric single crystals with langasite structure: $La_3Ga_5SiO_{14}$ (LGS), $La_3Ga_{5.5}Nb_{0.5}O_{14}$ (LGN) and $La_3Ga_{5.5}Ta_{0.5}O_{14}$ (LGT). II. Piezoelectric and elastic properties," *Journal of Crystal Growth*, vol. 216, no. 1, pp. 293–298, 2000.

[2] B. V. Mill and Y. V. Pisarevsky, "Langasite-type materials: from discovery to present state," in *Proceedings of IEEE International Frequency Control Symposium and Exhibition*, pp. 133–144, June 2000.

[3] T. Fukuda, H. Takeda, K. Shimamura et al., "Growth of new langasite single crystals for piezoelectric applications," in *Proceedings of the 11th IEEE International Symposium on Appliations of Ferroelectrics (ISAF-XI)*, pp. 315–319, August 1998.

[4] J. Bohm, R. B. Heimann, M. Hengst, R. Roewer, and J. Schindler, "Czochralski growth and characterization of piezoelectric single crystals with langasite structure: $La_3Ga_5SiO_{14}$ (LGS), $La_3Ga_{5.5}Nb_{0.5}O_{14}$ (LGN), and $La_3Ga_{5.5}Ta_{0.5}O_{14}$ (LGT). Part I," *Journal of Crystal Growth*, vol. 204, no. 1, pp. 128–136, 1999.

[5] H. Takeda, K. Shimamura, T. Kohno, and T. Fukuda, "Growth and characterization of $La_3Nb_{0.5}Ga_{5.5}O_{14}$ single crystals," *Journal of Crystal Growth*, vol. 169, no. 3, pp. 503–508, 1996.

[6] G. M. Kuzmicheva, E. N. Domoroschina, V. B. Rybakov, A. B. Dubovsky, and E. A. Tyunina, "A family of langasite: growth and structure," *Journal of Crystal Growth*, vol. 275, no. 1-2, pp. e715–e719, 2005.

[7] B. H. T. Chai, A. N. P. Bustamante, and M. C. Chou, "A new class of ordered langasite structure compounds," in *Proceedings of IEEE International Frequency Control Symposium and Exhibition*, pp. 163–168, June 2000.

[8] B. V. Mill', E. L. Belokoneva, and T. Fukuda, "New compounds with a $Ca_3Ga_2Ge_2O_{14}$-type structure: $A_3XY_3Z_2O_{14}$ (A = Ca, Sr, Ba, Pb; X = Sb, Nb, Ta; Y = Ga, Al, Fe, In; Z = Si, Ge)," *Russian Journal of Inorganic Chemistry*, vol. 43, no. 8, pp. 1168–1175, 1998.

[9] M. M. C. Chou, S. Jen, and B. H. T. Chai, "New ordered langasite structure compounds—crystal growth and preliminary investigation of the material properties," in *Proceedings of IEEE Ultrasonics Symposium*, pp. 225–230, October 2001.

[10] T. Karaki, R. Sato, M. Adachi, J. I. Kushibiki, and M. Arakawa, "Piezoelectric properties of $Ca_3NbGa_3Si_2O_{14}$ single crystal," *Japanese Journal of Applied Physics B*, vol. 43, no. 9, pp. 6721–6724, 2004.

[11] I. H. Jung, A. Yoshikawa, T. Fukuda, and K. H. Auh, "Growth and structure of $A_3NbGa_3Si_2O_{14}$ (A=Sr, Ca) compounds," *Journal of Alloys and Compounds*, vol. 339, no. 1-2, pp. 149–155, 2002.

[12] I. H. Jung, Y. H. Kang, K. B. Shim, A. Yoshikawa, T. Fukuda, and K. H. Auh, "Single crystal growth and characterizations of $A_3NbGa_3Si_2O_{14}$-type compounds for piezoelectric applications," *Japanese Journal of Applied Physics B*, vol. 40, no. 9, part 1, pp. 5706–5709, 2001.

[13] Z. Wang, D. Yuan, D. Xu et al., "Growth and dielectric properties of $Ca_3NbGa_3Si_2O_{14}$ crystals," *Journal of Alloys and Compounds*, vol. 370, no. 1-2, pp. 291–295, 2004.

[14] Z. Wang, X. Cheng, D. Yuan et al., "Crystal growth and properties of $Ca_3NbGa_3Si_2O_{14}$ single crystals," *Journal of Crystal Growth*, vol. 249, no. 1-2, pp. 240–244, 2003.

[15] Z. Wang, D. Yuan, A. Wei et al., "Growth and optical activity of $Ca_3NbGa_3Si_2O_{14}$ single crystal," *Applied Physics A*, vol. 78, no. 4, pp. 561–563, 2004.

[16] X. Shi, D. Yuan, X. Yin, S. Guo, X. Zhang, and Z. Li, "Crystal growth and dielectric, piezoelectric and elastic properties of $Ca_3NbGa_3Si_2O_{14}$ single crystal," *Journal of Crystal Growth*, vol. 293, no. 2, pp. 485–488, 2006.

[17] A. A. Kaminskii, B. V. Mill, E. L. Belokoneva, C. E. Sarkisov, T. Y. Pastukhova, and G. G. Khogzhabagyan, "Crystal structure and stimulated emission $La_3Ga_{5.5}Nb_{0.5}O_{14}$—$Nd^{3+}$," *Journal of Inorganic Materials*, vol. 20, pp. 2058–2061, 1984 (Russian).

[18] T. S. Chernaya, S. S. Kazantsev, V. N. Molchanov et al., "Crystal structure of $La_3Nb_{0.5}Ga_{5.5}O_{14}$ at 20 K," *Crystallography Reports*, vol. 51, no. 1, pp. 23–28, 2006.

[19] G. M. Kuz'micheva, O. Zakharko, E. A. Tyunina, V. B. Rybakov, E. N. Domoroshchina, and A. B. Dubovski, "Neutron diffraction and X-ray diffraction investigations of langasite crystals," *Crystallography Reports*, vol. 53, no. 6, pp. 989–994, 2008.

[20] G. M. Kuz'micheva, O. Zaharko, E. A. Tyunina et al., "Point defects in langatate crystals," *Crystallography Reports*, vol. 54, no. 2, pp. 279–282, 2009.

[21] I. A. Kaurova, G. M. Kuz'micheva, V. B. Rybakov, A. B. Dubovskii, and A. Cousson, "Composition, structural parameters, and color of langatate," *Inorganic Materials*, vol. 46, no. 9, pp. 988–993, 2010.

[22] G. M. Kuz'micheva, I. A. Kaurova, V. B. Rybakov et al., "The color of langatate crystals and its relationship with composition and optical properties," *Crystal Research and Technology*, vol. 47, no. 2, pp. 131–138, 2012.

[23] I. A. Kaurova, G. M. Kuz'micheva, V. B. Rybakov, A. B. Dubovsky, A. Cousson, and O. Zaharko, "Growth and structural investigations of $Ca_3NbGa_3Si_2O_{14}$ crystals," *Khimicheskaya Technologiya*, vol. 11, no. 10, pp. 585–591, 2010 (Russian).

[24] G. M. Kuz'micheva, E. A. Tyunina, E. N. Domoroshchina, V. B. Rybakov, and A. B. Dubovskii, "X-ray diffraction study of $La_3Ga_{5.5}Ta_{0.5}O_{14}$ and $La_3Ga_{5.5}Nb_{0.5}O_{14}$14 langasite-type single crystals," *Inorganic Materials*, vol. 41, no. 4, pp. 412–419, 2005.

[25] G. M. Sheldrick, "A short history of SHELX," *Acta Crystallographica A*, vol. 64, no. 1, pp. 112–122, 2007.

[26] A. A. Pugacheva, B. A. Maksimov, B. V. Mill' et al., "Growth and structure of $La_3Zr_{0.5}Ga_5Si_{0.5}O_{14}$ crystals," *Crystallography Reports*, vol. 49, no. 1, pp. 53–59, 2004.

[27] G. B. Bokiy, *Introduction to Crystal Chemistry*, MGU, Moscow, Russia, 1954.

[28] H. Kimura, S. Uda, O. Buzanov, X. Huang, and S. Koh, "The effect of growth atmosphere and Ir contamination on electric properties of $La_3Ta_{0.5}Ga_{5.5}O_{14}$ single crystal grown by the floating zone and Czochralski method," *Journal of Electroceramics*, vol. 20, no. 2, pp. 73–80, 2008.

[29] E. N. Domoroshchina, G. M. Kuz'micheva, V. B. Rybakov, A. B. Dubovsky, E. A. Tyunina, and S. Y. Stepanov, "Relationship between growth conditions, structure and optical properties of langasite crystals $La_3Ga_5SiO_{14}$," *Perspective Materials*, vol. 4, pp. 17–30, 2004 (Russian).

[30] O. A. Buzanov, E. V. Zabelina, and N. S. Kozlova, "Optical properties of lanthanum-gallium tantalate at different growth and post-growth treatment conditions," *Crystallography Reports*, vol. 52, no. 4, pp. 691–696, 2007.

[31] E. A. Tyunina, G. M. Kuz'micheva, O. Zaharko, and A. B. Dubovsky, "Effect of growth and post-growth treatment conditions on langasite crystals optical properties," *Vestnik MITHT*, vol. 5, no. 5, pp. 27–35, 2010 (Russian).

[32] M. Adachi, T. Funakawa, and T. Karaki, "Growth of substituted langasite-type $Ca_3NbGa_3Si_2O_{14}$ single crystals, and their dielectric, elastic and piezoelectric properties," in *Proceedings of the 13th IEEE International Symposium on Applications of Ferroelectronics (ISAF '02)*, pp. 411–414, June 2002.

7

Nanoscale Phenomena Occurring during Pyrolysis of *Salix viminalis* Wood

Aleksandra W. Cyganiuk,[1] Roman Klimkiewicz,[2] Andrzej Olejniczak,[1,3] Anna Kucińska,[1] and Jerzy P. Łukaszewicz[1]

[1] *Faculty of Chemistry, Nicholas Copernicus University, Ulica Gagarina 11, 87-100 Torun, Poland*
[2] *Institute of Low Temperature and Structure Research PAN, Ulica Okólna 2, 50-422 Wrocław, Poland*
[3] *Flerov Laboratory of Nuclear Reactions, Joint Institute for Nuclear Research, Dubna 141980, Russia*

Correspondence should be addressed to Jerzy P. Łukaszewicz; lukaszju@chem.uni.torun.pl

Academic Editor: Eric Guibal

Selective utilisation of unique properties of *Salix viminalis* wood enables preparation of materials of nanotechnologic properties. Thermal decomposition of lignin-cellulose organic matter results in the formation of a nanostructured porous carbon matrix (charcoal). Narrowed pore size distribution (PSD) in the subnanometer range allows to consider the charcoals as carbon molecular sieves (CMSs), which are capable of separating even chemically inert gases like neon, krypton, and nitrogen. High tolerance of *Salix viminalis* to heavy metal ions enables enriching living plant tissues with metal ions like lanthanum and manganese. Such ions may later form $LaMnO_3$ with parallel transformation of plant tissues (organic matter) to carbon matrix using a heat treatment. In this way, one gets a hybrid material: a porous carbon matrix with uniformly suspended nanocrystallites of $LaMoO_3$. The crystallites are in the catalytically active phase during the conversion of n-butanol to heptanone-4 with high yield and selectivity.

1. Introduction

Thermal treatment of biomass (including wood) is usually seen as an ordinary technological process. However, in some cases, this process may trigger some phenomena, which, doubtlessly, should be regarded as a part of widely understood nanotechnology, because of the properties of materials obtained during pyrolysis. The current research is aimed to prove that the terms "nanomaterial," "nanotechnology," "nanoscale" should not be exclusively considered in relation to artificial, synthetic materials obtained by sophisticated methods. We demonstrate on selected examples that nanoscale effects are achievable by particular treatments of natural products and/or by utilisation of phenomena, which occur in living plants. The presented examples relate to the plant called *Salix viminalis*. *Salix viminalis* is one of so-called short-rotation coppice [1], planted as an easily renewable source of energy [2]. There are almost no other cases of utilising *Salix viminalis* beside those related with energy or phytoremediation [3] of water and soil. Apart from its proved high tolerance to heavy metal ions, *Salix viminalis* grows very fast in the mild climate zone, yielding hard wood in large quantities per a single hectare of crop area [4]. The high agricultural yield of *Salix viminalis* may considerably reduce the cost of products obtained from this material, provided that proposed methods are implemented in the industrial use.

1.1. Carbon Molecular Sieves. Carbon molecular sieve (CMS) is often defined as a carbonaceous material of a narrowed pore size distribution (PSD). CMS is widely used for gas separation [7], purification of mixtures, or in catalytic processes [8]. The phenomenon of selective adsorption depends not only on the size of pores, but also on other properties of CMS, such as electron properties. Unlike the other known molecular sieving mineral materials [9], CMSs possess adequate chemical (pH) and thermal (in an inert atmosphere) stability and high hydrophobicity (if not chemically modified).

Number of procedures and precursors for preparing CMSs have been proposed and developed since the discovery of molecular sieving effect of Saran char in the late 1940s [10]. More recently, highly ordered microporous and mesoporous materials, like zeolites or MCM silicas, have proved to be useful molecular sieves [11]. Silicas, as well as other inorganic materials, may also serve as a template for fabrication of carbon porous adsorbents of an adjustable pore structure [12]. However, the application of proposed methods in mass production of CMSs remains complicated as it involves several production steps and requires hazardous reagents. Further more, the product yield of the complex fabrication methods seems to be rather low, and therefore these methods are hardly applicable for industrial aims. Nearly the same shortcomings relate to fabrication of carbon nanotubes (CNTs). In general, after opening and fragmentation, CNTs might be regarded as a carbon-type adsorbent of narrowed pores size distribution (PSD) [13]. Such narrowed PSDs were mentioned also for active carbons, for preparation of which, initial materials were wood shells of walnut [14] or palm fruit [15]. It is claimed that such fabricated active carbons possess pore structure typical for molecular sieves, but in many cases, the determined PSD is not narrowed and therefore being far from the ideal one [16]. Regarding the mentioned applications of solids of narrowed PSD (including CMSs) and the shortcomings of some already discovered fabrication ways, one has to state that there is an obvious need for an inexpensive carbon-type adsorbent, in which pores are really uniform and their size is below 1 nm, that is, somehow similar to size of molecules and atoms.

1.2. Formation of Metal Oxide Nanocrystallites in Carbon Matrix. Carbon-based materials are of a great interest in terms of their application in catalysis [17]. Such materials are catalysts, themselves but more frequently, they serve as a support for an actual catalytic phase, which often consists of metal/metal oxide clusters [18]. Laboratory practice exploits several routine ways to obtain carbon-based materials with metal/metal oxide clusters on the surface. The most classic methods are colloid deposition, electroplating, and ion exchange [19]. Currently available methods have not been applied for the deposition of clusters of perovskite-type oxides. Perovskite-type oxides are applied as oxygen reduction catalysts for fuel cells [20] or air-metal batteries [21]. Besides that, one of the newest trends in catalysis is the application of pure perovskite type oxides for the conversion of n-alcohols into aldehydes and/or ketones. Some catalysts are able to perform a secondary condensation of created aldehydes into ketones containing $2n$-1 carbon atoms in the aliphatic chain. The "n" mark denotes the number of C atoms in n-alcohol, which undergo the conversion. Lanthanum and manganese containing perovskite type oxides [22, 23] belong to the group of catalysts of potential catalytic activity towards ketonization of primary alcohols. Moreover, the application of carbon-type catalyst support enables utilisation of specific surfaces [24, 25] and catalytic [26, 27] properties of carbon. In addition, one may expect that some synergetic effects occur in the metal oxide-carbon [28, 29] system as often happens in case of other hybrid catalysts.

2. Experimental

2.1. Fabrication of Molecular Sieves. Harvested *Salix viminalis* rods are dried and grounded into shavings, about 1 cm long, then pyrolyzed. Carbonisation is carried out in two stages: (I) the preliminary stage—1 h at 600°C for expelling some volatile species (II) the secondary stage—1 h at arbitrarily chosen temperature (600, 700, 800, and 900°C) for expelling residual volatile fractions and the formation of nanopore-rich polycrystalline carbon matrix. Such carbons were denoted as SV. Two first numbers after the SV symbol like SV6171 describe primary carbonization (e.g., 6–600°C, 1-1 hour), while the two remaining numbers describe secondary carbonization (e.g., 7–700°C, 1-1 hour). Additionally activated carbon sample were prepared as well. Raw *Salix viminalis* wood was saturated for 48 h with $ZnCl_2$ solution (4 mole/dm^3). The saturation was followed by a standard two-step carbonization: preliminary (1 h at 600°C) and secondary (1 h at 600°C). Carbonized samples were intensively rinsed with water to eluate $ZnCl_2$ which is leaving cavities (pores) in carbon matrix. Such activated carbon was denoted as DSV.

2.2. Fabrication of Hybrid $LaMnO_3$/Carbon Catalysts. Recently, one of us proposed [30] a novel way of fabricating hybrid carbon-based metal oxide-containing materials. The novelty of the method consists in the exploitation of natural phenomenon of metal ion transportation in living plants. Metal ions, after introduction to transport-responsible tissues in a living plant, are transported to almost every cell, since *Salix viminalis* is highly tolerant to the presence of metal ions (including heavy metals) in its body. Freshly cut stems of *Salix viminalis* were immersed (vertical alignment) in a water solution containing equimolar quantities of $La(NO_3)_3$ and $Mn(NO_3)_2$ (example concentrations: 0.001 M, 0.01 M, and 0.1 M). The stems retained the ability of intensive capillary suction resulting in a gradual rise of the solution along the stems. After saturation with La^{3+} and Mn^{2+} ions, the stems were dried, diminished, and carbonized (600–800°C, a two-step procedure) in an inert gas atmosphere (N_2). First, the metal ion saturated wood was carbonized for 1 h at 600°C (N_2 atmosphere) in order to remove volatile species and to transform the wood (lignin-cellulose) matrix into carbon-based matrix. Next, heat treatment (1 h, N_2 flow) at the temperature of 800°C enabled preserving developed pore structure, and, what is the most important, it enabled the transformation of introduced metal ions into appropriate metal oxides.

2.3. Determining a Pore Structure. The pore structure and specific surface areas were determined using a widely accepted method, which utilises the phenomenon of low temperature adsorption of chemically neutral gases [31]. Authors applied nitrogen as an inert adsorptive. Nitrogen adsorption isotherms were recorded at the temperature of liquid nitrogen (−196°C) using Micromeritics ASAP 2010. The standard

TABLE 1: Weight content of main elements in selected carbonised plants. Results obtained using the EDS/EDX method at particular spots on the surface of carbon samples. Results selected according to the highest carbon content ever recorded for a particular carbon type.

Carbonized plant	Element content (% w/w)					
	C	O	Ca	Mg	K	Residual elements
Basket willow (wood) *Salix viminalis*	81.4	15.8	0.7	—	1.6	0.5
Elder (wood) *Sambucus nigra*	80.5	16.9	0.4	0.3	1.6	0.3
Brittle willow (wood) *Salix fragilis*	81.9	16.7	—	0.2	1.1	0.1
Common pine (wood) *Pinus sylvestris*	85.6	13.5	0.5	0.1	0.3	—
Pistachio tree (fruit shells) *Pistacia vera*	79.0	17.4	—	—	1.9	1.7
Plum tree (fruit stone) *Prunus domestica*	86.3	13.2	—	—	0.5	—
Norway maple (wood) *Acer platanoides*	84.2	14.3	0.4	0.3	0.6	0.2
Black poplar (wood) *Populus nigra*	79.5	19.0	—	0.5	0.9	0.1
Common birch (wood) *Betula pendula*	84.8	14.3	—	0.2	0.7	—

software provided by the manufacturer of ASAP 2010 was employed for the regression of primary obtained adsorption data (nitrogen adsorption versus relative partial pressure of the adsorptive). Two regression models were applied: BET [32] for the calculation of specific surface area of carbons and H-K method [33] for the calculation of pore size distribution (PSD) [34]. No changes were introduced to the standard calculation procedures offered by the commercial software [35]. All tested carbon samples were degassed in vacuum (0.133 Pa) at elevated temperature (250°C) for extended time (3 hours) before nitrogen adsorption measurements.

2.4. Gas Mixture Separation. Chromatographic tests were performed using a gas chromatograph Schimadzu GC-14B, supplied with a TCD detector kept at constant temperature of 110°C. Carrier gas (helium) was fed with 10, 15, 20, 25, 30, 40, and 50 mL/min intensities. The flow rate value was set, depending on the results of van Deemter optimisation procedure, performed prior to the separation tests. A glass chromatographic loop was 2.5 m long and its inner diameter was 2.6 mm. The separation tests were performed at several temperatures (70, 60, 50, 40, and 30°C) and consisted in recording of chromatograms for injected samples (pure gases, 2-component and 3-component mixtures). The chromatograms let us determine retention times and separation factors.

2.5. Catalytic Tests. The conversion of n-butanol was applied as a testing reaction. All runs were performed at the atmospheric pressure without addition of any carrier gases. A vertical quartz reactor of 10 mm inside diameter with a fixed bed containing 4 mL (0.77 g only) of catalyst of 0.6–1.2 mm particle size was used. The reactor was placed in a vertical pipe furnace. The tests were initiated at 300°C and conducted in a function of increasing temperature up to 480°C. Alcohol was introduced to the preheating zone at the top of the reactor, using a syringe infusion micropump at the variable flow rate of 1 to 3 mL/h. The products were analysed using gas chromatography.

2.6. Other. The morphology of the samples was investigated by means of an electron microscope (LEO 1430 VP, Electron Microscopy Ltd.) which was supplied with two detectors for secondary electrons (SEs), back-scattered electrons (BSEs), and X-ray elemental analyser (EDX). Such experimental setup enabled authors to determine the elemental composition of specific areas and display all details on scanning electron images (HRTEM (JEOL, JEM 2000 EX)). Apart from SEM observations, other investigations were performed as well.

3. Results and Discussion

The obtained carbon consists mainly of C, O, N, and H atoms (Table 1). Table 1 additionally presents original data obtained by the authors on the elemental composition of carbons produced by pyrolysis of other nonconventional kinds of wood. The results obtained for *Salix viminalis* are typical for wood originated chars which in ca. 80% (atomic content) consists of carbon. Such chars are also characteristic by high content of oxygen (ca. 15%) which influences severely the selection of a test gas for PSD determination that is practically excluding polar species like CO_2.

3.1. Formation of Uniform Nanopores in Carbon Matrix Obtained by Carbonisation of S. viminalis Wood. Figure 1

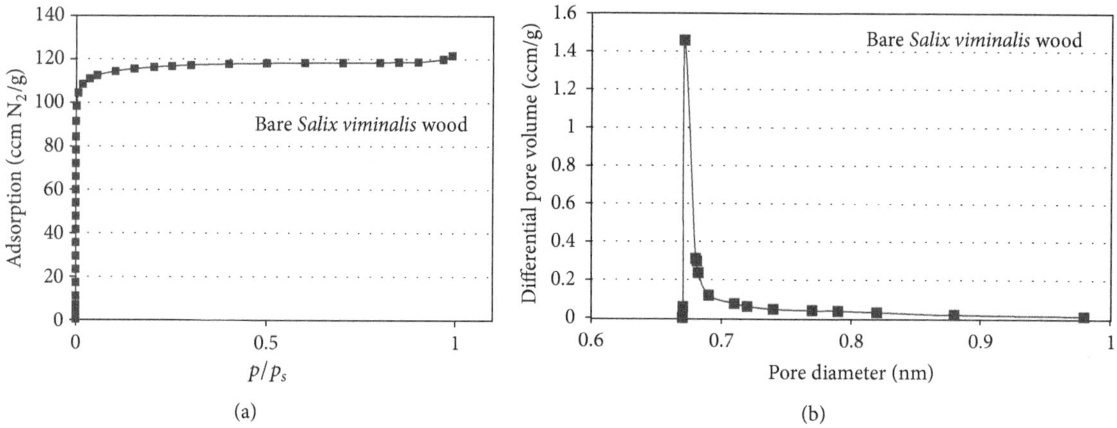

FIGURE 1: (a) Nitrogen adsorption isotherms recorded at −196°C for a virgin carbon sample SV6161 fabricated from *Salix viminalis* wood. (b) PSD functions determined from corresponding nitrogen adsorption isotherms.

depicts a nitrogen adsorption isotherm, recorded at −196°C for investigated carbonized *Salix viminalis* wood pieces. The determined nitrogen adsorption isotherms for such carbons were of the type I of the IUPAC classification, with the plateau reached for very low values of relative pressures (p/p_s) of the adsorbate. Such shape of isotherms allows to assume that the adsorbent is potentially a strictly microporous (nanoporous in fact) solid material; however, this assumption has to be proved by additional analyses. Regression of the adsorption data [33] pointed out very narrowed PSD which was the main aim of the performed research: inexpensive fabrication of nanosized molecular-sieve type carbons. The effective diameter of pores is definitely below 1 nm, what makes such produced carbons a unique and promising adsorbent for gas separation aims [36] (see later chapters). Beside gas separation, gas storage may be regarded as another possible field of application for these carbons. In the case of simple carbonization of *Salix viminalis* wood, specific surface area of raw carbons reaches the values of 300–400 m^2/g. Structure of investigated CMSs is depicted in the HRTEM micrograph (Figure 2). The carbon phase consists of numerous ordered domains, which are randomly oriented graphite crystallites. Similar concept of the structure of pyrolytic carbons is widely presented in the literature in forms of idealized diagrams. Differently oriented graphene sheets built slit-like cavities, which are considered as typical pores in partly graphitized pyrolytic carbons.

Just obtained raw pyrolytic carbons and/or organic precursors for their fabrication can be subjected to several procedures leading to the development of the total micropore volume and specific surface area (H_3PO_4 and CO_2 activation procedures). The activation treatments increase the total nitrogen uptake at p/p_s above 0.95, but the shape of the nitrogen adsorption isotherm and PSD remained unchanged. The same effect, that is, qualitatively unchanged PSD, has been found as a result of fabrication procedures performed at different carbonization temperatures. Up-to-date obtained results enable us to state that most of the mentioned activation methods lift the value of specific surface area above 1100 m^2/g. In general, activation procedures yielded carbons of a well-developed surface area and micropore volume. However, the previously mentioned very narrow PSD, observed for nonactivated carbons, was preserved in case of activated carbons.

In general, the PSD determination is a complex analysis, with results highly dependant on the kind of adsorptive, that is, a gas being physically adsorbed by the solid under investigation. The regression model of adsorption data plays a crucial role as well. Usually, Ar, N_2, and CO_2 are applied for this purpose. According to some studies [37], the usefulness of N_2 is limited due to its slow diffusion into micropores (linear diameter below 2 nm) at −196°C. Physical adsorption of CO_2 (at 0°C or 25°C) was proposed as new standard procedure for PSD determination [37] regardless of some evidently negative properties of CO_2.

(i) CO_2 is not chemically inert, particularly at room temperature (evident acidic properties which may lead to specific reaction with base centres on the surface).

(ii) CO_2 molecule is not a dipole but high negative charge is cumulated on both O atoms (specific molecular interactions may be expected).

(iii) CO_2 has a specific longitude shape ("long" 3-atom linear molecule may cause orientation issues).

Recently, some research suggested inapplicability of N_2 preferring CO_2, in contrast to other studies proving comparativeness of Ar, N_2, and CO_2 results in terms of PSD determination [38]. Theoretical simulation of Ar, N_2, and CO_2 adsorption [39–41] in a model pore structure proved particular usefulness of Ar as a probe gas. Low temperature adsorption of N_2 was let to determine PSDs only slightly differing from Ar-based PSDs. According to the studies, applicability of CO_2 in case of carbons of high content of heteroatoms (particularly oxygen) seems to be limited in case of the carbon surface with implanted surface functional groups, since numerically generated CO_2-based PSDs suffered from the presence of false pores (nonexisting in the assumed pore structure). Thus, despite some objections, N_2 adsorption at

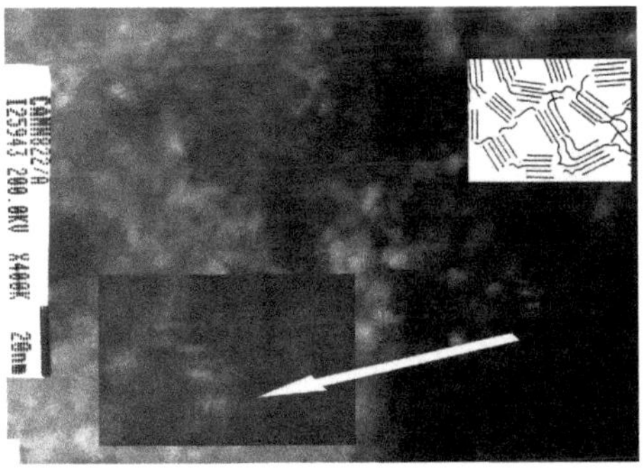

FIGURE 2: Structure of virgin carbon obtained from *Salix viminalis* wood SV6161 studied by HRTEM: randomly oriented graphite crystallites suspended in amorphous carbon phase. Historically, first theoretical model of carbon was elaborated in early fifties in 20th century [5]: inserted in the right upper corner. A novel model has been proposed recently [6].

−196°C may still serve as a standard experiment, providing reliable qualitative information on the run of PSD curves for real carbons, particularly those obtained by carbonisation of wood (high contents of polar surface oxygen and nitrogen groups).

3.2. Separation of Inert Gases by Means of Nanoporous CMS. In the light of the perveious statements, the obtained *Salix viminalis*-originated carbons possess molecular sieve-type pore structure, which is dominated by ultra-fine pores of size placed in the subnanometric range. Such discrete pore size distribution should be proved in practice despite determination of PSD functions presented earlier. Authors proposed a gas mixture separation (N_2 and Kr) in dynamic conditions. The gases differ from each other in terms of symmetry, Kr is supposed to be a sphere (no dipole or quadrupole moment), while N_2 is a linear molecule with differentiated size if measured along or perpendicular to the nitrogen-nitrogen triple bond. In some circumstances, quadrupole moment of N_2 molecule may play important role. The literature quotes similar kinetic diameter for nitrogen and krypton (N_2—0.364 nm and Kr—0.360 nm) [42], but some resources [43] suggest other diameters of N_2 molecule, depending on the molecule orientation, that is, whether measured perpendicular to N–N bond (0.300 nm). Separation of the slightly differing (geometrically) gases might be effective, provided an adsorbent contains pores of a size comparable to dimensions of the molecules under consideration. In practice, high content of nanopores (diameter below 1 nm) provides effective gas separation as in pressure swing adsorption (PSA) air separation method. In our study, a mixture of Kr and N_2 was injected into a chromatographic column packed with nanosized CMS, originated from *Salix viminalis* wood. Table 2 contains retention times, while Table 3 presents the data from Table 2 recalculated to separation factors for both gases.

Retention times for Kr are, in general, very long (minimum 8 minutes) at all investigated temperatures (30–70°C) or even longer than 20 minutes. Krypton atoms are retained, on average three times longer in the column filled with nanosized CMSs obtained from *Salix viminalis*. This may result from comparable dimensions of potential energy "well" in nanopores (calculated linear dimension below 1 nm) and the size of Kr atoms which leads to a relatively durable entrapment of the atoms in pores. Retention times in the same experimental conditions for N_2 are considerably shorter, ranging from 3 to 4.5 minutes and only slightly depending on the temperature of measurement conditions. The effect of shorter retention may be explained by several factors, one of which is a nonspherical shape of N_2 molecules. This conclusion seems to be probable since isosteric heat of adsorption Q_{ST} on microporous carbons (carbon molecular sieves) is very nearly equal for N_2 (23.9 ± 0.8 kJ/mol) and for Kr (22.9 ± 0.3 kJ/mol) [44]. Thus, natural tendency to prefer exothermic processes does apply to this case. Extensive studies [44] on the mechanism of adsorption on CMS led to the conclusion that two basic factors are essential for the overall process: (1) entering the pore (based on relent potential barrier at the pore entrance) and (2) diffusion along the pore (based on spot-to-spot hopping mechanism). However, author stated [44] that the rate-limiting process was the entry through the pore aperture, which may be considered as a "geometric factor," being in fact a certain potential barrier at the entrance. Both factors (1) and (2) are somehow summarized in the observed adsorption rate. If entering pores is a key factor as suggested and proven in several studies, even small differences in size and shape of adsorbing molecules highly influence adsorption-desorption kinetics. Such phenomenon is utilised in air separation for N_2 and O_2. The size of both biatomic molecules slightly differs (kinetic diameters: N_2—0.364 nm and O_2—0.346 nm), but oxygen desorption kinetics is faster than of nitrogen, and oxygen is released prior to nitrogen from pressurized chambers in the PSA method for air separation [45]. That enables efficient separation of the two basic components of air, which is commonly used in industry, over active carbon bed, provided its porosity is properly tailored (sub-nanoscale pores dominate). Moreover, adsorption/desorption activation energies can also favour faster adsorption/desorption of oxygen. Studies [46] on the adsorption of oxygen and nitrogen on CMSs proved differences in adsorption/desorption activation energies. The activation energy for nitrogen adsorption was (at 30–70°C) ca. 43 kJ/mol, while the corresponding value for oxygen was (0–40°C) ca. 35 kJ/mol. That resulted in a relatively fast adsorption/desorption of oxygen if compared to nitrogen.

We assume that a corresponding situation occurs in our study, causing fast adsorption/desorption of nitrogen (short retention times) and slowing the processes down in the case of krypton (long retention times). Thus, smaller molecules, like specifically oriented N_2, may relatively easily pass through the centre of pore entry (where the height of barriers is the lowest), even having low kinetic energy.

Weak dependency of nitrogen retention times on temperature if compared to Kr was observed. Bigger Kr atoms must possess high kinetic energy in order to overcome repulsing potential barrier at the pore entry. Therefore, the observed Kr retention times are more temperature dependent.

TABLE 2: Retention times recorded for N_2 and Kr separated on example carbon molecular sieves originated from *Salix viminalis* wood.

Gas	T (°C)									
	70		60		50		40		30	
	t_r	$\pm\Delta t_r$	t_r	$\pm\Delta t_r$	t_r	$\pm\Delta t_r$	t_r	$\pm\Delta t_r$	t_r	$\pm\Delta t_r$
Carbon SV6163										
Kr	1544	41.7	1770	23.2	2128	77.3	2654	47.4	3393	176.2
N_2	487	1.5	526	1.0	567	1.8	641	8.0	737	1.0
Carbon DSV6161										
Kr	886	7.8	1039	4.5	1285	64.5	1507	29.9	1758	15.3
N_2	386	2.7	412	2.7	459	1.6	504	0.8	551	4.6

t_r: retention time (s).

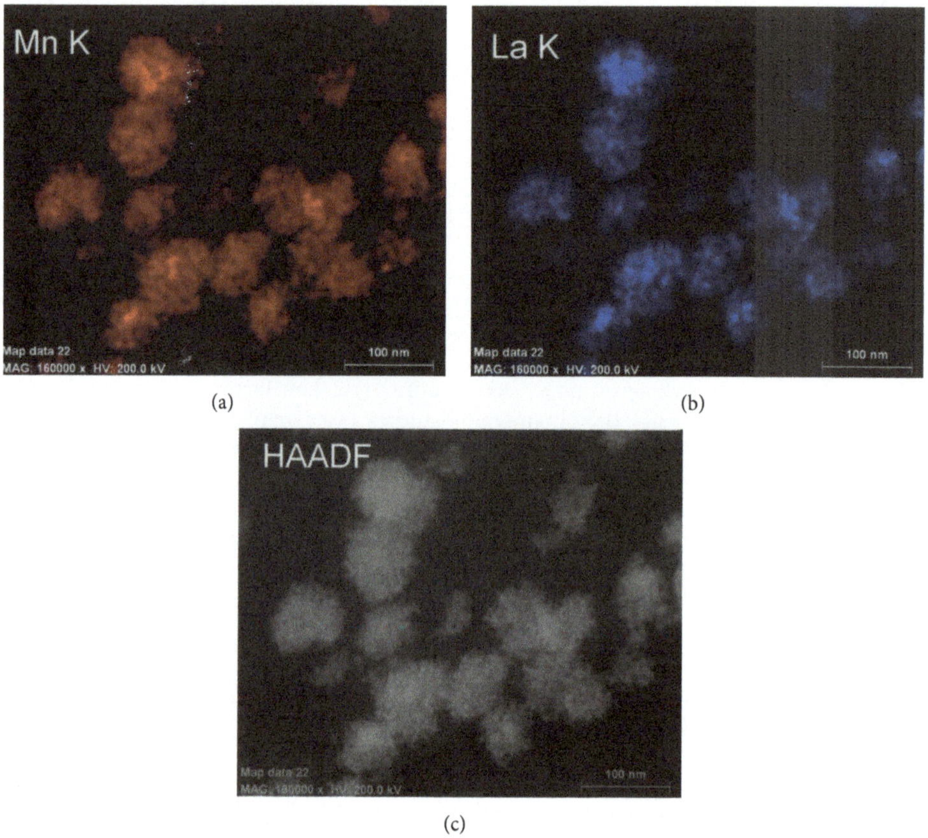

FIGURE 3: SEM-EDS/EDX images presenting common occurrence of La and Mn atoms in carbon sample obtained from *Salix viminalis* wood saturated with La^{3+} and Mn^{2+} ions prior to carbonization.

3.3. Formation of Nanoclusters of Complex Oxide ($LaMnO_3$ Perovskite) in Carbon Matrix, Obtained by Carbonization of S. viminalis Wood: Fabrication of a Catalyst for Ketonization of n-Alcohols. Figure 3 presents a series of HR-TEM images recorded for a carbon, obtained by carbonization of *Salix viminalis* wood, previously enriched (saturated) with La and Mn atoms. Short carbonisation times did not lead to the formation of $LaMnO_3$, crystals [47] in the carbon amorphous matrix. However, EDS/EDX maps show common occurrence of La and Mn atoms in particulars spots in the carbon matrix. Important changes are visible for samples, subjected to a heat treatment for 22 hours. This observation is consistent with other studies on the synthesis of pure $LaMnO_3$ [47] and formerly presented XRD diffraction studies, in which diffraction spectra were in perfect agreement with XRD spectrum recorded for $LaMnO_{3.15}$ [48].

Figure 4 shows that the extension of heat treatment time (22 hours) leads to the formation of crystalline inorganic nanoclusters inserted into carbon matrix, with apparent results of graphitization. Past XRD and XPS investigations [49] doubtlessly proved that clusters consist of $LaMnO_3$ with no significant excess of other lanthanum and manganese derivatives like single metal oxides (La_2O_3, Mn_2O_3, and/or MnO_2). Thus, there is a real background to state that

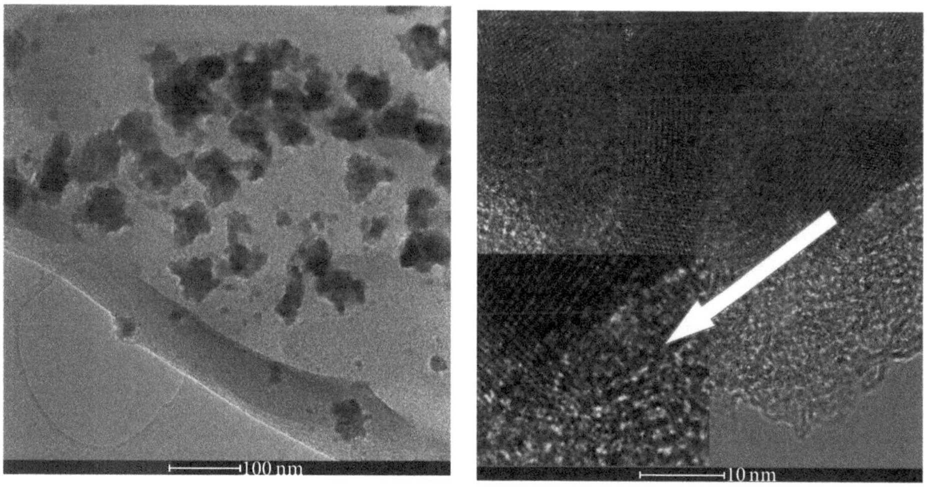

(a) (b)

FIGURE 4: HRTEM image presenting interface among $LaMnO_3$ crystallite and partly graphitized carbon matrix.

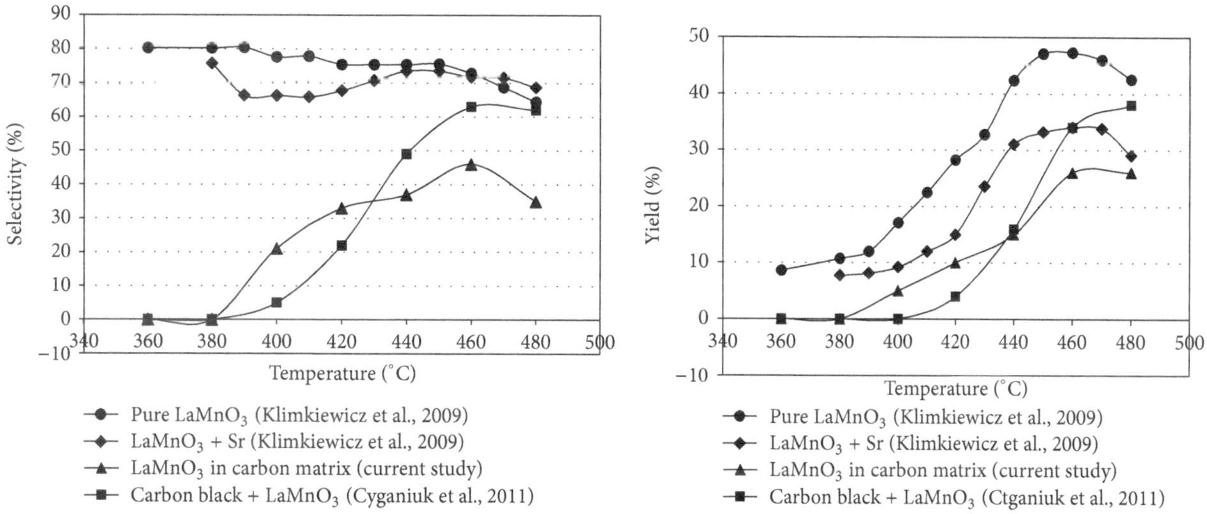

(a) (b)

FIGURE 5: Selectivity (a) and yield (b) in the catalytic conversion of n-butanol to heptanone-4 over different catalyst containing $LaMnO_3$.

TABLE 3: Retention coefficients determined for N_2 and Kr, separated on example *Salix viminalis* wood originated carbons (carrier gas flow 15 mL/min).

	Gas mixture Kr + N_2			
T (°C)	SV6163		DSV6161	
	R_s	$\pm\Delta R_s$	R_s	$\pm\Delta R_s$
70	4.90	0.34	7.91	0.68
60	4.88	0.23	9.00	0.84
50	5.04	0.73	9.23	2.61
40	4.91	0.67	10.56	1.26
30	6.23	1.60	10.78	0.99

during carbonisation of mixed metal ions in plant tissues, even complex metal oxides, like perovskite-type oxides, may participate at synthesis, in the form of nanocrystallites, whose diameter is ca. 10 nm. Such high dispersion of $LaMnO_3$ nanocrystallites results from uniform distribution of La^{3+} and Mn^{2+} ions in the living plant cells. During carbonisation, an organic matter transforms into a porous carbon matrix, while the metal cations react, yielding perovskite-type oxide crystallites, separated by layers of carbon.

Despite the performed impregnation of raw wood with metal cations (La^{3+} and Mn^{2+}), the obtained hybrid material ($LaMnO_3$/carbon) is still porous and resembles a perfect molecular sieve. The run of PSD functions for $LaMnO_3$/carbon samples is as narrow as for raw carbons, which do not contain inorganic phase. However, intensive enrichment of *S. viminalis* wood with La and Mn resulted in reduced specific surface area from 404 m^2/g (sample D0.01 LaMn 61822) to 235 m^2/g (sample D0.1 LaMn 61822) upon increasing concentration of impregnating solutions from 0.01 M to 0.10 M.

Some results of catalytic tests consisting in conversion of n-butanol are presented in Figure 5. One may state that the tested perovskite-type oxide $LaMnO_3$ in the form of nanocrystallites is an active catalyst, at higher temperatures but also for the formation of ketones (heptanone-4). In the same experimental conditions, pure carbon samples, that is, without $LaMnO_3$, are not catalytically active. Thus, the noticed catalytic activity results from the presence of real catalysts—$LaMnO_3$ nanocrystals. The material is less active (Figure 5) than pure $LaMnO_3$; however, the catalysts phase content is extremely low (ca. 0.30% atomic content of La and Mn atoms proved by XPS measurements). The tested catalysts show activity (yield and selectivity) comparable to other hybrid catalysts (carbon black—$LaMnO_3$) containing 10% of the perovskite-type oxide (by weight). Both key catalytic parameters, that is, selectivity and yield, are satisfactory at this stage of research. Even in extreme conditions, liquefied products were colourless without any yellowish shade. The shade would be a sign of undesired products like some resins and tars products.

4. Conclusions

Salix viminalis seems to be an ordinary woody plant, but it possesses unique features like organic tissues transforming into a durable strictly porous carbon matrix, resembling a perfect molecular sieve (very narrow PSD function). Its second useful feature is a high tolerance to heavy metals, what enables saturation of plants' living tissues with heavy metal ions. Carbonisation of metal saturated tissues leads to the formation of metal oxide (even complex oxides like $LaMnO_3$) crystallites suspended in the carbon matrix. Both pore size in carbon matrix and the size of metal oxide crystallites enable accounting such obtained materials to the group of nanomaterials: nanoporous carbon adsorbent and hybrid $LaMnO_3$ (nano-dispersed)/carbon catalyst. Sieving properties were proved by fast and efficient separation of N_2/Kr gas mixture over the nanoporous carbon adsorbent despite minimal differences in the dimensions of nitrogen molecule and krypton atom. The presence of highly dispersed $LaMnO_3$ phase on the carbon-type support enabled effective conversion of n-butanol into heptanone-4, despite the minute content the catalytically active phase.

Nomenclature

PSD:	Pore size distribution function
CMS:	Carbon molecular sieve
CNT:	Carbon nanotube
BET Surface Area:	Specific surface area determined by BET method (m^2/g)
HRTEM:	High resolution transmission electron microscopy
SEM:	Scanning electron microscopy
EDS/EDX:	Energy-dispersive X-ray spectroscopy (or XEDS).

References

[1] L. Christersson and L. Sennerby-Forsse, "The swedish programme for intensive shortrotation forests," *Biomass and Bioenergy*, vol. 6, no. 1-2, pp. 145–149, 1994.

[2] K. Stanczyk and M. Ludwik, "Energy forests. Energy Forest Development on Areas in Central—Eastern Europe, where the Agricultural Production is Uneconomical," Central Mining Institute, Katowice, Poland, 2003.

[3] M. Labrecque, T. I. Teodorescu, and S. Daigle, "Biomass productivity and wood energy of *Salix* species after 2 years growth in SRIC fertilized with wastewater sludge," *Biomass and Bioenergy*, vol. 12, no. 6, pp. 409–417, 1997.

[4] W. Sobczyk, "Plonowanie wierzby wiciowej w świetle badań naukowych," *Polityka Energetyczna*, vol. 10, no. 2, pp. 547–555, 2007.

[5] R. E. Franklin, "Crystallite growth in graphitizing and non-graphitizing carbons," *Proceedings of the Royal Society A*, vol. 209, pp. 196–218, 1951.

[6] P. J. F. Harris, "Fullerene-like models for microporous carbon," *Journal of Materials Science*, vol. 48, no. 2, pp. 565–577, 2013.

[7] A. I. Shirley and N. O. Lemcoff, "Air separation by carbon molecular sieves," *Adsorption*, vol. 8, no. 2, pp. 147–155, 2002.

[8] I. I. Ivanova, A. S. Kuznetsov, V. V. Yuschenko, and E. E. Knyazeva, "Design of composite micro/mesoporous molecular sieve catalysts," *Pure and Applied Chemistry*, vol. 76, no. 9, pp. 1647–1658, 2004.

[9] A. Corma, "From microporous to mesoporous molecular sieve materials and their use in catalysis," *Chemical Reviews*, vol. 97, no. 6, pp. 2373–2419, 1997.

[10] P. H. Emmett, "Adsorption and pore-size measurements on charcoals and whetlerites," *Chemical Reviews*, vol. 43, no. 1, pp. 69–148, 1948.

[11] V. Antochshuk and M. Jaroniec, "Simultaneous modification of mesopores and extraction of template molecules from MCM-41 with trialkylchlorosilanes," *Chemical Communications*, no. 23, pp. 2373–2374, 1999.

[12] Z. G. Shi, Y. Q. Feng, L. Xu, S. L. Da, and M. Zhang, "A template method to control the shape and porosity of carbon materials," *Carbon*, vol. 42, no. 8-9, pp. 1677–1682, 2004.

[13] Q. Wang, S. R. Challa, D. S. Sholl, and J. K. Johnson, "Quantum sieving in carbon nanotubes and zeolites," *Physical Review Letters*, vol. 82, no. 5, pp. 956–959, 1999.

[14] Z. Hu and E. F. Vansant, "Carbon molecular sieves produced from walnut shell," *Carbon*, vol. 33, no. 5, pp. 561–567, 1995.

[15] J. S. Tan and F. N. Ani, "Carbon molecular sieves produced from oil palm shell for air separation," *Separation and Purification Technology*, vol. 35, no. 1, pp. 47–54, 2004.

[16] S. Manocha, "Porous carbons," *Sadhana*, vol. 28, no. 1-2, pp. 335–348, 2003.

[17] P. Serp and J. L. Figueiredo, *Carbon Materials for Catalysis*, Wiley, Hoboken, NJ, USA, 2009.

[18] A. E. Aksoylu, M. Madalena, A. Freitas, M. F. R. Pereira, and J. L. Figueiredo, "Effects of different activated carbon supports and support modifications on the properties of Pt/AC catalysts," *Carbon*, vol. 39, no. 2, pp. 175–185, 2001.

[19] K. Yasuda and Y. Nishimura, "The deposition of ultrafine platinum particles on carbon black by surface ion exchange—increase in loading amount," *Materials Chemistry and Physics*, vol. 82, no. 3, pp. 921–928, 2003.

[20] S. P. Jiang, "Issues on development of (La,Sr)MnO_3 cathode for solid oxide fuel cells," *Journal of Power Sources*, vol. 124, no. 2, pp. 390–402, 2003.

[21] S. Imaizumi, K. Shimanoe, Y. Teraoka, N. Miura, and N. Yamazoe, "Preparation of carbon-supported perovskite-type oxides LaMn 1-$_y$,Fe$_y$O$_{3+\delta}$ based on reverse homogeneous precipitation method," *Journal of the Electrochemical Society*, vol. 151, no. 10, pp. A1559–A1564, 2004.

[22] I. Maurin, P. Barboux, Y. Lassailly, J. P. Boilot, F. Villain, and P. Dordor, "Charge-carrier localization on Mn surface sites in granular $LaMnO_{3+\delta}$ samples," *Journal of Solid State Chemistry*, vol. 160, no. 1, pp. 123–133, 2001.

[23] A. Barnabé, M. Gaudon, C. Bernard, C. Laberty, and B. Durand, "Low temperature synthesis and structural characterization of over-stoichiometric $LaMnO_{3+\delta}$ perovskites," *Materials Research Bulletin*, vol. 39, no. 4-5, pp. 725–735, 2004.

[24] S. Biniak, M. Walczyk, and G. S. Szymański, "Modified porous carbon materials as catalytic support for cathodic reduction of dioxygen," *Fuel Processing Technology*, vol. 79, no. 3, pp. 251–257, 2002.

[25] M. A. Montes-Morán, D. Suárez, J. A. Menéndez, and E. Fuente, "On the nature of basic sites on carbon surfaces: an overview," *Carbon*, vol. 42, no. 7, pp. 1219–1225, 2004.

[26] P. A. Gauden and M. Wiśniewski, "CO_2 sorption on substituted carbon materials. Computational chemistry studies," *Applied Surface Science*, vol. 253, no. 13, pp. 5726–5731, 2007.

[27] Z. J. Sui, J. H. Zhou, Y. C. Dai, and W. K. Yuan, "Oxidative dehydrogenation of propane over catalysts based on carbon nanofibers," *Catalysis Today*, vol. 106, no. 1–4, pp. 90–94, 2005.

[28] D. B. Tagiyev, G. O. Gasimov, and M. I. Rustamov, "Carbon deposits on the surface of CaO/SiO_2 as active catalysts for the oxidative dehydrogenation of ethylbenzene," *Catalysis Today*, vol. 102-103, pp. 197–202, 2005.

[29] R. Klimkiewicz and E. B. Drag, "Catalytic activity of carbonaceous deposits in zeolite from halloysite in alcohol conversions," *Journal of Physics and Chemistry of Solids*, vol. 65, no. 2-3, pp. 459–464, 2004.

[30] J. P. Łukaszewicz, R. Wesołowski, and A. Arcimowicz, Patent application. Patent Office of Poland, No. P 384122, 2007.

[31] Standard DIN 66134, "Bestimmung der Porengrößenverteilung und der spezifischen Oberfläche mesoporöser Feststoffe durch Stickstoffsorption; Verfahren nach Barrett, Joyner und Halenda (BJH)," DIN Deutsches Institut fuer Normung, Berlin, Germany, 1998.

[32] Standard ISO 9277, "Determination of the Specific Surface Area of Solids by Gas Adsorption Using the BET Method," International Organization for Standardization, Geneva, Switzerland, 1995.

[33] G. Horvath and K. Kawazoe, "Method for the calculation of effective pore size distribution molecular sieve carbon," *Journal of Chemical Engineering of Japan*, vol. 16, no. 6, pp. 470–475, 1983.

[34] N. A. Seaton, J. P. R. B. Walton, and N. quirke, "A new analysis method for the determination of the pore size distribution of porous carbons from nitrogen adsorption measurements," *Carbon*, vol. 27, no. 6, pp. 853–861, 1989.

[35] Operator's manual, "Micromeritics ASAP 2010 accelerated surface area and porosimetry system," Norcross, Micromeritics Instrument Corporation, 1995.

[36] B. Buczek, "Characterization of carbon molecular sieves based on nitrogen and carbon dioxide adsorption," *Chemical and Process Engineering*, vol. 31, no. 2, pp. 217–224, 2010.

[37] D. Lozano-Castelló, D. Cazorla-Amorós, and A. Linares-Solano, "Usefulness of CO_2 adsorption at 273 K for the characterization of porous carbons," *Carbon*, vol. 42, no. 7, pp. 1231–1236, 2004.

[38] J. Jagiello and M. Thommes, "Comparison of DFT characterization methods based on N_2, Ar, CO_2, and H_2 adsorption applied to carbons with various pore size distributions," *Carbon*, vol. 42, no. 7, pp. 1225–1229, 2004.

[39] S. Furmaniak, A. P. Terzyk, P. A. Gauden, P. J. F. Harris, and P. Kowalczyk, "Can carbon surface oxidation shift the pore size distribution curve calculated from Ar, N_2 and CO_2 adsorption isotherms? Simulation results for a realistic carbon model," *Journal of Physics Condensed Matter*, vol. 21, no. 31, Article ID 315005, 2009.

[40] S. Furmaniak, A. P. Terzyk, P. A. Gauden, P. J. F. Harris, and P. Kowalczyk, "The influence of carbon surface oxygen groups on Dubinin-Astakhov equation parameters calculated from CO_2 adsorption isotherm," *Journal of Physics Condensed Matter*, vol. 22, no. 8, Article ID 085003, 2010.

[41] A. P. Terzyk, S. Furmaniak, P. A. Gauden, P. J. F. Harris, J. Włoch, and P. Kowalczyk, "Hyper-parallel tempering Monte Carlo simulations of Ar adsorption in new models of microporous non-graphitizing activated carbon: effect of microporosity," *Journal of Physics Condensed Matter*, vol. 19, no. 40, Article ID 406208, 2007.

[42] C. R. Reid and K. M. Thomas, "Adsorption of gases on a carbon molecular sieve used for air separation: linear adsorptives as probes for kinetic selectivity," *Langmuir*, vol. 15, no. 9, pp. 3206–3218, 1999.

[43] K. S. W. Sing and R. T. Williams, "The use of molecular probes for the characterization of nanoporous adsorbents," *Particle and Particle Systems Characterization*, vol. 21, no. 2, pp. 71–79, 2004.

[44] W. A. Steele, "The physical interaction of gases with crystalline solids. I. Gas-solid energies and properties of isolated adsorbed atoms," *Surface Science*, vol. 36, no. 1, pp. 317–352, 1973.

[45] G. K. Step and M. V. Petrovichev, "Pressure swing adsorption for air separation and purification," *Chemical and Petroleum Engineering*, vol. 38, no. 3-4, pp. 154–158, 2002.

[46] C. R. Reid, I. P. O'koye, and K. M. Thomas, "Spherical adsorptives as probes for kinetic selectivity," *Langmuir*, vol. 14, no. 9, pp. 2415–2425, 1998.

[47] J. P. Łukaszewicz, S. Imaizumi, M. Yuasa, K. Shimanoe, and N. Yamazoe, "New approach towards preparation of efficient gas diffusion-type oxygen reduction electrode," *Journal of Materials Science*, vol. 41, no. 19, pp. 6215–6220, 2006.

[48] S. F. Miralaï, R. Avni, E. Francke, D. Morvan, J. Amouroux, and H. Nickel, "$LaMnO_3$ perovskite thin film deposition, from aqueous nitrate solutions of La and Mn, in a low-pressure plasma expanded through a nozzle (PETN)," *Thin Solid Films*, vol. 303, no. 1-2, pp. 17–26, 1997.

[49] A. Cyganiuk, R. Klimkiewicz, A. Olejniczak, and J. P. Łukaszewicz, "Biotechnological fabrication of $LaMnO_3$-carbon catalyst for n-butanol conversion to ketones," *Carbon*, vol. 48, no. 1, pp. 99–106, 2010.

8

Growth and Characterization of ULMA Single Crystals Doped with Ammonium Chloride

B. Sivasankari[1] and P. Selvarajan[2]

[1] Department of Physics, Kalasalingam University, Krishnankoil 626 190, India
[2] Department of Physics, Aditanar College of Arts and Science, Tiruchendur 628 216, India

Correspondence should be addressed to P. Selvarajan; pselvarajanphy@yahoo.co.in

Academic Editor: Steven Suib

Single crystals of pure and ammonium chloride-doped urea L-malic acid (ULMA) were grown by slow evaporation technique. Many interesting results on several properties of ammonium chloride impurity added to ULMA single crystals have been observed and studied. The different morphology of ammonium chloride-doped ULMA crystals was noticed and the effect of ammonium chloride doping on the growth, optical, and hardness properties has been investigated. The presence of functional groups has been estimated by FTIR analysis. The lattice parameters of the grown crystals were studied by single crystal X-ray diffraction technique. Powder X-ray diffraction studies confirm the diffraction planes of the grown crystals. The UV-visible spectrum shows the cutoff wavelength at 220 nm. A study on the thermal properties has also been carried out. The NLO property of the grown crystals was confirmed by SHG studies.

1. Introduction

Crystalline salts of amino acid complexes have recently attracted considerable interest among researchers due to its wide range of applications [1, 2]. Urea L-malic acid (ULMA) is one such material and few reports are available on ULMA crystal in the literature [3–7]. However, there is no report available, to the best knowledge of the authors, on the doping of this material. ULMA crystallizes in the monoclinic system with space group $P2_1$. The reported lattice parameters are $a = 9.0335(8)$ Å, $b = 6.9356(5)$ Å, $c = 6.8008(6)$ Å, and $\beta = 94.67(0)^\circ$ at 295 K [3, 4]. ULMA crystal has a wide transmission range in the visible region and could be used as a nonlinear optical (NLO) material. Microelectronics need low dielectric constant (ε_r) materials and ULMA crystal has low values of dielectric constant, and hence it can be used to reduce the RC delay, reduce power consumption, and reduce cross talk. It is reported that doped NLO materials have more advantages than undoped materials [8, 9]. Urea L-malic acid (ULMA) is an organic material, and if an inorganic material like ammonium chloride is used as the dopant, it is expected that the physical and chemical properties of the host crystal; namely, urea L-mailc acid may be altered and NLO properties of ULMA crystal may be improved, and hence in the present investigation, pure and ammonium chloride-doped urea L-malic acid (ULMA) single crystals were grown by slow evaporation technique. The grown crystals were characterized by FTIR, single crystal XRD and powder XRD studies, UV transmission, hardness, TG/DTA, and SHG studies. The results obtained from various studies of undoped and ammonium chloride-doped ULMA crystals are here reported and discussed.

2. Experimental Studies

2.1. Crystal Growth and Morphology. High-purity chemicals with deionized water were used to grow crystals by slow evaporation technique. Urea and L-malic acid were taken in 1 : 1 molar ratio, and ammonium chloride was added in three different weight percentages of 5 wt%, 10 wt%, and 15 wt% to the solution of urea L-malic acid. Uniform stirring was maintained and the temperature and volume were also kept constant as 307 K and 100 mL for all crystals to grow.

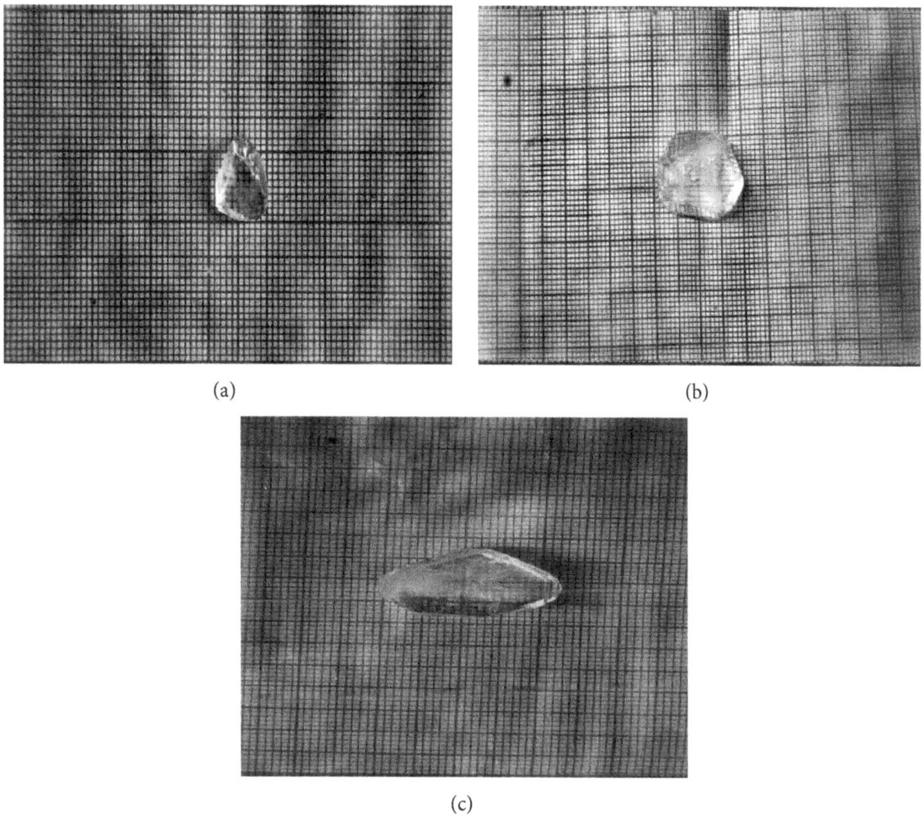

FIGURE 1: A photograph showing ammonium chloride doped ULMA crystals. (a) ULMA doped with ammonium chloride, (5 wt%), (b) ULMA doped with ammonium chloride, (10 wt%), (c) ULMA doped with ammonium chloride, (15 wt%).

Nuclei started to appear in the saturated solution after 15 days and crystals of size $20 \times 10 \times 7\,mm^3$ have been observed after the growth period of 35 days. From these grown crystals, defect-free crystals were selected and used for the studies. Figure 1 shows the as-grown crystals of ULMA doped with different weight percentages of ammonium chloride. In our study, crystallization indicates that dopant addition of different weight percentages of ammonium chloride to ULMA results in a change of crystal habit. Low concentrations of ammonium chloride as impurity were found not to affect the prismatic faces of pure ULMA. ULMA crystal grown from ammonium chloride of 5 wt% as dopant was found to have prismatic shape. With increasing the dopant percentage addition from 10 wt% to 15 wt%, extra faces with platy habit were found in the grown crystals. As reported in literature [10], it may be due to adsorption which led to changes in growth mechanism and habit modification. The supersaturation and stirring may be also found to affect the crystal habit.

2.2. Characterization Techniques. To analyse the presence of functional groups, Fourier Transform Infrared Spectrum (FTIR) was recorded using a Shimadzu spectrophotometer with KBr pellet technique in the range of 4000 to 500 cm^{-1}. Single crystal X-ray diffraction data have been collected using Enraf Nonius CAD-4 Diffractometer ($\lambda = 0.7107$ Å in $\omega/2\theta$ mode) with graphite monochromated Mo K_α radiation. Powder XRD was recorded by employing a powder X-ray diffractometer (PANalytical multipurpose diffractometer) with nickel filtered Cu K_α radiation ($\lambda = 1.54056$ Å). The UV-visible-IR transmission spectrum of ULMA was recorded on a SHIMADZU UV-240 IPC spectrophotometer in the range of 190–1100 nm. Single crystals of ULMA of thickness of 1 mm were used for this study. Vickers hardness measurements were carried out using ultramicrohardness tester fitted with a diamond indenter. The indentations were made using a Vickers pyramidal indenter for various loads from 25 to 200 g. Several trials of measurements were made on the prominent (010) face, and the average diagonal length was calculated for indentation of 5 s. Simultaneous TG and DTA were carried out using a thermal analyser SDT Q600 between 10 to 700°C at a heating rate of 20°C/min in nitrogen atmosphere [11, 12]. Preliminary experiment was undertaken to observe second-order nonlinear optical response, and the grown crystals were subjected to Kurtz powder test to find the nonlinear optical property [13]. The sample was illuminated with Nd : YAG laser using the first harmonics output of 1064 nm with pulse energy of 4.6 mJ. Urea was used for calibrating the SHG intensity.

3. Results and Discussion

3.1. FTIR Analysis. FTIR spectrum of the grown crystals is shown in Figure 2. By the interpretation of FTIR spectra, it is possible to show that certain functional groups are present

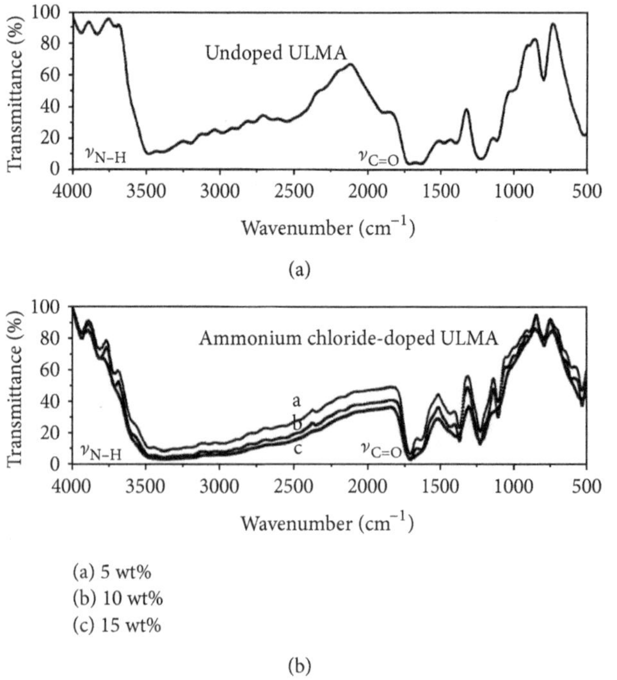

FIGURE 2: FTIR spectra of the grown undoped and ammonium chloride-doped ULMA samples.

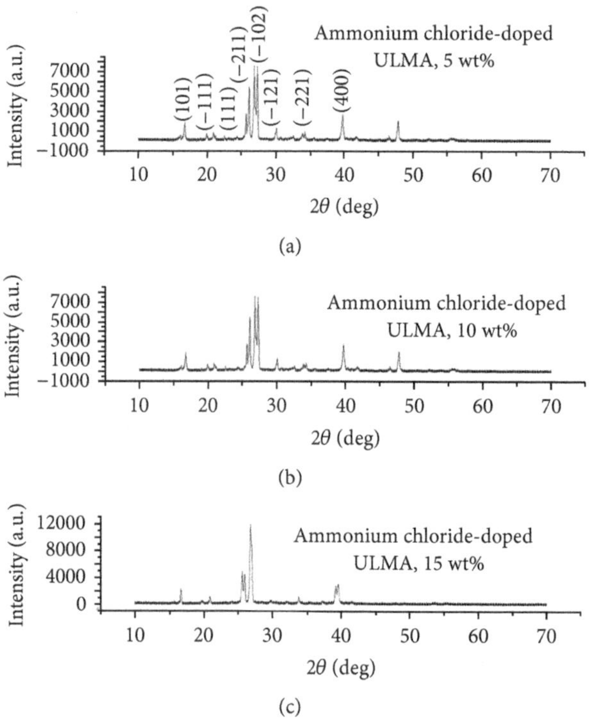

FIGURE 3: XRD patterns of the grown samples.

TABLE 1: Wave numbers of the absorption band maxima in the FTIR spectra of the grown samples.

Ammonium chloride concentration in the ULMA single crystal	Wave number and assignments, (cm^{-1})	
	Stretching vibration from amino groups	Carbonyl stretching vibrations
0 wt%	3489.77	1637.45
5 wt%	3459.75	1643.42
10 wt%	3468.79	1641.45
15 wt%	3474.73	1645.42

in the material. The results of FTIR analysis obtained in our present work for undoped ULMA single crystal were found to be in good agreement with the reported value [1, 6]. Intense band in the range of 3498–3250 cm^{-1} is due to stretching vibration from amino groups. An intense band in the range of 1645–1750 cm^{-1} is due to carbonyl stretching vibration. The FTIR peaks in ammonium chloride-doped ULMA crystals are wider than their corresponding peaks in undoped ULMA. The packing of ULMA in crystalline state is very peculiar due to hydrogen bonds as studied by de Matos Gomes et al. [7]. The assignments for the absorption peaks/bands of the FTIR spectra are given in accordance with the data reported in the literature [6, 14], and the frequency assignments for the absorption peaks corresponding to amino groups are provided in Table 1.

3.2. Single X-Ray Diffraction Studies. X-ray diffraction (XRD) was used for the identification of the crystal. The grown crystals were subjected to single crystal XRD studies, and the obtained values crystal cell parameters are given in the Table 2. It is observed from the results that the grown undoped and ammonium chloride-doped ULMA crystals crystallize in the monoclinic system and a small variation is noticed in the values of the lattice parameters of doped ULMA crystals as compared to those of undoped ULMA crystal. The obtained lattice parameters of pure LLMA crystal are observed to be in close agreement with data reported in the literature [3, 4].

3.3. Powder X-Ray Diffraction Studies. Single crystal XRD studies give the values of unit cell parameters of the samples. To confirm the obtained values of unit cell parameters, powder XRD studies were also carried out for the samples of this work. Also powder XRD studies will give ideas of diffraction planes of the crystals. Figure 3 shows the powder XRD spectra of the grown samples, and all the observed reflection peaks were indexed. The observed and calculated *d*-spacing values and the corresponding (hkl) planes are provided in Table 3. Using the powder XRD data, the lattice parameters were obtained with the help of UNITCELL software package, and it is found that the lattice parameters obtained from the powder XRD studies are very close to the those obtained from single crystal XRD studies. The slight shift of the reflection peaks in the powder XRD patterns of the doped samples indicates that the dopant has entered into the lattice of the host crystal.

3.4. Optical Transmission Studies. The UV-visible-IR transmission spectrum of the grown samples was shown in

TABLE 2: Unit cell parameters of the grown samples.

Unit cell parameters for pure ULMA	Unit cell parameters for ammonium chloride-doped ULMA crystals 5 wt%	10 wt%	15 wt%
$a = 9.045(3)$ Å	$a = 9.000(6)$ Å	$a = 9.000(5)$ Å	$a = 9.053(9)$ Å
$b = 6.923(5)$ Å	$b = 6.934(4)$ Å	$b = 6.915(5)$ Å	$b = 6.919(5)$ Å
$c = 6.814(4)$ Å	$c = 6.781(2)$ Å	$c = 6.727(3)$ Å	$c = 6.821(2)$ Å
$\alpha = \gamma = 90.00(0)^\circ$,	$\alpha = \gamma = 90.00(0)^\circ$,	$\alpha = \gamma = 90.00(0)^\circ$,	$\alpha = \gamma = 90.00(0)^\circ$,
$\beta = 94.88(1)^\circ$	$\beta = 94.61(1)^\circ$	$\beta = 94.60(1)^\circ$	$\beta = 94.59(1)^\circ$
$V = 425(1)$ Å^3	$V = 422(2)$ Å^3	$V = 418(3)$ Å^3	$V = 426(2)$ Å^3

TABLE 3: Theoretical and experimental d-values in Å, hkl values for prominent peaks powder XRD patterns of the grown samples.

Ammonium chloride-doped ULMA		Peak 101	−111	111	−211	−102	−121	−221	400
15 wt%	Theory d	5.312	4.502	4.241	3.425	3.322	2.922	2.566	2.118
	Calculated d	5.233	4.377	4.174	3.402	3.269	2.951	2.555	2.256
10 wt%	Theory d	5.251	4.467	4.244	3.322	3.211	2.938	2.534	1.986
	Calculated d	5.175	4.347	4.143	3.38	3.227	2.941	2.580	1.974
5 wt%	Theory d	5.297	4.452	4.245	3.322	3.264	2.972	2.525	2.024
	Calculated d	5.201	4.366	4.161	3.389	3.250	2.951	2.586	2.031

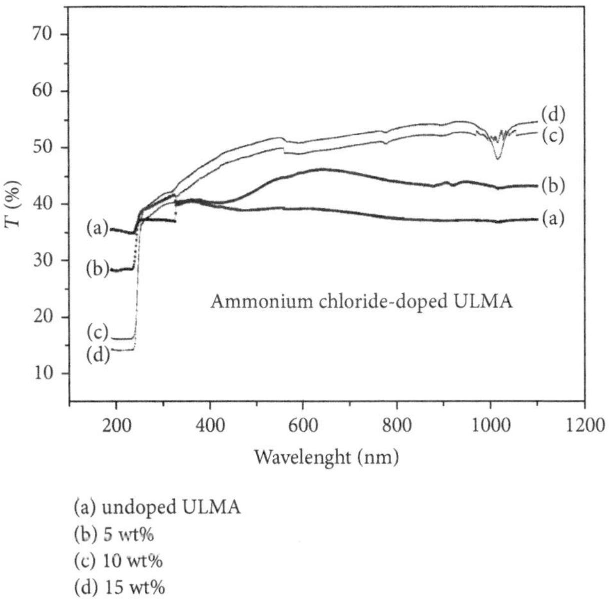

FIGURE 4: UV-visible-IR spectra of the grown samples

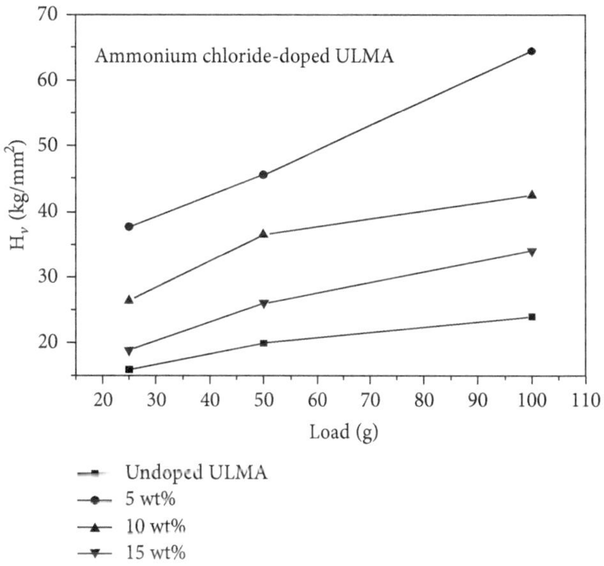

FIGURE 5: Variation of microhardness number with the applied load for pure and ammonium chloride-doped ULMA crystals.

Figure 4. For all optical applications in general and especially for SHG the material considered must be transparent in the wavelength region of interest. All crystals of undoped and ammonium chloride-doped ULMA can be utilized for second harmonic generation in visible wavelengths since the transparency is good in that range. An interesting result of increase in transmittance by the addition of ammonium chloride as impurity into ULMA crystal was observed. The transmittance increases by 45%, 52%, and 54% for the ammonium chloride-doped ULMA crystals with doping concentration of 5, 10, 15 wt% respectively, in the visible region, and the transmittance is comparatively low by around 40% for undoped ULMA crystal. This shows that the ammonium chloride as impurity has not destroyed the optical transparency of the crystal, and it is found to suppress the inclusions and improves the quality of ULMA crystal with higher transparency range. The lower cutoff which corresponds to the fundamental absorption occurs at 220 nm for undoped ULMA crystal. When ammonium chloride was added as dopant to ULMA, it is observed that lower cutoff wavelength slightly increased. The short cutoff

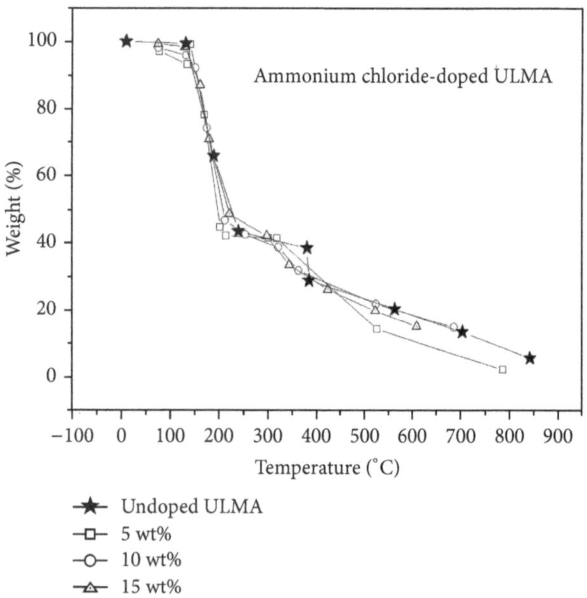

Figure 6: TGA curves for the grown samples.

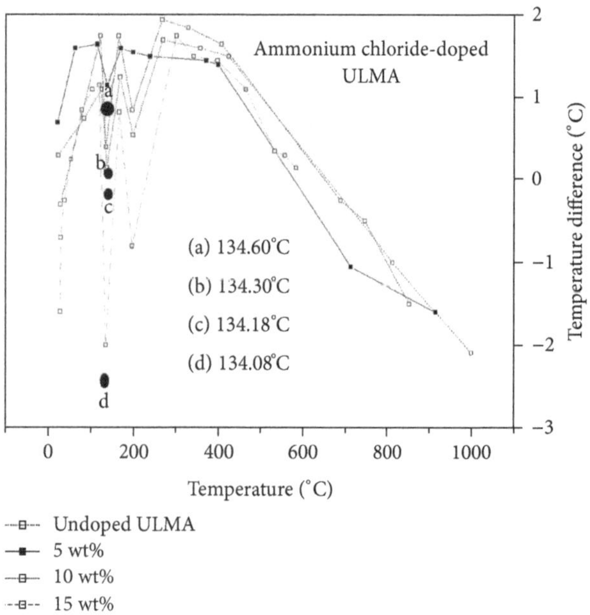

Figure 7: DTA curves for the grown samples.

wavelength facilitates the grown crystals of this work to be potential nonlinear optical materials for second harmonic generation. Absorption in the near ultraviolet region arises from electronic transitions associated within the samples. Using the formula $E_g = 1240/\lambda$ (nm) [15], the band gap is calculated to be 5.636 eV for all the grown samples.

3.5. Microhardness Studies. Vickers hardness number was evaluated from the relation $H_v = 1.8544\, P/d^2$ kg/mm^2, where H_v is Vickers hardness number, P is the indenter load in kilograms, and d is the diagonal length of the impression in millimetres [16]. Figure 5 shows the variation of microhardness values with applied load for the samples. The hardness value increases with increasing load. For loads above 200 g cracks developed on the surface of the crystal due to release of internal stress generated locally by indentation. It was observed that for low weight percentage concentration of ammonium chloride impurity, the microhardness is found to be more than double the times of pure ULMA single crystal. With the increase in higher impurity concentration, the strengthening decreases and the values of hardness are approaching those of pure ULMA crystal. This may be due to weakening the strength of bonds of the lattice of ULMA crystal as more weight percentage of dopants (ammonium chloride) are added. The results indicate that the mechanical properties could be increased with minimum selection of ammonium chloride as impurity.

3.6. TG and DTA Analysis. Figures 6 and 7 show the TG/DTA curves of pure and ammonium chloride-doped ULMA crystals. It is observed from the results that the melting point of ULMA crystal is slightly increased when low concentration of ammonium chloride was added as dopant. When the dopant concentration of ammonium chloride is more than 5 wt%, the decomposition point is found to be decreased, and these results are also confirmed from microhardness studies. For all grown samples we found that there is no loss of water molecules before thermal stability point and also no endothermic peaks before decomposition are recorded. The sharpness of the thermograms is also illustrative of the crystal purity. More than one endothermic peak may be due to complex formation. For all the grown samples, gradual weight loss is noted up to the melting point. The decomposition point for all the samples is observed in the temperature range of 193–196°C. The thermal stability is found to be more for all the samples and can be utilized in fabricating NLO devices. The obtained results are shown in Table 4.

3.7. Nonlinear Optical Test. The nonlinear optical (NLO) test was carried out for the samples and here an interesting result in SHG is observed. The second harmonic generation (SHG) is found to be appreciably increased by the addition of ammonium chloride as impurity into ULMA single crystal. The second harmonic generator wavelength was 532 nm. At 15 wt% of ammonium chloride as impurity into ULMA crystal, the SHG efficiency is found to be more than double the times of that of urea. The results of SHG efficiency of the grown samples are shown in Table 5. The dopant "ammonium chloride" is an inorganic material, and if it is added to the lattice of ULMA, it is possible that the dopant may neutralize the OH group of ULMA, and this might be the cause of the enhanced second harmonic generation (SHG) efficiency. If OH group of ULMA is neutralized by doping ammonium chloride, the electron delocalization will be more in the doped samples, and this leads to enhancement of NLO property in the ammonium chloride-doped ULMA sample [17].

TABLE 4: TG/DTA results.

	Compounds			
	Undoped ULMA	Ammonium chloride-doped ULMA		
		5 wt%	10 wt%	15 wt%
TG				
Stages of decomposition (°C)	132.02, 224.05, 386.32	135.02, 215.28, 320.00	133.30, 213.21, 322.16	130.09, 222.53, 298.80
Intial mass of the material subjected to analysis (mg)	3.2600	2.7500	2.9430	0.7609
Residue (mg)	0.0033	0.0381	0.0137	0.7609
DTA				
Irreversible endothermic transition temperatures (°C)	132.08, 194.72	134.60, 196.12	134.30, 195.98	132.18, 194.49
Melting point (°C)	132.08	134.60	134.30	132.18

TABLE 5: SHG efficiencies of the grown crystals.

Compounds	SHG efficiency
Pure ULMA crystal	0.34 times of urea
5 wt% of ammonium chloride added to ULMA	0.5 times of urea
10 wt% of ammonium chloride added to ULMA	1.32 times of urea
15 wt% of ammonium chloride added to ULMA	2.4 times of urea

4. Conclusion

ULMA crystals doped with ammonium chloride as dopant were grown and the morphology changes were noted. Grown crystals were characterized by single crystal and powder XRD studies and they confirmed that all the grown crystals belong to monoclinic system. Improvement in the optical transmission percentage and mechanical strength has been observed with ULMA crystal due to ammonium chloride as impurity addition. Hardness, thermal stability, and melting point were also determined for all the grown samples. The improved nonlinear optical properties of ammonium chloride-doped ULMA crystal proved it as a promising material for nonlinear application.

Acknowledgments

The authors would like to thank authorities of Aditanar College of Arts and Science, Tiruchendur, for the encouragement given to them to carry out the research work. Also they thank the staff members of M.K. University, Madurai, RRL, Trivandrum, St. Joseph's College, Trichy, Crescent Engineering College, Chennai, and CECRI, Karaikudi, for the characterization of the samples.

References

[1] A. S. J. Lucia Rose, P. Selvarajan, and S. Perumal, "Growth, structural, spectral, mechanical, thermal and dielectric characterization of phosphoric acid admixtured L-alanine (PLA) single crystals," *Spectrochimica Acta A*, vol. 81, pp. 270–275, 2011.

[2] S. A. Martin Britto Dhas, G. Bhagavannarayana, and S. Natarajan, "Growth and characterization of a new potential NLO material from the amino acid family-l-prolinium picrate," *Journal of Crystal Growth*, vol. 310, no. 15, pp. 3535–3539, 2008.

[3] V. K. Dixit, S. Vanishri, H. L. Bhat et al., "Crystal growth and characterization of a new nonlinear optical material: urea L-Malic acid," *Journal of Crystal Growth*, vol. 253, no. 1–4, pp. 460–466, 2003.

[4] S. Moitra and T. Kar, "Studies on the crystal growth and characterization of urea l-malic acid single crystals grown from different solvents," *Materials Letters*, vol. 62, no. 10-11, pp. 1609–1612, 2008.

[5] L. Zhu, J. Zhang, D. Chen et al., "Characteristics of an organic nonlinear optical material urea L-malic acid film," *Materials Letters*, vol. 60, no. 13-14, pp. 1740–1743, 2006.

[6] A. Deepthy, S. Vanishri, D. Ambika et al., "Photoacoustic investigations on thermal anisotropy in urea l-malic acid single crystals," *Materials Research Bulletin*, vol. 43, no. 7, pp. 1641–1648, 2008.

[7] E. de Matos Gomes, V. Venkataramanan, E. Nogueira et al., "Synthesis, crystal growth and characterization of a new nonlinear optical material—urea L-malic acid," *Synthetic Metals*, vol. 115, no. 1, pp. 225–228, 2000.

[8] P. Selvarajan, J. Glorium Arul Raj, and S. Perumal, "Characterization of pure and urea-doped γ-glycine single crystals grown by solution method," *Journal of Crystal Growth*, vol. 311, no. 15, pp. 3835–3840, 2009.

[9] C. Krishnan, P. Selvarajan, and S. Pari, "Synthesis, growth and studies of undoped and sodium chloride-doped Zinc Tris-thiourea Sulphate (ZTS) single crystals," *Current Applied Physics*, vol. 10, no. 2, pp. 664–669, 2010.

[10] R. J. Davey, J. W. Mullin, and M. J. L. Whiting, "Habit modification of succinic acid crystals grown from different solvents," *Journal of Crystal Growth*, vol. 58, no. 2, pp. 304–312, 1982.

[11] M. Senthil Pandian and P. Ramasamy, "Growth and characterization of solution-grown tetra glycine barium chloride (TGBC) single crystals," *Journal of Crystal Growth*, vol. 310, no. 10, pp. 2563–2568, 2008.

[12] I. Quasim, A. Firdous, B. Want, S. K. Khosa, and P. N. Kotru, "Single crystal growth and characterization of pure and sodium-modified copper tartrate," *Journal of Crystal Growth*, vol. 310, no. 24, pp. 5357–5363, 2008.

[13] K. Betzler, H. Hesse, and P. Loose, "Optical second harmonic generation in organic crystals: urea and ammonium-malate," *Journal of Molecular Structure*, vol. 47, pp. 393–396, 1978.

[14] C. J. Pouchert, *The Aldrich Library of Infrared Spectra*, Aldrich Chemical Company, Milwaukee, Wis, USA, 3rd edition, 1981.

[15] B. Sivasankari and P. Selvarajan, "Characterization of urea L-malic acid (ULMA) crystals grown in water, acetone and ethanol," *Journal of Experimental Sciences*, vol. 1, pp. 1–3, 2010.

[16] B. Helina, P. Selvarajan, and A. S. J. Lucia Rose, "Structural, optical, thermal, Mechanical and dielectrical characterizations of γ-glycine crystals grown in strontium chloride solution," *Physica Scripta*, vol. 85, pp. 055803–055808, 2012.

[17] P. V. Dhanaraj, C. K. Mahadevan, G. Bhagavannarayana, P. Ramasamy, and N. P. Rajesh, "Growth and characterization of KDP crystals with potassium carbonate as additive," *Journal of Crystal Growth*, vol. 310, no. 24, pp. 5341–5346, 2008.

Recent Developments in the Research of Splat Formation Process in Thermal Spraying

Kun Yang, Min Liu, Kesong Zhou, and Changguang Deng

Department of New Materials, Guangdong General Research Institute of Industrial Technology (Guangzhou Research Institute of Non-Ferrous Metals), No. 363 Changxing Road, Tianhe District, Guangzhou 510651, China

Correspondence should be addressed to Kun Yang; yangkun20021357@hotmail.com

Academic Editor: Iwan Kityk

Thermal spraying is a well-established surface modification technology which has been widely used in industrial applications. As the coating properties were mainly determined by the flattening nature of each splat, much attention has been increasingly paid to the study on the splat formation process of thermal sprayed particles. This paper is concerned with the development in the research of the splat formation process of the individual splat deposited by thermal spraying during the past few decades, including the experimental and numerical simulations up to today; some classical splashing models were also reviewed. As a simulation of the actual thermal spray process, the development of the flattening behavior of free falling droplet has been mentioned as well. On the basis of the current investigation, some recommendations for the future work have been advised.

1. Motivation to Study Splat Formation Process

"God made the bulk; the surface was invented by the devil" [1]. This comment indicates that the surfaces of parts are the location for many phenomena between the material and its environment, where all physical and chemical interactions and exchanges take place [2]. As most of the material damage and failure start from the surface of the material, beside the so-called nanotechnology, surface modification technology has been quickly developed. This method is a hopeful way to realize the sustainable human society in future, as it enables both high performance and recyclability of the material, through modifying the surface of a material by bringing physical, chemical, or biological characteristics different from the ones originally found on the surface of a material, which have been attracting a great deal of attention from various industries, as they present a way to get an entirely different material performance from the surface of materials merely through surface engineering techniques [3].

In general, thermal spraying is such a typical process that can provide thick coating ranging from 20 μm to several millimeters on the substrate. This method has been developed since the early 1910 when Dr. Schoop introduced the technique with a flame as the heat source and molten lead for feedstock materials [4], while both electrical (plasma or arc) or chemical means (combustion flame) were usually used as the source of energy for thermal spraying nowadays.

Using this technology, thick coatings can be deposited on the substrate over a large area at a high deposition rate as compared to other coating processes in a very short time; in addition, almost all kinds of materials feed in powder or wire form can be used as the feedstock, including metals, alloys, ceramics, plastics, cermets, and composites. Accordingly, it has been increasingly applied in various fields. Including mechanics, aeronautics, aerospace, chemistry and oil, electronic, military, automotive, medical, marine, and mining, and their development has continuously increased over the last decade [5–11], for example, typically used as thermal barrier coating, TBCs, in power plants; this method has also been widely used to protect the substrate or remanufacturing of the surface damaged products.

However, the controllability or reliability of the process has not been established yet until today. Hence, the question of how to improve the thermal-spray technology is quite worthwhile for paying more attention to it.

The coatings fabricated by thermal spraying are always built up on the roughen substrate by the impingement of the molten or semimolten particles, which flatten after impact and solidify quickly, forming a layer. At the beginning of the coating buildup, particles impact directly onto the substrate. The phenomena occurring at this stage determine the adhesion of the coating to the substrate. Following this, the next layer is deposited on the top of the first one, until a coating of the desired thickness is obtained [12]. As the individual splat is unit cell for the entire coating buildup, the flattening and solidification of the individual particle on the substrate surface is the fundamental process for the coating fabrication. Coating microstructure and corresponding properties, such as porosity and adhesion strength, depend strongly on the flattening nature of each splat [13]. Therefore, it is necessary to make clear of the splat formation mechanism of the thermal sprayed particles to establish the controlling way for the coating fabrication.

Consequently, splat characteristics can be recognized by measuring the in-fight condition and investigating on the final splat shape. The desired splat formation can be obtained by changing the spraying parameters and splat-substrate materials combinations and contact nature. By selecting the optimum conditions, thermal spray process can be controlled effectively.

In general, the deposition process of the sprayed particles can be divided into three steps as illustrated in Figure 1. Firstly, the feedstock material was accelerated and heated by the flame prior to impact onto the substrate, then the molten or half-molten particle with high velocity impacts onto the substrate; following this, the molten particle spreading on the substrate surface was driven by the dynamic impact pressure and inertial of the particle; finally, it solidified on the substrate surface with various shapes.

Although the feedstock powder size used in the spraying is always ranging from several tens to several hundreds micrometers, which might not be indentified with the human eyes, the splat flattening and solidification process should be determined by the joint effects of many factor which will be introduced in Section 2. On the other hand, due to the complex shape of the semimolten particles [14] on rough substrate [15, 16], most of the common studies are focused on a smoothed flat surface. According to controlling the spraying parameters and splat-substrate natures, many types of splats can be obtained. The investigation is always focused on two typical types of splats as shown in Figure 2. Splash splat with a center splat surrounded by a ring of fragments or numerous splash fingers connect with the central solidification core as shown in Figure 2(a), and disk-shaped splat with only a regular central disk as shown in Figure 2(b) can be commonly obtained on the flat substrate surface under designated conditions.

Various investigations on splat formation process have appeared in the literature during the past few decades, and many theoretical, analytical, and experimental studies have been conducted. However, as the current understandings seem still not applicable for all the splat formation problems, no definitive agreement has been achieved so far. With the purpose to understand the current status of the investigation on the splat formation process and control the actual thermal spraying more efficiency, this paper is devoted to giving an overview on the splat formation problem of the individual splat deposited by thermal spraying. Lastly, the prospect for the future study on this issue is given.

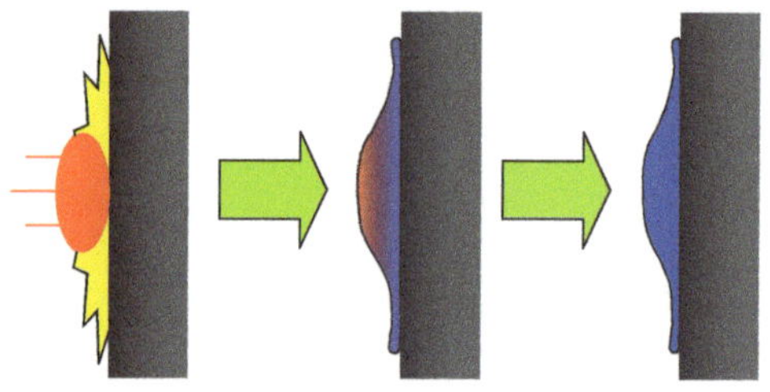

Figure 1: Scheme of the splat formation process of a thermal sprayed particle.

2. Splat Transition Behavior in Thermal Spraying

A transition phenomenon in a flattening behavior of the thermal sprayed particle on the flat substrate surface was firstly introduced in 1995 [17], which reported that when the substrate temperature was increased above one critical temperature, the splat shapes of most materials sprayed onto flat substrates underwent a transition from a distorted shape with splash to a disk-shaped splat. The transition temperature, T_t, as a critical substrate temperature over which more than 50% of splats are disk shaped, can be considered as a useful tool for the process control.

Figure 3 shows how the fraction of disk-shaped splats varies with the substrate temperature for Ni particles sprayed on an AISI304 stainless steel substrate, and how the transition temperature is defined [13, 17–19]. Actually, most of the feedstock materials sprayed onto the flat substrate exist such a transition behavior through increasing substrate temperature [22, 26–44]. It is assumed that the adsorption/desorption of adsorbates and condensates on the substrate surface [22, 29–33], surface characteristic change [33–36], and thermal contact resistance [37, 38], along with the wetting induced by the topography change might dominate the transition [45, 46]. It is quite interesting and practically meaningful that the transition in the splat shape of an individual particle corresponds well to the transition in the adhesion strength of the coating fabricated at corresponding temperature conditions [13, 47].

On the other hand, controlling the ambient pressure is another convenient way to optimize the actual coating fabrication process. The spraying is conducted in protective environments using vacuum chambers normally back filled with a protective gas at low pressure. This method has been attracting a wide attention because coating particles can be accelerated to much higher speeds without oxidation to metal, and the reaction of the coating materials with plasma operating gases during spraying can be controlled. However, only few attentions have been paid to the splat formation process under reduced pressure conditions. Despite this, existence of a similar transition was indicated recently as indicated in Figure 4 [48, 49], that is, the transition from splash

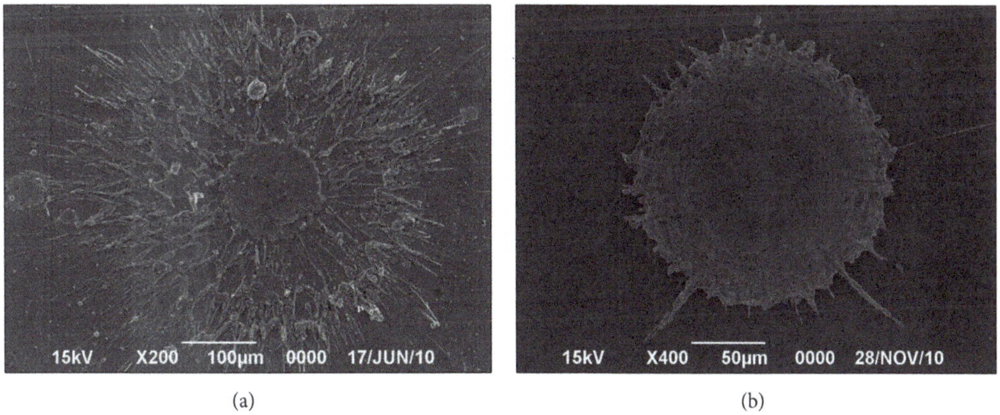

FIGURE 2: Typical splat morphology: (a) splash splat and (b) disk-shaped splat.

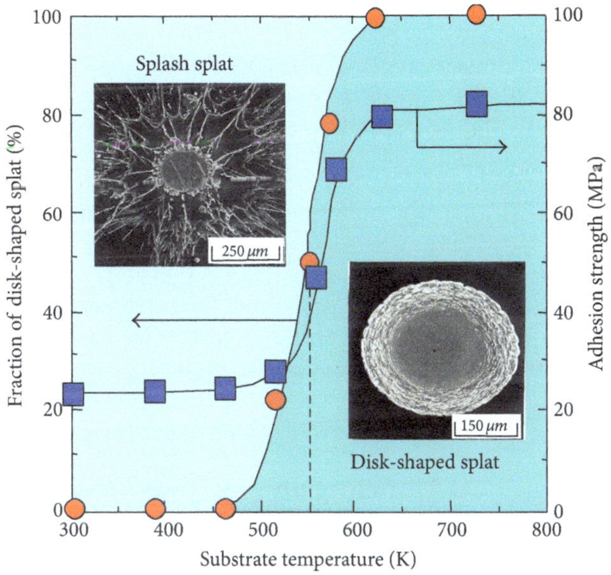

FIGURE 3: Dependence both of fraction of disk-shaped splat and coating adhesion strength on substrate temperature [13, 17–19].

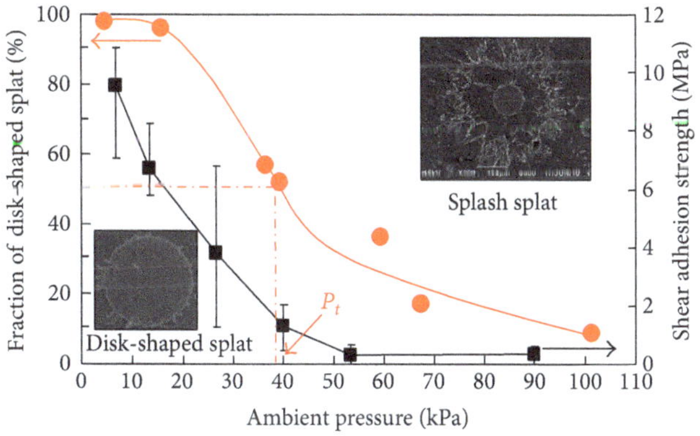

FIGURE 4: Dependence both of fraction of disk-shaped splat and coating adhesion strength on ambient pressure [20, 21].

splat to disk-shaped one occurred by reducing the ambient pressure [18, 20, 21, 48–51]. Fukumoto et al. systematically investigated the effect of desorption of adsorbates on particle flattening behavior in low pressure plasma spraying, while the substrate temperature being kept at room temperature [18, 19]. Transition pressure, P_t, was defined as the critical ambient pressure at which half of the splats were disk type. The transition pressure is characteristic of the surface: no chemical modification and topography change of the surface occurs when pressure is reduced, and desorption is the only possible physical change taking place. However, it is not fully understood yet. Similarly, the coating adhesion strength has a close relation with the shapes of the depositions through changing ambient pressure [21].

To summarize, splat shapes changed transitionally from splash type to disk-shaped one as increasing substrate temperature or reducing ambient pressure. The coating adhesion strength has a close relation with the splat shape. Generally, three different bonding mechanisms involved in thermally sprayed coating: chemical bonding, physical bonding, and mechanical bonding [52–56]. Mechanical bonding usually prevails for most coatings. As the individual splat is the unit cell for the coating buildup, if all the splats adhesive well with the substrate, favorable adhesion strength can be expected for the coatings. On the contrary, poor adhesion strength might exist between the coating and substrate surface. Accordingly, the central disk-like part of the splat should be good adhesive with the substrate, because of that the dynamic impact pressure of the in-flight particle prior to the impinging should be perpendicular to the substrate surface. When the splashing or fragmentation of the splat occurred, the flight direction of the splashing fingers should parallel to the substrate surface, no dynamic pressure might have an effect on the splashing fingers. Hence, few interlocking might be formed. In other words, the existence of splash finger might be the weak point for the adhesive between splat and substrate surface.

3. Experimental Investigation on Splat Formation Process

The initial stage of the splat flattening behavior always focused on the spray parameters [26, 44, 57–60]. A group of spraying parameters were systematically investigated by Li et al. [58]; the splat flattening behavior reveals that the particle's kinetic energy and its molten status are critical in determining the splat morphology. Salimijazi et al. [44] also noted that the in-flight particle temperature and velocity at the point of impact have a significant effect on the final morphology of the splat. On the other hand, the influence of arc current, operating gas flow, and spraying distance on the velocity and temperature of sprayed particles was also measured in the previous study [59, 60]. However, their results proposed that there was no significant change in velocity and temperature with these process parameters.

Instead of the particle oriented factors, feedstock materials on the splat formation process was also evaluated [27, 61–66]; the particle physical properties such as thermal conductivity, surface tension, latent heat, and oxide formation free energy were considered. The influence of substrate materials [61, 67, 68], in particular, the influence of thermal conductivity for different substrate materials was strongly mentioned. It was found that the smaller the thermal conductivity of the substrate materials, the easier the splat deposited on the substrate as disk-shaped splat. Consequently, the splashing phenomenon was influenced considerably by the initial solidification of the spreading droplet onto flat substrate surface. While for the same substrate material, the state of substrate surface such as oxidation state [26, 27, 39, 69], surface roughness and topography modification through substrate polishing or grit blasting [70–72], heating [33–36, 44, 73], laser treatment [33, 42], and micropattern [74–76] was deeply investigated.

For the splat morphology transition phenomena, most studies have focused on the effect of surface roughness or topography change on splat morphology, and little attention was paid to the surface absorbates until Li et al. [22, 77] initially simulated the surface adsorbates/condensates by a prelaid organic substance. They proposed an evaporated-gas model on splat splashing and, as a result, the transition temperature was regarded as creating the desorption process on the surface. In further research by Jiang et al. [31], it was shown that the ZrO_2 splat morphology changes progressively with the decrease in the surface adsorbates/condensates content. Fukumoto et al. also proposed the effect of desorption of adsorbates and condensates on the splat collected under reduced ambient pressure [18, 19]. Meanwhile, Hyland et al. [52, 78–80] studied the role of surface chemistry on the splat formation process by the immersion of the substrate in boiling water to grow specific types of oxide/oxyhydroxide on the surface. Their results suggested that the splat morphology was more strongly influenced by water release from the dehydration of oxyhydroxide to oxide than by the simple presence of the oxide layer on the substrate surface.

It has been proposed that the wetting of droplet to substrate surface may have an effect on the droplet flattening [18, 19, 26, 27, 33, 45, 46, 62, 81, 82]. The degree of wetting is always described by the contact angle (θ) at which the liquid-vapor interface meets the solid-liquid interface [83]. The contact property of the splat with substrate surface strongly depends on the wetting of the liquid metal to the solid substrate; Suganuma reported that the standard energy of formation of the oxide layer from the metal can be closely related to the wetting of the molten metal on the oxide substrate [84], while the wetting of a liquid metal relative to a solid oxide depends on the thermodynamics of the oxide material; that is, the more thermodynamically unstable the oxide, the easier the wetting [85], which may affect strongly the splat formation process.

Numerous observations have been conducted through changing spraying parameters and powder/substrate combinations. However, thermal spraying is a complex and short-period process; it is difficult to clarify the flattening behavior of the thermal sprayed particles directly with preventing technology. Most of the previous investigations were carried out by observing the geometry shape of the splat but without knowing the actual flattening and solidification process, because of the difficulties in performing high resolution studies of the splat-substrate interface. To overcome this difficulty, the pore and grain distribution at cross-section of the individual splat was observed by the assistance of focus ion beam (FIB) and transmission electron microscopy (TEM) [12, 52, 86–90]. The results proposed that different microstructures could be found under designated conditions, the substrate chemistry, roughness, and temperature all play an important role in the splat formation process, but the substrate temperature effect is the most prominent of these.

4. Numerical Modeling on Splat Formation Process

In the actual thermal spraying process, the diameters of the feedstock powders were always within several tens micrometers, while the splat flattening and solidification finished in a very short time in several microseconds order. It is very difficult to observe the splat formation process with the prevailing technology. Instead, some numerical models have been carried out extensively for verifying the possible influencing factors on the splat formation process [38, 80, 91–105].

The first numerical model of a liquid drop splashing onto a flat surface was proposed by Harlow and Shannon [92]. The full Navier-Stocks equations were solved in cylindrical coordinates in order to investigate the splash of a liquid drop onto a flat plate. The model was mainly applicable for the early stage of droplet impact when kinetic energy was dominant, because the effects of surface tension and viscosity were neglected to simplify the model. Another simple model using a simple model of two-dimensional radial flow of a spherical droplet was proposed by Madejski [93]. More physical characteristics of the droplet were considered in his model; the degree of flattening of a droplet depended upon the Weber, Reynolds, and Peclet numbers, and upon the freezing constant.

Since then, more and more scientists focused on the numerical simulation to investigate the splat flattening and solidification process. Firstly, Sobolev and Guilemany [94–100] theoretically studied the effects of many factors on splat flattening and solidification. Their theoretical results agree well with the experimental observations. A splat formation model has been developed based on the classic nucleation theory accounting for heterogeneous nucleation kinetic and crystal growth to predict the nucleation temperature and grain size distribution [101], the influence of the substrate material, interfacial thermal contact resistance, and wettability on the nucleation, grain size distribution, and rapid solidification process has been investigated. Another integrated model for the interaction between melt flow and nonequilibrium solidification in thermal spraying was developed by Wang et al. [102]. The effects of velocity profiles, droplet size, initial substrate and droplet temperature, and initial impinging velocity on rapid solidification have been investigated. In order to clarify the micropore formation mechanism, a theoretical model was proposed by H. Fukanuma [103], his study indicated that the impinging velocity, the ambient gas pressure, the particle diameter, and the molten material viscosity contribute to producing porosity.

In recent years, 3D mathematical models of acceleration, heating, and deformation of the particles have been developed to predict the particle transport and splatting on impact with a flat substrate were discussed. A 3D model for the splatting of high-velocity oxyfuel (HVOF) sprayed polymer particles on rough surfaces has been also developed recently [106]; the results indicated that an increase in the magnitude of the mean substrate roughness promoted splat instability and formation of radial fingers, while an increase in general surface roughness may result in a lower spreading ratio of thermally sprayed polymer particles. Three-dimensional molecular dynamics simulation with considering the influence of the impact parameters to clarify at an atomic level the flattening process of a high-temperature droplet impacting a substrate at high speed [104], the flattening ratio increases in proportion to the droplet velocity and the droplet diameter in their study. In addition, the deposition behavior of sprayed particles on inclined substrate was also numerically studied [105].

5. Splashing Models for Thermal Sprayed Particles

The splashing phenomenon is frequently found during thermal spraying process. As reviewed above, in order to verify the mechanism of splashing occurrence of splat deposited on flat substrate, much attention has been increasingly paid in the past few decades. Some classical splashing models have been reported up to date; however, they still have not been fully agreed yet.

5.1. Adsorbed Gas Induced Splashing Model. One of the most widely mentioned models proposed is evaporated-gas-induced splashing model as proposed in Figure 5 [22]. The substrate surface is covered with several physical adsorbed condensates prior to the droplet impact onto the flat substrate, the most common condensate on which is water from moisture [22, 31]. Once the molten droplet was thermally sprayed onto the substrate, desorption of adsorbed water on the substrate surface occurred to form a gas barrier between the splat and substrate due to the intensively rapid heating of the substrate surface [30, 107]. The adsorbed gas induced pressure increased rapidly with the progress of splat flattening process, while the dynamic impact pressure spread out and dissipated quickly with splat flattening. Once the gas-induced pressure became larger than the impact pressure, the flowing fluid would be pushed up or thrown away in jetting by inertial of flowing fluid [22] and deposited as splash splat with a core central zone surrounded by the ring of debris.

However, some phenomena still cannot be explained by this assumption. For example, when the substrate was preheated up to one critical temperature, most of the adsorbed gas condensation has been removed completely [32, 108], with the continuing increase of substrate temperature above this critical temperature, there might be no significant change of surface adsorbates, while the initial dynamic impact pressure is a constant value due to the no change of spraying parameters; however, more and more disk-shaped splats deposited on the substrate surface instead of splash ones with higher temperature, so how does the transition of splat shape occur? Furthermore, even when the substrate was kept at a constant temperature, that is to say, there was no adsorption/desorption change on the substrate surface, different splat morphologies were obtained through changing the powder materials [61–64]. Therefore, some new criterions should have an effect on the splat formation.

5.2. Substrate Melting-Induced Splashing Model. As proposed by the previous researchers, thermal sprayed coating was always mechanical bonded with the substrate [52–56]; however, the substrate melting at the splat-substrate interface during splat collection process was also reported recently with both experimental and numerical simulation [23, 78, 101, 109–113].

Li et al. [23] reported that impacting of a molten droplet with a melting point much higher than the substrate resulted in melting of the substrate around the impact area, which might induce the splashing occurrence. Kitahara and Hasui [109] reported that substrate melting occurred when nickel, chromium, molybdenum, and tungsten were plasma-sprayed onto aluminum and mild steel substrates. Steffens et al. [110] indicated that substrate melting strongly depended on the droplet thermophysical properties, contact temperature, and solidification time. Increasing the substrate or droplet temperature or spraying on a substrate with a low value of thermal conductivity enhanced the degree of melting. Zhang et al. [101] also detected an intermetallic layer between the molybdenum splat and the mild steel substrate and suggested that substrate melting occurred. Melting of the substrate, along with intermixing and diffusion between substrate and splat materials, was found when the nickel-chromium powder was thermally sprayed onto the flat stainless steel 304 substrate, but not for the aluminum substrate [78, 111–113].

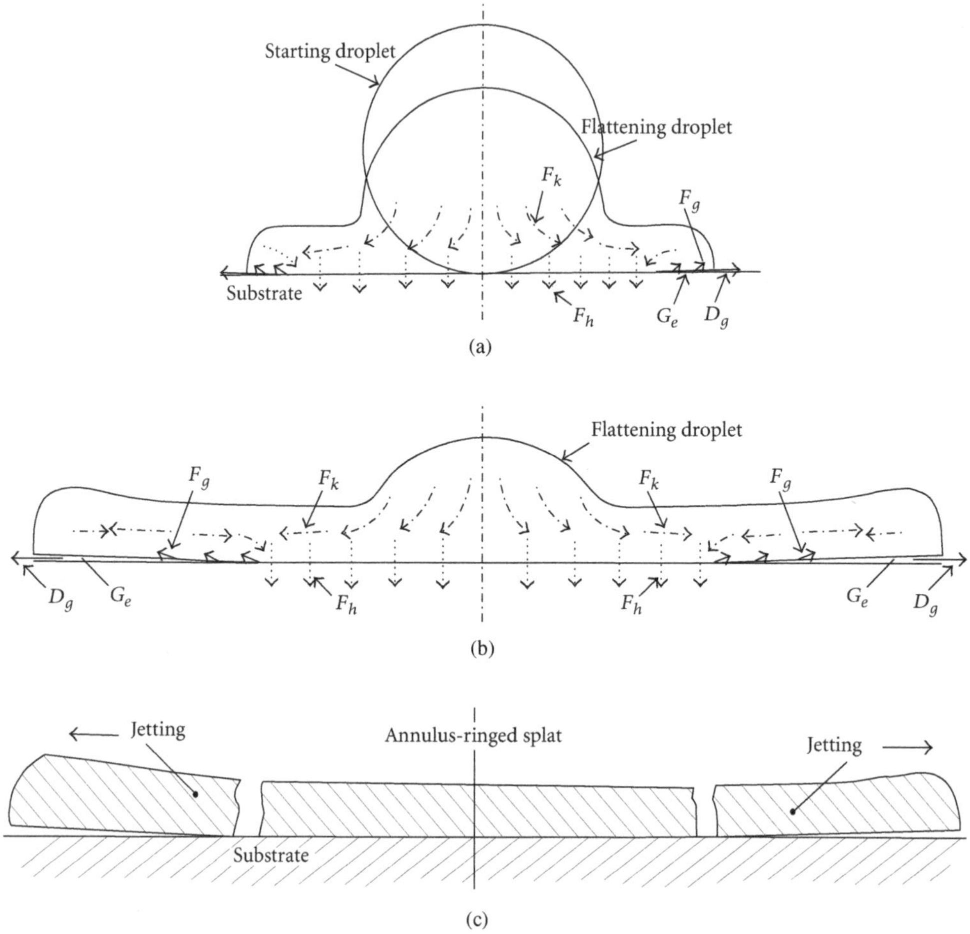

FIGURE 5: Evaporated-gas-induced splashing model [22]. (a) The early stage of flattening droplet; (b) formation of gas cushion and escape of the accumulated gas, (c) formation of fluid jetting. F_k: kinetic energy driven flow; F_g: evaporated gas driven flow; F_h: heat flow; G_e: evaporated gas; D_g: gas escape direction.

Substrate melting-induced splashing was reported by Li et al., on the basis of experimental observation as indicated in Figure 6 [23]. It was proposed that once the melting point of a molten droplet was much higher than the substrate, the substrate surface temperature would rise continuously due to the heat transfer from molten droplet to cold substrate, which would cause local melting of the substrate around the impact area. Due to a high dynamic pressure of the in-flight prior to the impinging, the melted substrate material would be extruded to flow along the solid surface below the liquid droplet, resulting in the formation of a crater around the impact point. The significant change of fluid flow direction led to the detaching of fluid from the substrate. When the dynamic pressure downward was not high enough to suppress the detachment of the fluid from the solid surface as the crater became large, the ejection of fluid occurred owing to the kinetic inertia of the flowing liquid. This led to the ejecting of droplet material in the form of splashing.

5.3. K Model. The Sommerfeld number, K, based on the study of water and ethanol droplet, has been often mentioned recently [114, 115], which was described as the following equations:

$$K = \mathrm{We}^{1/2}\mathrm{Re}^{1/4},$$
$$\mathrm{We} = \frac{\rho d v^2}{\gamma}, \qquad (1)$$
$$\mathrm{Re} = \frac{\rho d v}{\eta}.$$

We and Re were the Weber and Reynolds number. Here, ρ ($\mathrm{kg{\cdot}m^{-3}}$) is the droplet density; d (m) is the droplet diameter; v ($\mathrm{m{\cdot}s^{-1}}$) is the mean droplet velocity; γ ($\mathrm{N{\cdot}s{\cdot}m^{-2}}$) is dynamic viscosity of the droplet; η ($\mathrm{N{\cdot}m^{-1}}$) is the surface tension of the droplet, respectively.

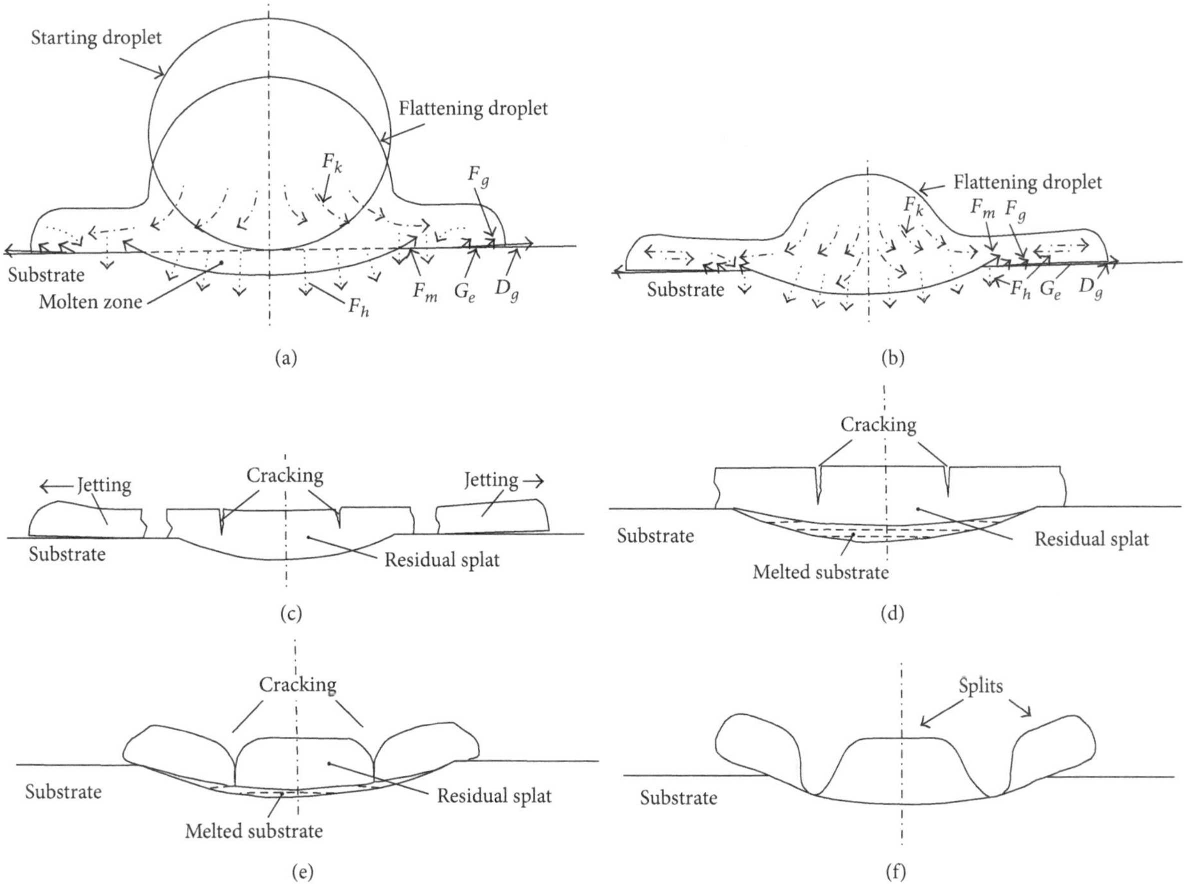

FIGURE 6: Substrate melting-induced splashing model [23]. F_k indicates the flow direction of the spreading fluid driven by kinetic energy. F_h is the heat transfer direction from the droplet to the substrate. (a) and (b) indicate the early stages of flattening, resulting in the melting of the substrate by the impacting droplet. F_m indicates the flow direction of the spreading fluid resulting from the crater, which is formed by the melted substrate surface region. F_g indicates the direction of the evaporated gas-induced force. When flattening completes (c), the splashing occurs, driven by the inertial force of the rapidly flowing fluid along direction F_m. Cracking occurs as the surface layer of the splat is solidified, as shown in (c) and (d). The displacement by floating of the cracked splits on low-melting-point liquid film results in the formation of splitting of the residual central splat (e) and the final splat (f).

The criteria for describing the nonsolidified liquid droplet on flat substrate surface were defined as [114]

if $K < 3$, splat rebounds;

if $3 < K < 57.7$, it results in deposition;

if $K > 57.7$, it induces splashing.

Therefore, the value of 57.7 could be treated as the critical value as K_c the predict the splashing occurrence or not. The K value of thermal sprayed particles can be precisely calculated through first measuring the particle temperature and velocity, and then introducing the measured temperature and velocity into each physical constant, such as density, viscosity, and surface tension. On the basis of the calculation, these critical values seem not applicable for actual thermal spraying conditions under some experimental conditions. These K values are significantly larger than that of K_c regardless of the particle size, or the spray distance [18, 19]. That is to say, regular sprayed particle has enough driving force for the splashing. However, the disk-shaped splat always appeared on the hot substrate, or the lowered ambient pressure in the actual thermal spraying. Escure et al. also proposed that, significant splashing occurred even if the K was smaller than the critical value of 57.7 when alumina particles were thermally sprayed onto the stainless steel substrate [116]. While in Li's study, nearly all the plasma sprayed particles yielded a K from several hundreds to thousands, which was quite higher than the critical value, but the splats deposited by those particles on preheated substrate always displayed a disk shape without any splashing [117].

Actually, the K model is independent of the particle properties. However, some of the transitional change in splat morphology under designated conditions cannot be explained by the traditional K. The splat transition phenomenon was reportedly realized just through controlling substrate conditions with no change of the in-flight particles. Consequently, a new independent criterion for splashing, the splashing parameter on flattening K_f, was defined by

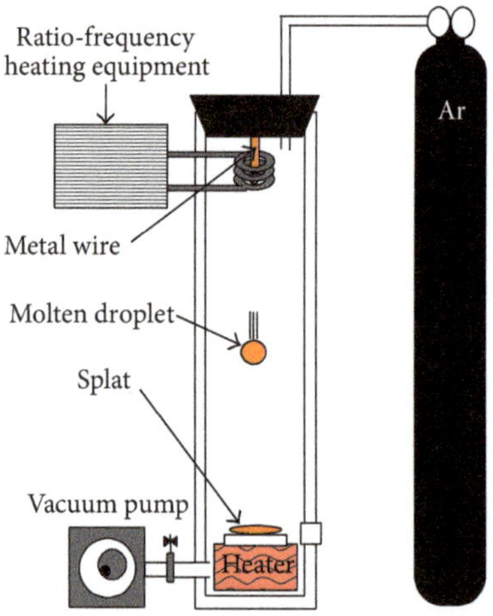

FIGURE 7: Scheme of free falling experimental apparatus [24].

Fukumoto et al. [24, 59, 118], by integrating the substrate temperature effect and the particle conditions:

$$K_f = 0.5a^{1.25}\text{Re}^{-0.3}K. \quad (2)$$

Here, a is defined as the ratio of the maximum flattening velocity to the impact velocity of the particle and it is influenced by the substrate temperature. However, the detailed physical meaning of the above expression is still not clearly defined.

5.4. Solidification Parameter Model. A splat shape predicting model was proposed by Dhiman and Chandra et al. [40, 53, 119]. The solidification parameter () which was defined as the ratio of the solid layer thickness to splat thickness was used to predict the splat morphology in a thermal spray process as illustrated in the following:

$$\Theta = \frac{s}{h} = \frac{s/D_0}{h/D_0} = \frac{s^*}{h^*}. \quad (3)$$

Here, D_0: droplet diameter before impact; h: splat thickness; h^*: dimensionless splat thickness ($= h/D_0$); s: thickness of droplet's solidified layer; s^*: dimensionless solid layer thickness ($= s/D_0$).

The criteria for predicting the final splat shape of thermal sprayed particles on flat substrate surface were defined as follows:

(1) If the solid layer is very thin ($s \ll h$, or $\Theta \ll 1$), which might has no effect on spreading, the splat would spread into a thin sheet liquid and rupture internally.

(2) Once the solid layer grows by a significant amount during spreading ($s \sim 0.1h$ to $0.3h$, or $h \sim 0.1$–0.3), it would restrain the splat from spreading too far and becoming thin enough to rupture. In this case, a disk-shaped splat will be produced.

(3) If the solidification is very rapid ($s \sim h$ or $\Theta \sim 1$), the solid layer will obstruct the outward spreading liquid.

6. Development in Research of Free Falling Droplet

Recently, free falling experiments, in which a metal droplet fell freely and impinged on a flat substrate, was conducted to simulate the thermal spray process [24, 25, 37, 37, 120–125]. Figure 7 shows the schematic of the experimental apparatus [24], in which millimeter-sized metallic droplet was heated and melted by radio-frequency heating equipment prior to the falling, the substrate temperature and ambient pressure can be controlled.

Fukumoto et al. observed the final shapes of the free falling Cu droplet collected on AISI304 substrate through controlling substrate temperature and ambient pressure. The droplet shapes were summarized in Figure 8 [25]. In their experiment, the chamber pressure was kept at atmospheric pressure when varied the substrate temperature, while no substrate heating was conducted when varied the ambient pressure. From the figure, the droplet shapes changed transitionally from splash type to disk-shaped one both by increasing substrate and reducing ambient pressure. In other words, the droplet shape has a similar transition tendency with the thermal spraying process. Accordingly, the investigation on the flattening behavior of thermal sprayed particles through observation on the free falling droplet should be reasonable and meaningful. The relating factors, such as thermal contact resistance and the droplet velocity at impact, were investigated carefully. According to their studies, the thermal contact resistance was determined by the area of contact between the molten droplet and rough solid substrate [120]. Furthermore, the number of fingers formed around the droplet increased with impact velocity [120, 121].

Thermal history at droplet-substrate interface during the droplet spreading and solidification process was proposed in the recent literature. A hole was drilled at the center of substrate, and through this hole, J-type thermocouple was inserted. The thermo-electromotive force was converted to digital signal and recorded by a data logging system [37, 122, 123]. Fukumoto et al. measured the interface temperatures varied with spreading time after the molten droplet impacted onto the flat substrate, and experimentally controlled the dominations of both substrate temperature and ambient pressure [49, 123]. Their results indicated that the substrate temperature and ambient pressure had an equivalent effect on the solidification rate of the spreading droplet and that faster solidification could be achieved on the substrate holding at high temperature and/or locating under reduced ambient pressure. Similar results had been proposed when some kinds of micrometer-sized particles were thermally sprayed onto flat substrate surfaces [30, 37–39]. In a word, the cooling rule of the droplet is opposite to the theory of Newtonian cooling [126], but the mechanism is still not clear at this moment.

7. Prospect for the Future Study

Although various works on the splat formation process have been investigated during the past few decades, some of the splat formation mechanisms are still not clarified. The key

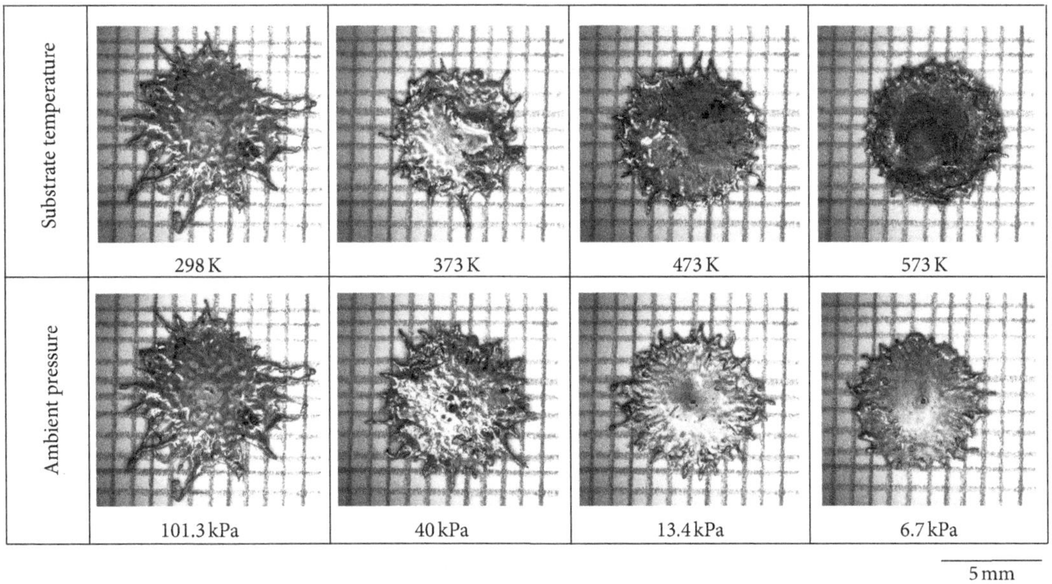

FIGURE 8: Dependence of splat morphologies on the substrate temperature and ambient pressure in the deposition chamber [25].

issues to be answered include why or how does the disk-shaped splat appear instead of splashing occurrence and what is the essential of the flattening problem? The author advises that continuing efforts should be devoted to making a deeper understanding on the splat formation process.

(1) Lots of observations on the splat morphologies, for example, the splat geometry shapes and pore distributions, have been widely conducted. However, it is still necessary to characterize the splat-substrate interface at high resolution. Thus, transmission electron microscope (TEM) is advised to observe the solidification structure at the interface in more detail after cutting by focus ion beam (FIB). In particular, the difference of the structures between disk-shaped splat and splash splat collected under designations should be compared systematically.

(2) As the splat formation process always finishes in several to several tens microseconds, and the feedstock size is about only several tens micrometers, it is quite difficult to observe the splat formation process directly with the prevailing technology; thus, a numerical simulation has been widely employed to clarify the flattening and solidification process. However, not enough efforts through controlling ambient pressure have been carried out yet. Therefore, some contributions to this aspect are strongly recommended.

(3) The role of wetting on the flattening behavior of thermal sprayed particles has been proposed in some literature, but most of the studies were focused on the static wetting, which was a thermal equilibrium process and dominated by the surface energy of the molten droplet. During the actual splat formation process, the splat boundary was in motion until the flattening finished, so it is dynamic wetting of substrate by molten droplet. Does the dynamic wetting have an effect on the splat formation process in only several microseconds? If so, how does the dynamic wetting affect the splat flattening process?

(4) Equivalent transition behavior of the splat geometry shaping from splash to disk-shaped ones was found through both increasing substrate temperature and reducing ambient pressure; nevertheless, there existed different microstructure and solidification behaviors. Consequently, it is quite necessary to make clear what the dominant factors in different experimental designations are.

(5) Much attention has been paid to the flattening behavior of thermal sprayed particles under reduced ambient pressure. However, it is quite difficult to get enough coarse depositions under low pressure conditions. Is this a technical problem or do any special phenomena occur? One assumption is that complete fragmentations occur once the molten droplet impacts onto flat substrate surface and that the fragmentations are deposited as smaller disk-shaped splats. However, no such deposition process can be proved through observing the final splat shape. Consequently, some technical improvements to collect more splats under reduced ambient pressure should be conducted, and more attention should be paid to the impinging behavior under low pressure designations.

(6) Free falling experiment was proved as a useful method to simulate the actual thermal spraying process. The previous results indicated that the flattening pattern of the molten droplet was decided just after collision onto solid surface, which was earlier enough to

finalization of the flattening. But we cannot explain how they decide so quickly? And how the influencing factors affect this decision? Therefore, intensive research is needed.

(7) Coatings are built up from the individual particles deposited on the substrate or the previous deposited splat; therefore, the study of splat formation process was not only from the point of view of scientific interest, but also from that of technical consequences as well. Consequently, the investigation on the splat problem should be concerned on the materials with industrial background, for example, materials for new generation thermal barrier coatings (TBCs) and wear coatings.

(8) Coating properties depend strongly on the single splat shape, because the individual splat is the unit cell for the coating buildup, but the previous study was mainly focused on the adhesion strength between the coating and substrate. The coating adhesion is the most fundamental characteristic for the entire coating, a coating will never protect a substrate if it does not adhere to it; however, the dependence of the other coating properties on splat shape should be paid a special attention to.

8. Summary

With the purpose to understand the research development of the splat formation process in thermal spraying, this paper reviewed the investigation activities on this aspect during the past few decades, the transition behavior of splat shape changing from splash type to disk-shaped one through controlling substrate temperature or ambient pressure, and the contributions of experimental and numerical modeling to the splat formation process, followed by the review of some classical splashing models. The research development of free falling experiment was also mentioned. Although numerous literature has been published, there are still some problems to be answered on this issue. The intensive research for this issue is still needed.

Acknowledgments

K. Yang would like to acknowledge Professor Fukumoto Masahiro (Toyohashi University of Technology, Japan) for the opportunity to study in his laboratory and for his assistance and discussions throughout the period of this work. This paper was partially supported both by the National Program on Key Basic Research Project (973 Program, 2012CB625100) and by a Grant from the Ph.D. Programs Foundation of Guangzhou Research Institute of Nonferrous Metals.

References

[1] U. Hofer, "Self-trapping of electrons at surfaces," *Science*, vol. 279, no. 5348, pp. 190–191, 1998.

[2] M. Cardon, "How (proper) blast cleaning averted a naval conflict," *The Shot Peening*, vol. 20, no. 2, p. 38, 2006.

[3] Y. Arata, "Advanced new technology for thermal spray," in *Proceedings of the International Symposium on Advanced Thermal Spraying Technology and Allied Coatings*, pp. 1–8, Osaka, Japan, 1988.

[4] P. Fauchais, A. Vardelle, and B. Dussoubs, "Quo Vadis thermal spraying?" *Journal of Thermal Spray Technology*, vol. 10, no. 1, pp. 44–66, 2001.

[5] R. A. Miller, "Current status of thermal barrier coatings—an overview," *Surface and Coatings Technology*, vol. 30, no. 1, pp. 1–11, 1987.

[6] G. Barbezat, "Advanced thermal spray technology and coating for lightweight engine blocks for the automotive industry," *Surface and Coatings Technology*, vol. 200, no. 5-6, pp. 1990–1993, 2005.

[7] J. Matějíček, P. Chráska, and J. Linke, "Thermal spray coatings for fusion applications—review," *Journal of Thermal Spray Technology*, vol. 16, no. 1, pp. 64–83, 2007.

[8] M. L. Thorpe, "Thermal spray. Industry in transition," *Advanced Materials and Processes*, vol. 143, no. 5, pp. 50–61, 1993.

[9] F. Kassabji, G. Jacq, and J. P. Durand, "Thermal spray application for the next millennium a business development perspective," in *Proceedings of the 15th International Thermal Spray Conference*, pp. 1677–1680, May 1998.

[10] M. Ducos and J. P. Durand, "Thermal coatings in Europe: a business prospective," in *Proceedings of the International Thremal Spray Conferences*, pp. 1267–1271, May 2001.

[11] M. Fukumoto, S. Y. Hwang, C. J. Li, K. Tani A, Nakahira, and M. K. A. Khor, "The current status of thermal spraying in Asia," *Journal of Thermal Spray Technology*, vol. 17, no. 1, pp. 5–13, 2008.

[12] G. Bolelli, K. Sabiruddin, L. Lusvarghi, E. Gualtieri, S. Valeri, and P. P. Bandyopadhyay, "FIB assisted study of plasma sprayed splat-substrate interfaces: NiAl-stainless steel and alumina-NiAl combinations," *Surface and Coatings Technology*, vol. 205, no. 2, pp. 363–371, 2010.

[13] M. Fukumoto, H. Hayashi, and T. Yokoyama, "Relationship between particle's splat pattern and coating adhesive strength of HVOF sprayed Cu-alloy," *Journal of Japan Thermal Spraying Society*, vol. 2, no. 3, pp. 149–156, 1995.

[14] H. Ageorges and P. Fauchais, "Plasma spraying of stainless-steel particles coated with an alumina shell," *Thin Solid Films*, vol. 370, no. 1, pp. 213–222, 2000.

[15] P. Fauchais, M. Vardell, A. Vardell, and L. Bianchi, "Plasma spray: study of the coating generation," *Ceramics International*, vol. 22, no. 4, pp. 295–303, 1996.

[16] P. Fauchais, M. Vardell, A. Vardell, L. Bianchi, and A. C. Leger, "Parameters controlling the generation and properties of plasma sprayed zirconia coatings," *Plasma Chemistry and Plasma Processing*, vol. 16, no. 1, pp. s99–s125, 1996.

[17] M. Fukumoto, S. Katoh, and I. Okane, "Splat behavior of plasma sprayed particles on flat substrate surface," in *Proceedings of the 14th International Thermal Spray Conference*, pp. 353–358, 1995.

[18] M. Fukumoto, M. Shiiba, H. Kaji, and T. Yasui, "Three-dimensional transition map of flattening behavior in the thermal spray process," *Pure and Applied Chemistry*, vol. 77, no. 2, pp. 429–442, 2005.

[19] M. Fukumoto, Y. Tanaka, and E. Nishioka, "Flattening problem of thermal sprayed particles," *Materials Science Forum*, vol. 449–452, no. 2, pp. 1309–1312, 2004.

[20] S. Sampath and H. Herman, "Rapid solidification and microstructure development during plasma spray deposition," *Journal of Thermal Spray Technology*, vol. 5, no. 4, pp. 445–456, 1996.

[21] M. Fukumoto, K. Yang, M. Yamada, and T. Yasui, "Splat formation mechanism in thermal spraying," *Journal of Solid Mechanics and Material Engineering*, vol. 5, no. 12, pp. 1001–1014, 2011.

[22] C. J. Li and J. L. Li, "Evaporated-gas-induced splashing model for splat formation during plasma spraying," *Surface and Coatings Technology*, vol. 184, no. 1, pp. 13–23, 2004.

[23] C. J. Li, C. X. Li, G. J. Yang, and Y. Y. Wang, "Examination of substrate surface melting-induced splashing during splat formation in plasma spraying," *Journal of Thermal Spray Technology*, vol. 15, no. 4, pp. 717–724, 2006.

[24] M. Fukumoto, E. Nishioka, and T. Matsubara, "Flattening and solidification behavior of a metal droplet on a flat substrate surface held at various temperatures," *Surface and Coatings Technology*, vol. 120-121, pp. 131–137, 1999.

[25] M. Fukumoto, K. Yang, K. Tanaka, T. Usami, T. Yasui, and M. Yamada, "Effect of substrate temperature and ambient pressure on heat transfer at interface between molten droplet and substrate surface," *Journal of Thermal Spray Technology*, vol. 20, no. 1-2, pp. 48–58, 2011.

[26] L. Bianchi, A. Grimaud, F. Blein, P. Lucchese, and P. Fauchais, "Comparison of plasma-sprayed alumina coatings by RF and DC plasma spraying," *Journal of Thermal Spray Technology*, vol. 4, no. 1, pp. 59–66, 1995.

[27] M. Vardelle, A. Vardelle, A. C. Leger, P. Fauchais, and D. Gobin, "Influence of particle parameters at impact on splat formation and solidification in plasma spraying processes," *Journal of Thermal Spray Technology*, vol. 4, no. 1, pp. 50–58, 1995.

[28] J. M. Houben, "Future development in thermal spraying," in *Proceedings of the 2nd National Conference on Thermal Spray*, pp. 1–19, 1984.

[29] S. Brunauer, P. H. Emmett, and E. Teller, "Adsorption of gases in multimolecular layers," *Journal of the American Chemical Society*, vol. 60, no. 2, pp. 309–319, 1938.

[30] A. McDonald, M. Lamontagne, C. Moreau, and S. Chandra, "Impact of plasma-sprayed metal particles on hot and cold glass surfaces," *Thin Solid Films*, vol. 514, no. 1-2, pp. 212–222, 2006.

[31] X. Jiang, Y. Wan, H. Herman, and S. Sampath, "Role of condensates and adsorbates on substrate surface on fragmentation of impinging molten droplets during thermal spray," *Thin Solid Films*, vol. 385, no. 1-2, pp. 132–141, 2001.

[32] I. A. Polunina, A. A. Isirikyan, K. E. Polounine, and S. S. Mikhailova, "Water influence on the surfactant adsorption on TiO_2," *Colloids and Surfaces A*, vol. 160, no. 2, pp. 141–146, 1999.

[33] M. Fukumoto, H. Nagai, and T. Yasui, "Influence of surface character change of substrate due to heating on flattening behavior of thermal sprayed particles," *Journal of Thermal Spray Technology*, vol. 15, no. 4, pp. 759–764, 2006.

[34] J. Cedelle, M. Vardelle, and P. Fauchais, "Influence of stainless steel substrate preheating on surface topography and on millimeter- and micrometer-sized splat formation," *Surface and Coatings Technology*, vol. 201, no. 3-4, pp. 1373–1382, 2006.

[35] M. Fukumoto, I. Ohgitani, M. Shiiba, and T. Yasui, "Effect of substrate surface change by heating on transition in flattening behavior of thermal sprayed particles," in *Proceedings of the International Thermal Spray Conference*, pp. 246–251, May 2004.

[36] M. Fukumoto, I. Ohgitani, H. Nagai, and T. Yasui, "Effect of substrate surface change by heating on flattening behavior of thermal sprayed particles," in *Proceedings of the International Thermal Spraying Conference*, 2005.

[37] Y. Heichal and S. Chandra, "Predicting thermal contact resistance between molten metal droplets and a solid surface," *Journal of Heat Transfer*, vol. 127, no. 11, pp. 1269–1275, 2005.

[38] A. McDonald, C. Moreau, and S. Chandra, "Thermal contact resistance between plasma-sprayed particles and flat surfaces," *International Journal of Heat and Mass Transfer*, vol. 50, no. 9-10, pp. 1737–1749, 2007.

[39] A. McDonald, C. Moreau, and S. Chandra, "Effect of substrate oxidation on spreading of plasma-sprayed nickel on stainless steel," *Surface and Coatings Technology*, vol. 202, no. 1, pp. 23–33, 2007.

[40] R. Dhiman, A. G. McDonald, and S. Chandra, "Predicting splat morphology in a thermal spray process," *Surface and Coatings Technology*, vol. 201, no. 18, pp. 7789–7801, 2007.

[41] D. K. Christoulis, D. I. Pantelis, N. de Dave-Fabrègue, F. Borit, V. Guipont, and M. Jeandin, "Effect of substrate temperature and roughness on the solidification of copper plasma sprayed droplets," *Materials Science and Engineering A*, vol. 485, no. 1-2, pp. 119–129, 2008.

[42] S. Costil, H. Liao, A. Gammoudi, and C. Coddet, "Influence of surface laser cleaning combined with substrate preheating on the splat morphology," *Journal of Thermal Spray Technology*, vol. 14, no. 1, pp. 31–38, 2005.

[43] M. F. Morks, Y. Tsunekawa, M. Okumiya, and M. A. Shoeib, "Splat morphology and microstructure of plasma sprayed cast iron with different preheat substrate temperatures," *Journal of Thermal Spray Technology*, vol. 11, no. 2, pp. 226–232, 2002.

[44] H. R. Salimijazi, L. Pershin, T. W. Coyle et al., "Effect of droplet characteristics and substrate surface topography on the final morphology of plasma-sprayed zirconia single splats," *Journal of Thermal Spray Technology*, vol. 16, no. 2, pp. 291–299, 2007.

[45] R. N. Wenzel, "Resistance of solid surfaces to wetting by water," *Industrial and Engineering Chemistry*, vol. 28, no. 8, pp. 988–994, 1936.

[46] T. Uelzen and J. Müller, "Wettability enhancement by rough surfaces generated by thin film technology," *Thin Solid Films*, vol. 434, no. 1-2, pp. 311–315, 2003.

[47] V. Pershin, M. Lufitha, S. Chandra, and J. Mostaghimi, "Effect of substrate temperature on adhesion strength of plasma-sprayed nickel coatings," *Journal of Thermal Spray Technology*, vol. 12, no. 3, pp. 370–376, 2003.

[48] K. Yang, K. Tomita, M. Fukumoto, M. Yamada, and T. Yasui, "Effect of ambient pressure on flattening behavior of thermal sprayed particles," *Journal of Thermal Spray Technology*, vol. 18, no. 4, pp. 510–518, 2009.

[49] K. Yang, K. Tanaka, T. Usami, and M. Fukumoto, "Role of ambient pressure on splat formation and coating adhesion strength during thermal spraying process," *Quarterly Journal of the Japan Welding Society*, vol. 29, no. 3, pp. 151s–155s, 2011.

[50] M. F. Morks, Y. Tsunekawa, M. Okumiya, and M. A. Shoeib, "Splat microstructure of plasma sprayed cast iron with different chamber pressures," *Journal of Thermal Spray Technology*, vol. 12, no. 2, pp. 282–289, 2003.

[51] K. Yang, Y. Ebisuno, K. Tanaka, M. Fukumoto, T. Yasui, and M. Yamada, "Verification of the flattening behavior of thermal-sprayed particles and free-falling droplets through controlling ambient pressure," *Surface and Coatings Technology*, vol. 205, no. 13-14, pp. 3816–3823, 2011.

[52] S. Brossard, P. R. Munroe, A. T. T. Tran, and M. M. Hyland, "Study of the effects of surface chemistry on splat formation for plasma sprayed NiCr onto stainless steel substrates," *Surface and Coatings Technology*, vol. 204, no. 9-10, pp. 1599–1607, 2010.

[53] S. Chandra and P. Fauchais, "Formation of solid splats during thermal spray deposition," *Journal of Thermal Spray Technology*, vol. 18, no. 2, pp. 148–180, 2009.

[54] Y. Z. Xing and C. J. Li, "Bonding characteristics of a plasma-sprayed Yttria-stabilized zirconia splat on a high-temperature substrate," in *Proceedings of the 4th Asian Thermal Spray Conference*, pp. 285–288, 2009.

[55] P. Fauchais, "Understanding plasma spraying," *Journal of Physics D*, vol. 37, no. 9, pp. R86–R108, 2004.

[56] M. Dorfman, "Thermal spray basics," *Advanced Materials and Processes*, vol. 160, pp. 47–50, 2002.

[57] G. Montavon, S. Sampath, C. C. Berndt, H. Herman, and C. Coddet, "Effects of vacuum plasma spray processing parameters on splat morphology," *Journal of Thermal Spray Technology*, vol. 4, no. 1, pp. 67–74, 1995.

[58] L. Li, A. Vaidya, S. Sampath, H. Xiong, and L. Zheng, "Particle characterization and splat formation of plasma sprayed zirconia," *Journal of Thermal Spray Technology*, vol. 15, no. 1, pp. 97–105, 2006.

[59] M. Fukumoto, E. Nishioka, and T. Nishiyama, "The criterion for splashing in flattening of thermal sprayed particles onto flat substrate surface," *Surface and Coatings Technology*, vol. 161, no. 2-3, pp. 103–110, 2002.

[60] M. Fukumoto, T. Yamaguchi, M. Yamada, and T. Yasui, "Splash splat to disk splat transition behavior in plasma-sprayed metallic materials," *Journal of Thermal Spray Technology*, vol. 16, no. 5-6, pp. 905–912, 2007.

[61] C. J. Li, J. L. Li, W. B. Wang, A. Ohmori, and K. Tani, "Effect of particle-substrate materials combinations on morphology of plasma sprayed splats," in *Proceedings of the 15th International Thermal Spray Conference*, pp. 481–487, May 1998.

[62] Y. Tanaka and M. Fukumoto, "Investigation of dominating factors on flattening behavior of plasma sprayed ceramic particles," *Surface and Coatings Technology*, vol. 120-121, pp. 124–130, 1999.

[63] M. Fukumoto, S. Kato, M. Ohwatari, and Y. Huang, "Splatting and solidification behavior of plasma sprayed metallic powders impinging on flat SUS304 substrate," *Journal of the Japan Institute of Metals*, vol. 59, pp. 1178–1184, 1995.

[64] Y. Tanaka and M. Fukumoto, "Dominant factor in the behavior of a particular flat ceramic-coated particles," *Quarterly Journal of the Japan Welding Society*, vol. 17, no. 2, pp. 259–265, 1999 (Japanese).

[65] M. Fukumoto and Y. Huang, "Flattening mechanism in thermal sprayed nickel particle impinging on flat substrate surface," *Journal of Thermal Spray Technology*, vol. 8, no. 3, pp. 427–432, 1999.

[66] Y. Tanaka and M. Fukumoto, "Effects of solidification and wetting on flattening behavior of thermal sprayed ceramic particles," *Quarterly Journal of the Japan Welding Society*, vol. 18, no. 1, pp. 26–32, 2000 (Japanese).

[67] Y. Huang, M. Ohwatari, and M. Fukumoto, "Effect of substrate material on flattening behavior of plasma sprayed Ni particles," in *Proceedings of the 6th International Symposium*, pp. 731–736, JWS, 1996 (Japanese).

[68] Y. Huang, M. Ohwatari, and M. Fukumoto, "Flattening behavior of plasma sprayed Ni particles on various substrate materials," *Journal of the Japan Institute of Metals*, vol. 60, no. 12, pp. 1215–1221, 1996 (Japanese).

[69] M. F. Bahbou and P. Nylén, "On-line measurement of plasma-sprayed Ni-particles during impact on a Ti-surface: influence of surface oxidation," *Journal of Thermal Spray Technology*, vol. 16, no. 4, pp. 506–511, 2007.

[70] M. Fukumoto, I. Ohgitani, and T. Yasui, "Effect of substrate surface change on flattening behaviour of thermal sprayed particles," *Materials Transactions*, vol. 45, no. 6, pp. 1869–1873, 2004.

[71] S. Amada, K. Imagawa, and S. Aoki, "Splat profile of impinging droplets on rough substrates: influence of surface roughness," *Surface and Coatings Technology*, vol. 154, no. 1, pp. 27–33, 2002.

[72] M. Qu and A. Gouldstone, "On the role of bubbles in metallic splat nanopores and adhesion," *Journal of Thermal Spray Technology*, vol. 17, no. 4, pp. 486–494, 2008.

[73] C. Moreau, P. Gougeon, and M. Lamontagne, "Influence of substrate preparation on the flattening and cooling of plasma-sprayed particles," *Journal of Thermal Spray Technology*, vol. 4, no. 1, pp. 25–33, 1995.

[74] K. Shinoda, A. Yamada, M. Kambara, Y. Kojima, and T. Yoshida, "Deformation of alumina droplets on micro-patterned substrates under plasma spraying conditions," *Journal of Thermal Spray Technology*, vol. 16, no. 2, pp. 300–305, 2007.

[75] H. B. Parizi, L. Rosenzweig, J. Mostaghimi et al., "Numerical simulation of droplet impact on patterned surfaces," *Journal of Thermal Spray Technology*, vol. 16, no. 5-6, pp. 713–721, 2007.

[76] A. McDonald, S. Chandra, and C. Moreau, "Photographing impact of plasma-sprayed particles on rough substrates," *Journal of Materials Science*, vol. 43, no. 13, pp. 4631–4643, 2008.

[77] C. J. Li, J. L. Li, and W. B. Wang, "The effect of substrate preheating and surface organic covering on splat formation," in *Proceedings of the 15th International Thermal Spray Conference*, pp. 473–480, May 1998.

[78] A. T. T. Tran, *Effect of surface chemistry on splat formation during plasma spraying [Ph.D. thesis]*, University of Auckland, 2010.

[79] A. T. T. Tran and M. M. Hyland, "The role of substrate surface chemistry on splat formation during plasma spray deposition by experiments and simulations," *Journal of Thermal Spray Technology*, vol. 19, no. 1-2, pp. 11–23, 2010.

[80] A. T. T. Tran, M. M. Hyland, T. Qiu, B. Withy, and B. J. James, "Effects of surface chemistry on splat formation during plasma spraying," *Journal of Thermal Spray Technology*, vol. 17, no. 5-6, pp. 637–645, 2008.

[81] Y. Tanaka and M. Fukumoto, "Influence of solidification and wetting on flattening behavior of plasma sprayed ceramic particles," *International Journal of Materials and Product Technology*, vol. 1, pp. 518–523, 2001.

[82] H. Li, S. Costil, H. L. Liao, and C. Coddet, "Role of the laser surface preparation on the adhesion of Ni-5%Al coatings deposited using the PROTAL process," *Journal of Thermal Spray Technology*, vol. 15, no. 2, pp. 191–197, 2006.

[83] E. G. Shafrin and W. A. Zisman, "Constitutive relations in the wetting of low energy surfaces and the theory of the retraction method of preparing monolayers," *Journal of Physical Chemistry*, vol. 64, no. 5, pp. 519–524, 1960.

[84] K. Suganuma, "Interfacial bond strength and interfacial energy metal-ceramic," *Bulletin of the Japan Institute of Metals*, vol. 29, no. 11, pp. 882–887, 1990 (Japanese).

[85] K. Nogi, N. Iwamoto, and K. Ogin, "Mechanism of wetting of ceramics by molten metal," *Bulletin of the Japan Institute of Metals*, vol. 31, no. 4, pp. 278–281, 1992 (Japanese).

[86] K. Yang, M. Fukumoto, T. Yasui, and M. Yamada, "Role of substrate temperature on microstructure formation in plasma-sprayed splats," *Surface and Coatings Technology*, vol. 214, no. 1, pp. 138–143, 2013.

[87] T. Chraskaa and A. H. King, "Effect of different substrate conditions upon interface with plasma sprayed zirconia—a TEM study," *Surface and Coatings Technology*, vol. 157, no. 2-3, pp. 238–246, 2002.

[88] T. Chraskaa and A. H. King, "Transmission electron microscopy study of rapid solidification of plasma sprayed zirconia—part II. Interfaces and subsequent splat solidification," *Thin Solid Films*, vol. 397, no. 1-2, pp. 40–48, 2001.

[89] H. Li, K. A. Khor, and P. Cheang, "Thermal sprayed hydroxyapatite splats: nanostructures, pore formation mechanisms and TEM characterization," *Biomaterials*, vol. 25, no. 17, pp. 3463–3471, 2004.

[90] A. T. T. Tran, M. M. Hyland, K. Shinoda, and S. Sampath, "Influence of substrate surface conditions on the deposition and spreading of molten droplets," *Thin Solid Films*, vol. 519, no. 8, pp. 2445–2456, 2011.

[91] M. Ivosevic, R. A. Cairncross, and R. Knight, "3D predictions of thermally sprayed polymer splats: modeling particle acceleration, heating and deformation on impact with a flat substrate," *International Journal of Heat and Mass Transfer*, vol. 49, no. 19-20, pp. 3285–3297, 2006.

[92] F. H. Harlow and J. P. Shannon, "The splash of a liquid drop," *Journal of Applied Physics*, vol. 38, no. 10, pp. 3855–3866, 1967.

[93] J. Madjeski, "Solidification of droplets on a cold surface," *International Journal of Heat and Mass Transfer*, vol. 19, no. 9, pp. 1009–1013, 1976.

[94] V. V. Sobolev and J. M. Guilemany, "Flattening of droplets and formation of splats in thermal spraying: a review of recent work—part 1," *Journal of Thermal Spray Technology*, vol. 8, no. 1, pp. 87–101, 1999.

[95] V. V. Sobolev and J. M. Guilemany, "Effect of oxidation on droplet flattening and splat-substrate interaction in thermal spraying," *Journal of Thermal Spray Technology*, vol. 8, no. 4, pp. 523–530, 1999.

[96] V. V. Sobolev and J. M. Guilemany, "Effect of substrate deformation on droplet flattening in thermal spraying," *Materials Letters*, vol. 35, no. 5-6, pp. 324–328, 1998.

[97] V. V. Sobolev and J. M. Guilemany, "Effect of wave processes on splat formation during thermal spraying," *Materials Letters*, vol. 42, no. 5, pp. 321–325, 2000.

[98] V. V. Sobolev, "Formation of splat morphology during thermal spraying," *Materials Letters*, vol. 36, no. 1–4, pp. 123–127, 1998.

[99] V. V. Sobolev and J. M. Guilemany, "Formation of splats during thermal spraying of composite powder particles," *Materials Letters*, vol. 42, no. 1-2, pp. 46–51, 2000.

[100] V. V. Sobolev and J. M. Guilemany, "Influence of wetting and surface effects on splat formation during thermal spraying," *Materials Letters*, vol. 37, no. 3, pp. 132–137, 1998.

[101] H. Zhang, X. Y. Wang, L. L. Zheng, and S. Sampath, "Numerical simulation of nucleation, solidification, and microstructure formation in thermal spraying," *International Journal of Heat and Mass Transfer*, vol. 47, no. 10-11, pp. 2191–2203, 2004.

[102] X. Y. Wang, H. Zhang, L. L. Zheng, and S. Sampath, "An integrated model for interaction between melt flow and non-equilibrium solidification in thermal spraying," *International Journal of Heat and Mass Transfer*, vol. 45, no. 11, pp. 2289–2301, 2002.

[103] H. Fukanuma, "Porosity formation and flattening model of an impinging molten particle in thermal spray coatings," *Journal of Thermal Spray Technology*, vol. 3, no. 1, pp. 33–44, 1994.

[104] J. Shimizu, E. Ohmura, Y. Kobayashi, S. Kiyoshima, and H. Eda, "Molecular dynamics simulation of flattening process of a high-temperature, high-speed droplet—influence of impact parameters," *Journal of Thermal Spray Technology*, vol. 16, no. 5-6, pp. 722–728, 2007.

[105] C. W. Kang, H. W. Ng, and S. C. M. Yu, "Plasma spray deposition on inclined substrates: simulations and experiments," *Journal of Thermal Spray Technology*, vol. 16, no. 2, pp. 261–274, 2007.

[106] M. Ivosevic, V. Gupta, J. A. Baldoni, R. A. Cairncross, T. E. Twardowski, and R. Knight, "Effect of substrate roughness on splatting behavior of HVOF sprayed polymer particles: modeling and experiments," *Journal of Thermal Spray Technology*, vol. 15, no. 4, pp. 725–730, 2006.

[107] M. Pasandideh-Fard, V. Pershin, S. Chandra, and J. Mostaghimi, "Splat shapes in a thermal spray coating process: simulations and experiments," *Journal of Thermal Spray Technology*, vol. 11, no. 2, pp. 206–217, 2002.

[108] A. V. Kiselev and A. V. Uvarov, "Infrared spectra and electron spin resonance of aluminium, silicon and titanium oxides and of adsorbed substances," *Surface Science*, vol. 6, no. 4, pp. 399–421, 1967.

[109] S. Kitahara and A. Hasui, "A study of the bonding mechanism of sprayed coatings," *Journal of Vacuum Science & Technology*, vol. 11, no. 4, pp. 747–753, 1974.

[110] H. D. Steffens, B. Wielage, and J. Drozak, "Interface phenomena and bonding mechanism of thermally-sprayed metal and ceramic composites," *Surface and Coatings Technology*, vol. 45, no. 1–3, pp. 299–308, 1991.

[111] S. Brossard, P. R. Munroe, A. Tran, and M. M. Hyland, "Study of the splat-substrate interface for a NiCr coating plasma sprayed onto polished aluminum and stainless steel substrates," *Journal of Thermal Spray Technology*, vol. 19, no. 1-2, pp. 24–30, 2010.

[112] S. Brossard, P. R. Munroe, A. T. T. Tran, and M. M. Hyland, "Study of the microstructure of NiCr splats plasma sprayed on to stainless steel substrates by TEM," *Surface and Coatings Technology*, vol. 204, no. 9-10, pp. 1608–1615, 2010.

[113] A. T. T. Tran, S. Brossard, M. M. Hyland, B. J. James, and P. Munroe, "Evidence of substrate melting of NiCr particles on stainless steel substrate by experimental observation and simulations," *Plasma Chemistry and Plasma Processing*, vol. 29, no. 6, pp. 475–495, 2009.

[114] C. Mundo, M. Sommerfeld, and C. Tropea, "Droplet-wall collisions: experimental studies of the deformation and breakup process," *International Journal of Multiphase Flow*, vol. 21, no. 2, pp. 151–173, 1995.

[115] P. Fauchais, M. Fukumoto, A. Vardelle, and M. Vardelle, "Knowledge concerning splat formation: an invited review," *Journal of Thermal Spray Technology*, vol. 13, no. 3, pp. 337–360, 2004.

[116] C. Escure, M. Vardelle, A. Vardelle, and P. Fauchais, "Visualization of the impact of drops on a substrate in plasma spraying: deposition and splashing modes," in *Proceedings of the International Thremal Spray Conferences*, pp. 805–812, May 2001.

[117] H. Li, S. Costil, H. L. Liao, C. J. Li, M. Planche, and C. Coddet, "Effects of surface conditions on the flattening behavior of

plasma sprayed Cu splats," *Surface and Coatings Technology*, vol. 200, no. 18-19, pp. 5435–5446, 2006.

[118] E. Nishioka, *Study on active control method for thermal plasma spray process [Ph.D. thesis]*, Toyohashi University of Technology, 2001.

[119] A. McDonald, M. Xue, S. Chandra, J. Mostaghimi, and C. Moreau, "Modeling fragmentation of plasma-sprayed particles impacting on a solid surface at room temperature," *Comptes Rendus—Mecanique*, vol. 335, no. 5-6, pp. 351–356, 2007.

[120] S. D. Aziz and S. Chandra, "Impact, recoil and splashing of molten metal droplets," *International Journal of Heat and Mass Transfer*, vol. 43, no. 16, pp. 2841–2857, 2000.

[121] Y. Tanaka, S. Yoshida, and R. Kawase, "Effect of impact velocity and preheated substrate temperature on flattening and solidifying behavior of free falling metal droplet," *Journal of Japan Thermal Spray Society*, vol. 47, pp. 54–60, 2010.

[122] Y. Heichal, S. Chandra, and E. Bordatchev, "A fast-response thin film thermocouple to measure rapid surface temperature changes," *Experimental Thermal and Fluid Science Journal*, vol. 30, no. 2, pp. 153–159, 2005.

[123] M. Pasandideh-Fard, S. Chandra, and J. Mostaghimi, "A three-dimensional model of droplet impact and solidification," *International Journal of Heat and Mass Transfer*, vol. 45, no. 11, pp. 2229–2242, 2002.

[124] R. G. Azar, Z. Yang, S. Chandra, and J. Mostaghimi, "Impact of molten metal droplets on the tip of a pin projecting from a flat surface," *International Journal of Heat and Fluid Flow*, vol. 26, no. 2, pp. 334–347, 2005.

[125] M. Pasandideh-Fard, R. Bhola, S. Chandra, and J. Mostaghimi, "Deposition of tin droplets on a steel plate: simulations and experiments," *International Journal of Heat and Mass Transfer*, vol. 41, no. 19, pp. 2929–2945, 1998.

[126] R. H. S. Winterton, "Newton's law of cooling," *Contemporary Physics*, vol. 40, no. 3, pp. 205–212, 1999.

10

A Model of Numerical Calculation of Conductivity for III-V MBE Epilayers Using a Hall Device

Andrzej Wolkenberg[1,2]

[1] *Institute of Electron Technology, Aleja Lotnikow 32/46, 02-668 Warsaw, Poland*
[2] *Institute of Materials Engineering, Czestochowa University of Technology, Aleja Armii Krajowej 19, 42-200 Czestochowa, Poland*

Correspondence should be addressed to Andrzej Wolkenberg; wolkenberg@neostrada.pl

Academic Editor: Ram Katiyar

An electrical conduction versus temperature model using a Hall device was developed. In the case of InAs, InGaAs, and GaAs MBE epilayers, the prediction agrees well with the experimental results. Herein, we explain here how these calculated fractions of total conductivity describe the measured values. The method allows for the calculation of the carrier concentration and mobility of each component of a multicarrier system. The extracted concentrations are used to characterise the different components of charge transport in the active layer. The conductance values G [S] of these components of charge transport were obtained. Also the scattering events for the investigated samples are presented. The analysis of the experimental results for three semiconductor compositions and different concentrations demonstrates the utility of our method in comparing the conductance of each component of the multilayered system as a function of temperature.

1. Models

1.1. Two Layer Model. We take as an example a physical two-layer model according of thin InAs MBE layers. After etching step by step Grange et al. [1] obtained a 1 μm InAs MBE layer with the concentration profile shown in Figure 2 (▼). Near the interface with the GaAs substrate, the researchers measured the concentration to be as high as $n = 1 \cdot 10^{18}/\text{cm}^3$.

This result is in agreement with the TEM image presented in Figure 1 obtained by Fawcett et al. [2]. We proposed that the MBE layer of InAs consists of two parts:

(i) a strongly defected layer measuring t_1 in thickness,

(ii) a correctly formed layer measuring $t_2 - t_1$ in thickness, where t_2 is the metallurgical thickness.

We present in Figure 2 data regarding the concentration values obtained by many authors. The following equation is used to relate the experimental concentration data with the InAs layer thickness:

$$n = \frac{n_1 \cdot t_1 + n_2 \cdot (t_2 - t_1)}{t_2}, \qquad (1)$$

where $n_1 = 2 \cdot 10^{19}\,\text{cm}^{-3}$, $t_1 = 18.2 \cdot 10^{-8}\,\text{cm}$, and $n_2 = 1 \cdot 10^{15}\,\text{cm}^{-3}$.

Therefore one can convince one's self that the two-layer model may be practically used in our calculation.

1.2. Calculation Background. Electron and hole concentrations are derived by the numerical solution of the neutrality equation:

$$n_e + x_{e,h} = p + N_{D1}^{+} + N_{D2}^{+} - N_{A}^{-}, \qquad (2)$$

where $n = n_e + x_{e,h}$ is the total conductivity electron concentration,

$$x_{e,h} = x_D - \sqrt{\frac{x_A N_V}{2}} \exp\left[-\frac{E_{X,A}}{kT}\right] \left[\text{cm}^{-3}\right], \qquad (3)$$

$x_{e,h}$ is the net concentration, x_D is the concentration of the donor-like states, x_A is the concentration of the acceptor like state, N_V is the effective density of states in the valence band, $E_{X,A}$ is the ionisation energy of the acceptor-like states, k is the Boltzmann constant, p is the hole concentration, N_{D1}^{+} is the concentration of shallow donors with activation energy

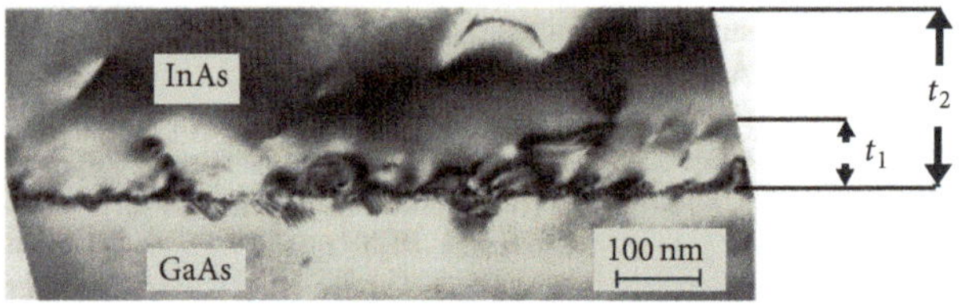

FIGURE 1: Two-layer pattern of MBE InAs layer after Fawcett et al. [2] and our interpretation: t_1—thin interface layer measuring 1.82 nm in thickness with high charge carrier concentration n_1 and high FWHM value; $(t_2 - t_1)$—thick bulk layer whose thickness is the difference between the metallurgical thickness t_2—and defected layer thickness t_1 with mean charge carrier concentration n_2.

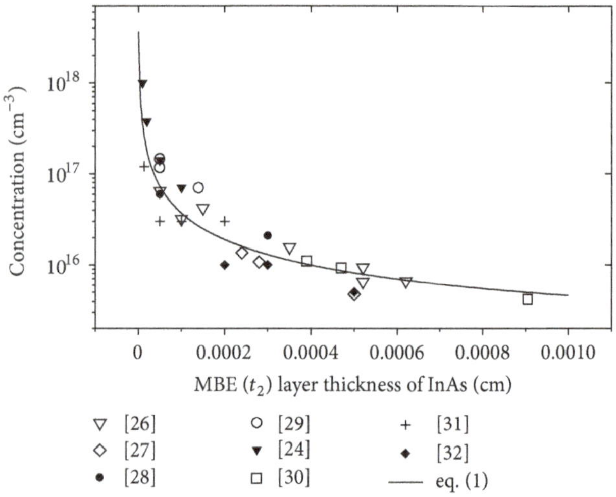

FIGURE 2: Charge carrier concentration versus thickness of t_2 layer of InAs deposited at 500°C on GaAs substrate.

E_{D1}, N_{D2}^{+} is the concentration of deep donors with activation energy E_{D2}, and N_A^- is the concentration of acceptors with activation energy E_A.

To obtain the mobility, we numerically integrate the equation

$$\frac{1}{\tau_{n,h}} = \frac{1}{\tau_{\text{ii}(n,h)}} + \frac{1}{\tau_{\text{dp}(n,h)}} + \frac{1}{\tau_{\text{pie}(n,h)}} + \frac{1}{\tau_{\text{po}(n,h)}} + \frac{1}{\tau_{\text{sc}(n,h)}} + \frac{1}{\tau_{\text{dis}(n,h)}} \left[\text{s}^{-1}\right], \quad (4)$$

where the number of used τ values depends on values presented in Tables 4, 8, and 12 and the scatterings times τ are presented in Table 1.

2. Introduction

Structural inaccuracy, especially in layered structures, is one factor that affects how the magnitude of a magnetic field influences measured charge transport properties [3]. While studying the transport properties of HgCdTe layers, Gold and Nelson [4] reported on the influence of different charge

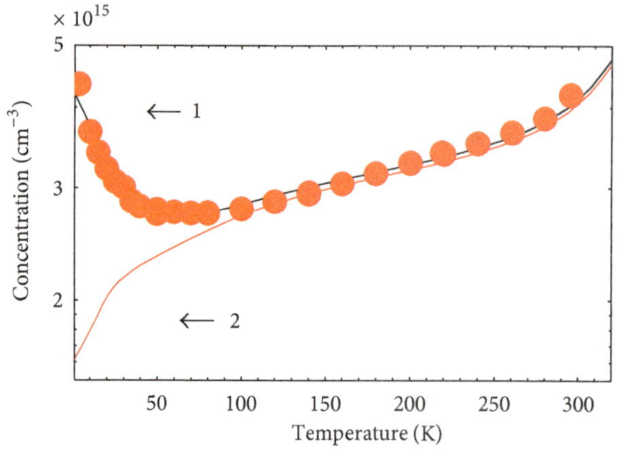

FIGURE 3: Concentration versus temperature n-InAs 9.0 μm; $N_A = 8.05 \cdot 10^{15}$ cm^{-3}, $N_{D1} = 1.26 \cdot 10^{16}$ cm^{-3}; $N_{D2} = 8.5 \cdot 10^{15}$ cm^{-3}; $E_A = 0.005$ eV; $E_{D1} = 0.001$ eV; $E_{D2} = 0.13$ eV; $x_A = 4.2 \cdot 10^{15}$ cm^{-3}; $x_D = 1.2 \cdot 10^{15}$ cm^{-3}; $E_{x_A} = 0.0008$ eV; red bullets: n_H experimental data, (1) black line: n calculated curve; (2) red line: n_H calculated curve [3].

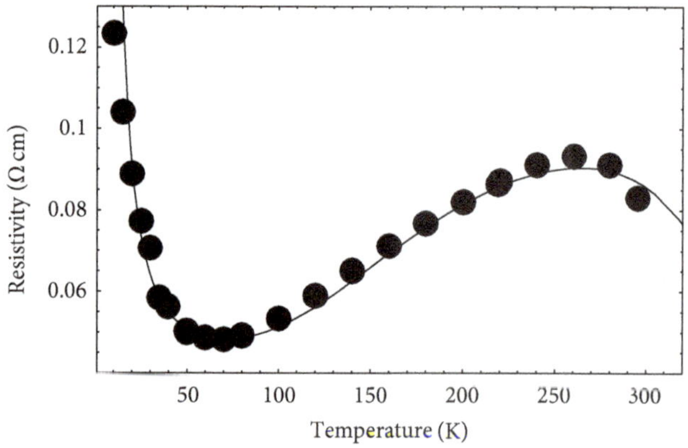

FIGURE 4: The resistivity versus temperature semiconductor parameters as in Figure 3, with μ calculated by scattering on ionized impurities, acoustic phonons with deformation potential, acoustic phonons with piezoelectric potential, polar optical phonons and space charge with $N_S \cdot Q = 0.03$, • • • ρ experimental points, black line: ρ calculated curve (e—electron charge).

carriers using Hall characterisation. An exhaustive review concerning these phenomena was undertaken by Meyer et al. [5]. Such disorder was first described and interpreted by the so-called mobility spectrum by Beck and Anderson [6] and many papers have focused on this problem [7–20]. For the proper characterisation of a multicarrier conduction system, the number of independent measurements must match the number of desired physical quantities. In actuality, the complete characterisation of multicarrier conduction is extremely complex. Even for two-carrier conduction, one must make four independent measurements to calculate the individual carrier concentration and mobility of each component [21]. To this end, other methods, such as the maximum entropy approach [22] or the application of Bryan's algorithm [23], must be considered. The concept most often used in the

TABLE 1: Scattering mechanisms used in calculations.

Semiconductor	Scattering on ionized donors and acceptors	Scattering on acoustic phonons (deformation potential)	Scattering on acoustic phonons (piezoelectric potential)	Scattering on optics polar phonons	Scattering on space charge	Scattering on dislocations
InAs	$\tau_{ii} = \frac{a}{b}$; $a =$ $5.332 \cdot 10^7 \varepsilon_r^2 m_{ef,(n,p)}^{0,5} eps^{1.25}$ $b =$ $N_I Z^2 \left[\ln(1+y) - \frac{y}{(1+y)}\right]$ $y =$ $\frac{3.145 \cdot 10^{36} m_{ef,(n,p)} \varepsilon_r T}{n, p} eps$ $N_I = n + 2 \cdot p$ $Z = 1$	$\tau_{dp} = \frac{3.012 \cdot 10^{-12}}{T \cdot m_{ef,(n,p)}^{1.5} eps^{0.5}}$	$\tau_{pie} = \frac{9.502 \cdot 10^{-8} \sqrt{eps/m_{ef,(n,p)}} T^2}{\left[(h_{14})^2((3/C_l) + (4/C_t))\right]}$	$\tau_{po} = \frac{0.02781 \cdot fT \cdot eps^{0.5}}{\text{stdi} \cdot m_{ef,(n,p)}^{0.5}}$ $TD = 337 - 9.1156(T - 4.75555)^{1.1179}$ $\times \exp\left[-\frac{T - 11.196}{15.781}\right]$ $fT1 = 1 - \frac{0.5841TD}{T} + 0.292\left(\frac{TD}{T}\right)^2$ $-0.037164\left[\frac{TD}{T}\right]^3 + 0.0012\left[\frac{TD}{T}\right]^4$ $fT2 = 8^{-1}(3\sqrt{\pi})\sqrt{\frac{TD}{T}}$; $fT3 = \text{iF}[T < 84, fT2, fT1]$ $fT = \left[\exp\left(\frac{TD}{T}\right) - 1\right] fT3$ $\text{stdi} = \frac{1}{\varepsilon_{opt}\varepsilon_0} - \frac{1}{\varepsilon_r \varepsilon_0}$	$\tau_{SC} = \frac{\sqrt{m_{ef(n,p)} \cdot 9.108 \cdot 10^{-28}}}{N_s Q \cdot \sqrt{eps}}$	
$In_{0.53}Ga_{0.47}As$	$\tau_{ii} = \frac{a}{b}$; $a = 5.332 \cdot 10^7 \varepsilon_r^2 m_{ef,(n,p)}^{0.5} eps^{2.8}$ $b =$ $N_I Z^2 \left[\ln(1+y) - \frac{y}{(1+y)}\right]$ $y =$ $\frac{3.145 \cdot 10^{36} m_{ef,(n,p)} \varepsilon_r T}{n, p} eps$ $N_I = n + 2 \cdot p$ $Z = 1$	$\tau_{dp} = \frac{3.012 \cdot 10^{-12}}{T \cdot m_{ef,(n,p)}^{1.5} eps^{0.5}}$	$\tau_{pie} = \frac{9.502 \cdot 10^{-8} \sqrt{eps/m_{ef,(n,p)}} T^2}{\left[(h_{14})^2((3/C_l) + (4/C_t))\right]}$	$\tau_{po} = \frac{0.02781 \cdot fT \cdot eps^{0.5}}{\text{stdi} \cdot m_{ef,(n,p)}^{0.5}}$ $TD = 358.44 - 9.1156(T - 4.75555)^{1.1179}$ $\times \exp\left[-\frac{T - 11.196}{15.781}\right]$ $fT1 = 1 - \frac{0.5841TD}{T} + 0.292\left(\frac{TD}{T}\right)^2$ $-0.037164\left[\frac{TD}{T}\right]^3 - 0.0012\left[\frac{TD}{T}\right]^4$ $fT2 = 8^{-1}(3\sqrt{\pi})\sqrt{\frac{TD}{T}}$; $fT3 = \text{iF}[T < 84, fT2, fT1]$ $fT = \left[\exp\left(\frac{TD}{T}\right) - 1\right] fT3$ $\text{stdi} = \frac{1}{\varepsilon_{opt}\varepsilon_0} - \frac{1}{\varepsilon_r \varepsilon_0}$	$\tau_{SC} = \frac{\sqrt{m_{ef(n,p)} \cdot 9.108 \cdot 10^{-28}}}{N_s Q \cdot \sqrt{eps}}$	$\tau_{disl} = \frac{9.082 \cdot 10^2 \sqrt{m_{ef(n,p)}}}{T^{0.44}\sqrt{k}}$

Table 1: Continued.

Semiconductor	Scattering on ionized donors and acceptors	Scattering on acoustic phonons (deformation potential)	Scattering on acoustic phonons (piezoelectric potential)	Scattering on optics polar phonons	Scattering on space charge	Scattering on dislocations
GaAs	$\tau_{ii} = \frac{a}{b}$; $a = 5.332 \cdot 10^7 \varepsilon_r^2 m_{ef,(n,p)}^{0.5} eps^{1.91}$ $b =$ $N_I Z^2 \left[\ln(1+y) - \frac{y}{(1+y)}\right]$ $y =$ $\frac{3.145 \cdot 10^{36} m_{ef,(n,p)} \varepsilon_r T}{n, p} eps$ $N_I = n + 2 \cdot p$ $Z = 1$			$\tau_{po} = \frac{0.02781 \cdot fT \cdot eps^{0.35}}{\text{stdi} \cdot m_{ef,(n,p)}^{0.5}}$ $TD = 358.44 - 9.1156(T - 4.75555)^{1.1179} \times \exp\left[-\frac{T - 11.196}{15.781}\right]$ $fT1 = 1 - \frac{0.5841TD}{T} + 0.292\left(\frac{TD}{T}\right)^2 - 0.037164\left[\frac{TD}{T}\right]^3 + 0.0012\left[\frac{TD}{T}\right]^4$ $fT2 = 8^{-1}(3\sqrt{\pi})\sqrt{\frac{TD}{T}}$; $fT3 = \text{iF}[T < 84, fT2, fT1]$ $fT = [\exp\left(\frac{TD}{T}\right) - 1] fT3$ $\text{stdi} = \frac{1}{\varepsilon_{opt}\varepsilon_0} - \frac{1}{\varepsilon_r \varepsilon_0}$	$\tau_{SC} = \frac{\sqrt{m_{ef(n,p)} \cdot 9.108 \cdot 10^{-28}}}{N_s Q \cdot \sqrt{eps}}$	

TABLE 2: Main properties of the investigated semiconductors at 300 K.

Investigated semiconductor	Energy gap [eV]	Intrinsic concentration [cm^{-3}]	Average distance between impurities R_d [Å]	Donor state Bohr radius a_B [Å]	R_d/a_B	Donor state ionisation energy E_B [meV]
InAs	0.354	$\sim 1\cdot 10^{15}$	384.5	367	1.05	1.36
$In_{0.53}Ga_{0.47}As$	0.743	$8.5\cdot 10^{11}$	1224.6	180	6.8	2.8
GaAs	1.43	$2.2\cdot 10^{6}$	809.5	104	7.8	5.25

TABLE 3: Growth conditions of InAs layers on (001) GaAs substrates.

No. of procedures	Thickness [μm]	Relation between V/III	Substrate temperature [°C]
92	4	5	440
298	5.6	4.5	450
301	4.7	3.7	500
330	9.05	5.8	506

TABLE 4: Scattering mechanism involved in calculation of InAs samples properties.

No.	Parameter	InAs 330 9.0 μm	InAs 301 4.7 μm	InAs 298 5.6 μm	InAs 292 4 μm
1	Mobility [cm^2/V · s]	τ_{ii}, τ_p, τ_{po}, τ_{pie}, τ_{sc}	τ_{ii}, τ_{dp}, τ_{po}, τ_{pie}, τ_{sc}, τ_{disl}	τ_{ii}, τ_{disl}	τ_{ii}, τ_{disl}

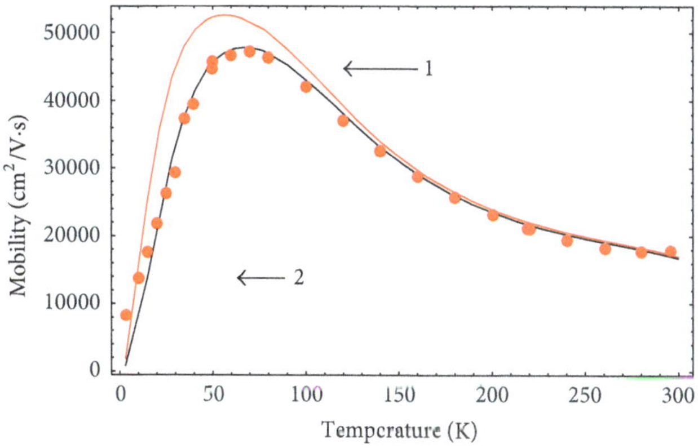

FIGURE 5: Mobility versus temperature calculated using the parameters presented in Figure 3 [3] • • • μ_H experimental points, (1) red line: μ_H calculated curve (see Section 1—Models); (2) (black line) μ calculated curve (see Section 1—Models).

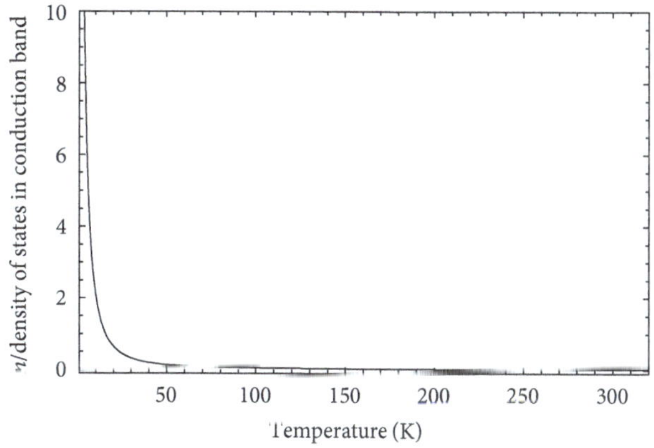

FIGURE 6: The ratio of electron concentration to density of states in the conduction band versus temperature for the structure shown in Figure 1.

characterisation of mobility spectra has been described in many papers by Antoszewski and collaborators [12, 24, 25]. The main features of the above-mentioned methods are as follows.

(1) The Hall voltage of the investigated semiconductor is not proportionally dependent on the magnetic field.

(2) Such behaviour in the conductivity process results in charge carriers (electrons and holes) with different mobilities.

(3) The mentioned authors [6, 12, 24, 25] used different mathematical approaches to obtain mobility spectra from mobility and conductivity data.

In cases in which this concept is applicable, the calculations are supported by the conductivity tensor dependence on the magnetic field [6]:

$$\sigma(H) \equiv \sigma_{xx}(H) + i\sigma_{xy}(H) = \int_{-\infty}^{\infty} \frac{s(\mu)(1 + i\mu H)}{1 + (\mu H)^2} d\mu, \quad (5)$$

where $s(\mu) = S^+(\mu) \rightarrow \mu > 0$; or $s(\mu) = S^-(|\mu|) \rightarrow \mu < 0$ and $S^+(\mu)$ and $S^-(\mu)$ are a pair of conductivity density functions.

Equation (5) allows for the treatment of the nonlinear magnetic field dependence on the Hall effect [10]. Many experimental results have been analysed using this method [8–12, 24, 25]. Antoszewski et al. concluded that a major advantage of mobility spectrum analysis (MSA) and quantitative mobility spectrum analysis (QMSA) over other methods is that these procedures are inherent not arbitrary that is, no prior assumptions are required [24, 25]. These investigators suggested that their improved (QMSA) method has sufficient reliability, versatility, and sensitivity for their proposed analysis which is fully computer-automated, using magnetic field dependence data as input.

Our approach accounts for several sources of charge carriers as components contribution to the total conductivity of a semiconductor layer [3]. These sources form the parts used to solve the neutrality equation, and the solution is obtained by numerical computer calculation. The values of

Table 5: Conductance components for InAs samples at temperatures 10 K and 70 K.

No.	Parameter	10 K		70 K			
		InAs 9 μm	InAs 4.7 μm	InAs 9 μm	InAs 4.7 μm	InAs 5.6 μm	InAs 4 μm
1	$n = n_e + x_e + n_{D2}$ [cm^{-3}]	$3.8 \cdot 10^{15}$	$7.9 \cdot 10^{15}$	$2.7 \cdot 10^{15}$	$6.62 \cdot 10^{15}$	$4.52 \cdot 10^{16}$	$4.085 \cdot 10^{17}$
2	μ_n [cm^2/V · s]	8000	3600	49000	33500	17350	*
3	σ_n [1/Ω cm]	4.9	4.56	21.17	35.53	125.63	*
4	σ_n/σ_n	1	1	1	1	1	*
5	n_e [cm^{-3}]	$1.7 \cdot 10^{15}$	$2.01 \cdot 10^{15}$	$5.03 \cdot 10^{15}$	$5.04 \cdot 10^{15}$	$1.495 \cdot 10^{16}$	$1.43 \cdot 10^{16}$
6	μ_{n_e} [cm^2/V · s]	7653	3320.15	49000	33153	23972.1	*
7	σ_{n_e} [1/Ω cm]	2.07	1.07	21.87	26.76	57.41	*
8	σ_{n_e}/σ_n	0.426	0.234	1.03	0.753	0.46	*
9	x_e [cm^{-3}]	$2.17 \cdot 10^{15}$	$5.94 \cdot 10^{15}$	$-1.06 \cdot 10^{14}$	$5.64 \cdot 10^{15}$	$3.023 \cdot 10^{16}$	$3.94 \cdot 10^{17}$
10	μ_{x_e} [cm^2/V · s]	19465	2286	−1914	7536.31	28202.4	*
11	σ_{x_e} [1/Ω cm]	6.76	2.175	0.033	6.81	136.58	*
12	σ_{x_e}/σ_n	1.4	0.48	0.0015	0.192	1.087	*
13	n_{D2} [cm^{-3}]	$-4.63 \cdot 10^{14}$	$-2.45 \cdot 10^{14}$	$-2.13 \cdot 10^{14}$	$-1.69 \cdot 10^{14}$	$-8.66 \cdot 10^{14}$	*
14	μ_{D2}[cm^2/V · s]	−100	−472.22	−2740	−1623.56	28518.2	*
15	σ_{D2} [1/Ω cm]	0.0074	0.0185	0.093	0.044	−3.96	*
16	σ_{D2}/σ_n	0.0015	0.00406	0.0044	0.0124	0.0315	*
17	Σn [cm^{-3}]	$3.407 \cdot 10^{15}$	$7.705 \cdot 10^{15}$	$4.7 \cdot 10^{15}$	$1.05 \cdot 10^{16}$	$4.43 \cdot 10^{16}$	$4.083 \cdot 10^{17}$
18	G_n [S]	$4.41 \cdot 10^{-3}$	$2.14 \cdot 10^{-3}$	$1.9 \cdot 10^{-2}$	$1.67 \cdot 10^{-2}$	$7.03 \cdot 10^{-2}$	*
19	G_{n_e} [S]	$1.86 \cdot 10^{-3}$	$5.03 \cdot 10^{-4}$	$1.97 \cdot 10^{-2}$	$1.26 \cdot 10^{-2}$	$3.2 \cdot 10^{-2}$	*
20	G_{x_e} [S]	$6.0 \cdot 10^{-3}$	$1.02 \cdot 10^{-3}$	$2.97 \cdot 10^{-5}$	$3.2 \cdot 10^{-3}$	$7.65 \cdot 10^{-2}$	*
21	G_{D2} [S]	$6.66 \cdot 10^{-6}$	$8.7 \cdot 10^{-6}$	$8.37 \cdot 10^{-5}$	$2.068 \cdot 10^{-5}$	$-2.2 \cdot 10^{-3}$	*
22	$\Sigma G_{n_e+x_e+n_{D2}}$ [S]	$7.87 \cdot 10^{-3}$	$1.53 \cdot 10^{-3}$	$1.98 \cdot 10^{-2}$	$1.58 \cdot 10^{-2}$	$10.63 \cdot 10^{-2}$	*
23	$\Sigma G_{n_e+x_e+n_{D2}}/G_n$	1.78	0.71	1.042	0.95	1.5	*

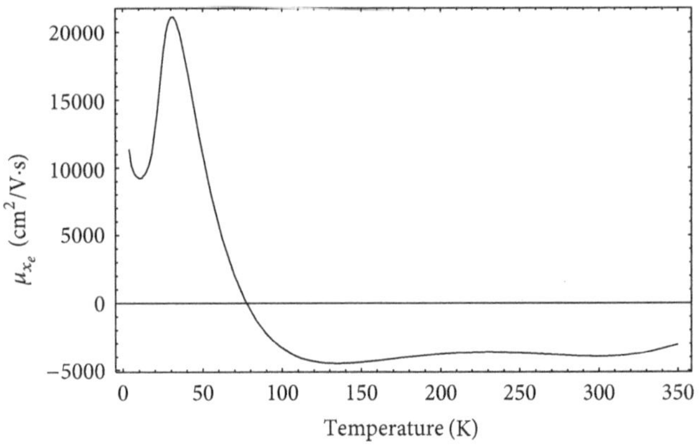

Figure 7: Mobility μ_{x_e} versus temperature, using the same calculation parameters used in Figures 3–5; negative values indicate hole behaviour (see Section 1—Models).

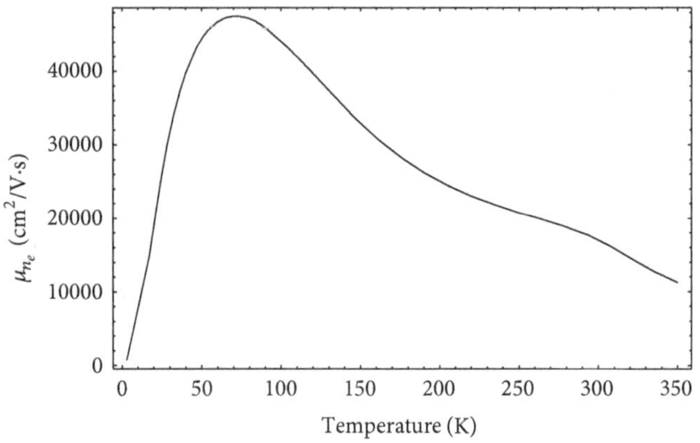

Figure 8: The electron mobility μ_{n_e} versus temperature, using the same calculation parameters used in Figures 3–5 (see Section 1—Models).

such charge carrier sources are chosen such that the calculated concentration curve approximates the experimental data [3, 26–32], as demonstrated for our samples (Table 2) by the example in Figure 3. We believe that the measured conductivity is affected by the presence of different donors in the semiconductor layers. Our model contains four charge carrier sources: two donors—one shallow (N_{D1}), located under the conductivity band (N_{D1}), and one deeper (N_{D2}); one unknown X, which in our approach, can be donor-like or acceptor-like, depending on the temperature; and one acceptor (N_A) as the source of holes. The elements N_{D1}, N_{D2}, and N_A are located in the bulk of the epitaxial layer while X is located at the interface between the substrate and the semiconductor within the first atomic layers (for

TABLE 6: Conductance components for InAs samples at 200 K.

No.	Parameter	200 K			
		InAs 9 μm	InAs 4.7 μm	InAs 5.6 μm	InAs 4 μm
1	$n = n_e + x_e + n_{D2}$ [cm^{-3}]	$3.28 \cdot 10^{15}$	$7.86 \cdot 10^{15}$	$4.68 \cdot 10^{16}$	$4.055 \cdot 10^{17}$
2	μ_n [cm^2/V $\cdot$ s]	24000	21900	16600	8700
3	σ_n [1/Ω cm]	12.6	27.576	124.46	564.
4	σ_n/σ_n	1	1	1	1
5	n_e [cm^{-3}]	$4.0 \cdot 10^{15}$	$8.12 \cdot 10^{15}$	$3.41 \cdot 10^{16}$	$9.88 \cdot 10^{16}$
6	μ_{n_e} [cm^2/V $\cdot$ s]	23570	29849.3	17016.3	10221
7	σ_{n_e} [1/Ω cm]	15.65	27.12	92.96	162.
8	σ_{n_e}/σ_n	1.24	0.983	0.75	0.286
9	x_e [cm^{-3}]	$-7.5 \cdot 10^{14}$	$-2.49 \cdot 10^{14}$	$1.276 \cdot 10^{16}$	$3.07 \cdot 10^{17}$
10	μ_{x_e} [cm^2/V $\cdot$ s]	-4765	-601.37	5692.03	31734
11	σ_{x_e} [1/Ω cm]	0.57	0.024	11.63	1559
12	σ_{x_e}/σ_n	0.045	0.00087	0.093	2.76
13	n_{D2} [cm^{-3}]	$-9.0 \cdot 10^{13}$	$5.88 \cdot 10^{14}$	$-1.54 \cdot 10^{15}$	*
14	μ_{D2} [cm^2/V $\cdot$ s]	-621	1506.67	5758.1	*
15	σ_{D2} [1/Ω cm]	0.009	0.142	-1.42	*
16	σ_{D2}/σ_n	0.0007	0.00515	-0.0011	*
17	Σn [cm^{-3}]	$3.16 \cdot 10^{-15}$	$8.46 \cdot 10^{15}$	$4.5 \cdot 10^{16}$	$4.058 \cdot 10^{17}$
18	G_n [S]	$1.134 \cdot 10^{-2}$	$1.3 \cdot 10^{-2}$	$6.97 \cdot 10^{-2}$	$22.56 \cdot 10^{-2}$
19	G_{n_e} [S]	$1.41 \cdot 10^{-2}$	$1.27 \cdot 10^{-2}$	$5.2 \cdot 10^{-2}$	$6.48 \cdot 10^{-2}$
20	G_{x_e} [S]	$5.13 \cdot 10^{-4}$	$1.13 \cdot 10^{-5}$	$6.5 \cdot 10^{-3}$	$62.36 \cdot 10^{-2}$
21	G_{D2} [S]	$8.1 \cdot 10^{-6}$	$6.67 \cdot 10^{-5}$	$-7.95 \cdot 10^{-4}$	*
22	$\Sigma G_{n_e+x_e+n_{D2}}$ [S]	$1.45 \cdot 10^{-2}$	$1.28 \cdot 10^{-2}$	$5.77 \cdot 10^{-2}$	$68.84 \cdot 10^{-2}$
23	$\Sigma G_{n_e+x_e+n_{D2}}/G_n$	1.28	0.98	0.83	3.05

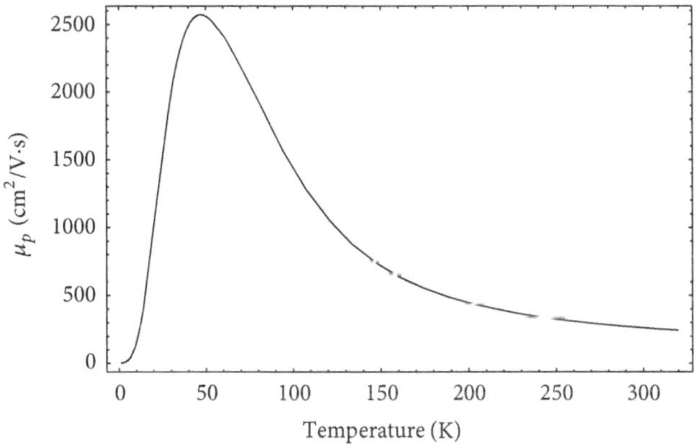

FIGURE 9: The hole mobility μ_p versus temperature, using the same calculation parameters used in Figures 3–5 (see Section 1—Models).

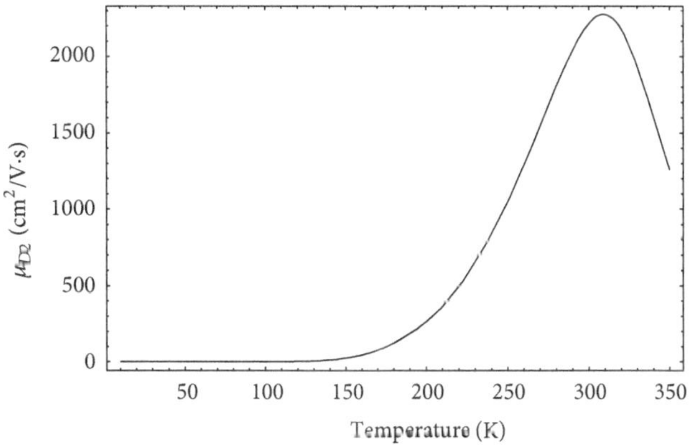

FIGURE 10: The electron mobility μ_{D2} versus temperature, using the same calculation parameters used in Figures 3–5 (see Section 1—Models).

an explanation of this concept, see Section 1—Models). We numerically solve the neutrality equation, thereby combining the values of N_{D1}, N_{D2}, X, and N_A together with their energies of activation, until we obtain the curves $n(T)$ or $n_H(T)$ lying as closely as possible to the experimental data at temperatures of 4–300 K. With values for $n(T)$ and $n_H(T)$ and using well-known physical relations (see Section 1—Models), we can calculate all transport properties needed to determine the conductance values. The negative values of the calculated transport parameters show that at the appropriate temperatures, charge carriers show hole-like properties. We suppose that our approach does not lose the physical interpretation of each step. The results were carefully analysed to ensure that the numerical calculations provided an accurate physical model of the studied effect.

TABLE 7: Conductance components for InAs samples at 300 K.

No.	Parameter	300 K InAs 9 μm	InAs 4.7 μm	InAs 5.6 μm	InAs 4 μm
1	$n = n_e + x_e + n_{D2}$ [cm^{-3}]	4.18 · 1015	$9.62 \cdot 10^{15}$	$4.97 \cdot 10^{16}$	$4.09 \cdot 10^{17}$
2	μ_n [cm^2/V · s]	16000	15800	14580	8400
3	σ_n [1/Ω cm]	10.7	24.35	116.08	550
4	σ_n/σ_n	1	1	1	1
5	n_e [cm^{-3}]	$5.62 \cdot 10^{15}$	$8.13 \cdot 10^{15}$	$4.27 \cdot 10^{16}$	$1.45 \cdot 10^{17}$
6	μ_{n_e} [cm^2/V · s]	16240	15611	14565.7	9090
7	σ_{n_e} [1/Ω cm]	14.6	20.33	99.64	211
8	σ_{n_e}/σ_n	1.36	0.835	0.858	0.38
9	x_e [cm^{-3}]	$-1.43 \cdot 10^{15}$	$-7.56 \cdot 10^{14}$	$7.073 \cdot 10^{15}$	$2.64 \cdot 10^{17}$
10	μ_{x_e} [cm^2/V · s]	−4430	−1109.67	2275.6	16535
11	σ_{x_e} [1/Ω cm]	1.01	0.134	2.58	698
12	σ_{x_e}/σ_n	0.095	0.0055	0.022	1.27
13	n_{D2} [cm^{-3}]	$4.5 \cdot 10^{14}$	$1.862 \cdot 10^{15}$	$-1.51 \cdot 10^{15}$	*
14	μ_{D2} [cm^2/V · s]	1965	3636.22	2302.22	*
15	σ_{D2} [1/Ω cm]	0.14	1.085	−0.56	*
16	σ_{D2}/σ_n	0.013	0.0445	−0.048	*
17	Σn [cm^{-3}]	$4.64 \cdot 10^{15}$	$9.24 \cdot 10^{15}$	$4.83 \cdot 10^{16}$	$4.09 \cdot 10^{17}$
18	G_n [S]	$9.63 \cdot 10^{-3}$	$1.145 \cdot 10^{-2}$	$6.5 \cdot 10^{-2}$	$22.0 \cdot 10^{-2}$
19	G_{n_e} [S]	$1.314 \cdot 10^{-2}$	$9.55 \cdot 10^{-3}$	$5.58 \cdot 10^{-2}$	$8.44 \cdot 10^{-2}$
20	G_{x_e} [S]	$9.09 \cdot 10^{-4}$	$6.3 \cdot 10^{-5}$	$1.44 \cdot 10^{-3}$	$27.92 \cdot 10^{-2}$
21	G_{D2} [S]	$1.26 \cdot 10^{-4}$	$5.1 \cdot 10^{-4}$	$3.14 \cdot 10^{-4}$	*
22	$\Sigma G_{n_e+x_e+n_{D2}}$ [S]	$1.417 \cdot 10^{-2}$	$1.012 \cdot 10^{-2}$	$5.75 \cdot 10^{-2}$	$36.36 \cdot 10^{-2}$
23	$\Sigma G_{n_e+x_e+n_{D2}}/G_n$	1.47	0.88	0.88	1.65

TABLE 8: Scattering mechanism involved in calculation of InGaAs samples properties.

No.	Parameter	$In_{0.53}Ga_{0.47}As$ 7 μm	$In_{0.53}Ga_{0.47}As$ 1.0 μm	$In_{0.53}Ga_{0.47}As$ 1.1 μm	$In_{0.53}Ga_{0.47}As$ 1.0 μm
1	Mobility [cm^2/V · s]	$\mu_{ii}, \mu_{dp}, \mu_{po}, \mu_{pie}$	$\mu_{ii}, \mu_{po}, \mu_{sc}$	$\mu_{ii}, \mu_{po}, \mu_{sc}$	$\mu_{ii}, \mu_{dp}, \mu_{po}, \mu_{pie}, \mu_{sc}$

3. Experimental

The procedure for epitaxial layers of InAs is described as an example. The layers were obtained by molecular beam epitaxy (MBE) in a Riber 32P reactor on GaAs (001) substrates. The preparation conditions are presented in Table 3.

The layers exhibited very smooth surfaces and good crystallographic quality (RHEED results) and electrical properties (Hall measurements). The electrical properties are measured "*ex situ*" by the Hall effect (measurements currents of 10^{-4} A–10^{-3} A in a 0.6 T magnetic field). The Hall device had a square-shaped configuration with dimensions of 5 × 5 mm. The ohmic contacts of the device were made of In 99% + Sn 1% alloy in a Bio-Rad Micro-science-Division (RC2400 Alloying Furnace). The linearity of the contacts characteristics were verified before each Hall effect measurement. All layers were intentionally undoped. The measured n-concentration of the wafers resulted from technical factors of preparation. $In_{0.53}Ga_{0.47}As$ layers were deposited by MBE on SI-InP and GaAs as well as SI-GaAs. We carried out the calculations as described previously [3] for a few samples of InAs, $In_{0.53}Ga_{0.47}As$ and GaAs with different charge carrier concentrations [3, 26–32]. The procedure consisted of the following steps:

(1) numerical solution of the neutrality equation,

(2) numerical calculation of the mobility of each component of the neutrality equation versus temperature;

(3) after determining the values for concentration and mobility versus temperature, we could obtain the conductivity values.

To validate the accuracy of the calculation, we compared the resistivity versus temperature, that is, calculated the curve using experimental data as in Figure 3 (resistivity was measured without a magnetic field; see Figure 4).

We present a more detailed example of the calculation for an InAs sample measuring 9.0 μm thick (see Tables 2, 4, 5, 6, and 7, and all figures) [28]. The InAs samples have the lowest E_G (Table 2). The results are presented in Figures 3, 4, 5, 6, 7, 8, 9, and 10 with the parameters used to solve the neutrality

TABLE 9: Conductance components for InGaAs samples at temperatures of 10 K and 70 K.

No.	Parameter	10 K		70 K		
		$In_{0.53}Ga_{0.47}As$ 7 μm	$In_{0.53}Ga_{0.47}As$ 7 μm	$In_{0.53}Ga_{0.47}As$ 1.0 μm	$In_{0.53}Ga_{0.47}As$ 1.1 μm	$In_{0.53}Ga_{0.47}As$ 1.0 μm
1	$n = n_e + x_e + n_{D2}$ [cm^{-3}]	$1 \cdot 10^{13}$	$1.2 \cdot 10^{13}$	$2.66 \cdot 10^{15}$	$1.095 \cdot 10^{17}$	$1.8 \cdot 10^{19}$
2	μ_n [cm^2/V · s]	6000	38000	48835	4086.1	1770.3
3	σ_n [1/Ω cm]	0.01	0.073	20.8	71.6	50.98
4	σ_n/σ_n	1	1	1	1	1
5	n_e [cm^{-3}]	$9.5 \cdot 10^{13}$	$1.5 \cdot 10^{13}$	$2.4 \cdot 10^{15}$	*	*
6	μ_{n_e} [cm^2/V · s]	815	37300	98000	*	*
7	σ_{n_e} [1/Ω cm]	0.0124	0.09	37.67	*	*
8	σ_{n_e}/σ_n	1.24	1.22	1.8	*	*
9	x_e [cm^{-3}]	$2.0 \cdot 10^{12}$	0.0	$4.92 \cdot 10^{14}$	*	*
10	μ_{x_e} [cm^2/V · s]	−730	−7650	−78000	*	*
11	σ_{x_e} [1/Ω cm]	−0.00023	0.0	−6.15	*	*
12	σ_{x_e}/σ_n	−0.023	*	−0.295	*	*
13	n_{D2} [cm^{-3}]	0.0	$8 \cdot 10^{9}$	$3.45 \cdot 10^{7}$	*	*
14	μ_{D2} [cm^2/V · s]	−720	−6710	−53100	*	*
15	σ_{D2} [1/Ω cm]	*	−0.000009	$-2.93 \cdot 10^{-7}$	*	*
16	σ_{D2}/σ_n	*	0.000123	$-1.41 \cdot 10^{-8}$	*	*
17	Σn [cm^{-3}]	$9.7 \cdot 10^{13}$	$1.5 \cdot 10^{13}$	$2.892 \cdot 10^{15}$	*	*
18	G_n [S]	$7 \cdot 10^{-6}$	$5.11 \cdot 10^{-5}$	0.00208	$7.88 \cdot 10^{-3}$	$5.098 \cdot 10^{-3}$
19	G_{n_e} [S]	$8.68 \cdot 10^{-6}$	$6.3 \cdot 10^{-5}$	0.0038	*	*
20	G_{x_e} [S]	$-1.61 \cdot 10^{-7}$	0	$-6.15 \cdot 10^{-4}$	*	*
21	G_{D2} [S]	*	*	*	*	*
22	$\Sigma G_{n_e+x_e+n_{D2}}$ [S]	$8.52 \cdot 10^{-6}$	$6.3 \cdot 10^{-5}$	0.003185	*	*
23	$\Sigma G_{n_e+x_e+n_{D2}}/G_n$	1.22	1.235	1.53	*	*

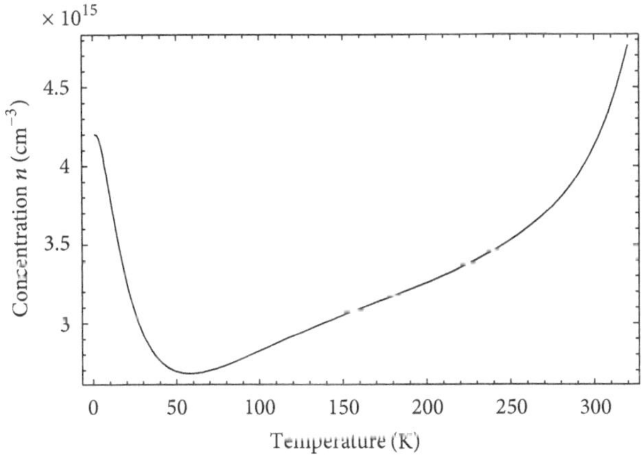

FIGURE 11: The n concentration versus temperature, using the same calculation parameters used in Figures 3–5 (see Section 1—Models).

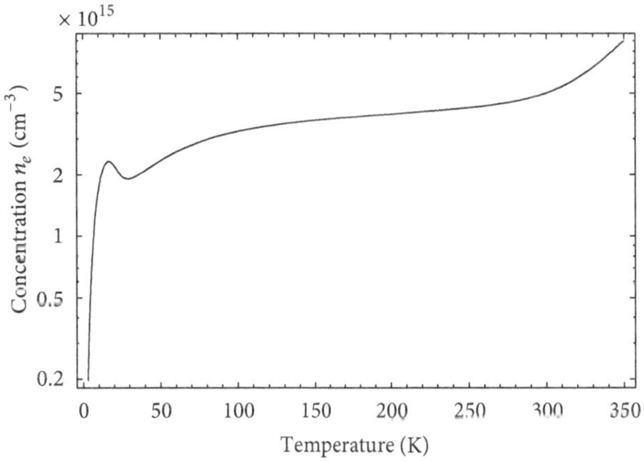

FIGURE 12: The n_e concentration versus temperature, using the same calculation parameters used in Figures 3–5; (see Section 1—Models).

equation (according to [3, 32]). In addition to the results presented in [28], we provide the Hall curves calculated by using the dependence $r_H = \langle \tau_{n,h}^2 \rangle / \langle \tau_{n,h} \rangle^2$ and the calculated values of n, n_e, x_e and n_{D2} versus temperature in Figures 11, 12, 13, and 14.

Figure 3 presents the results of solving the neutrality equation [3, 28, 32] for the InAs sample. In this case, we used the metallurgical thickness of the epitaxial layer, which allowed us to calculate the conformity between the thickness values used to solve for the concentration and resistivity versus temperature. At low temperatures, there was an increase in concentration (Figure 3), which was probably due to the presence of impurity band conduction. The sample has an additional donor concentration N_{D2} at 0.13 eV below

TABLE 10: Conductance components for InGaAs samples at 200 K.

No.	Parameter	200 K $In_{0.53}Ga_{0.47}As$ 7 μm	 $In_{0.53}Ga_{0.47}As$ 1.0 μm	 $In_{0.53}Ga_{0.47}As$ 1.1 μm	 $In_{0.53}Ga_{0.47}As$ 1 μm
1	$n = n_e + x_e + n_{D2}$ [cm^{-3}]	$4.0 \cdot 10^{13}$	$4.65 \cdot 10^{15}$	$1.04 \cdot 10^{17}$	$1.799 \cdot 10^{19}$
2	μ_n [cm^2/V · s]	19000	12810	8514	2010.5
3	σ_n [1/Ω cm]	0.122	9.54	141.85	5794
4	σ_n/σ_n	1	1	1	1
5	n_e [cm^{-3}]	$1.95 \cdot 10^{14}$	$8.0 \cdot 10^{15}$	$8.42 \cdot 10^{16}$	$1.2 \cdot 10^{18}$
6	μ_{n_e} [cm^2/V · s]	18370	14400	8842	3045
7	σ_{n_e} [1/Ω cm]	0.574	18.45	119.26	585
8	σ_{n_e}/σ_n	4.7	1.93	0.84	0.1
9	x_e [cm^{-3}]	$-1.85 \cdot 10^{14}$	$-2.7 \cdot 10^{15}$	$2.53 \cdot 10^{16}$	$1.68 \cdot 10^{19}$
10	μ_{x_e} [cm^2/V · s]	−13400	−6320	1643.3	35440
11	σ_{x_e} [1/Ω cm]	0.4	2.73	6.66	95381
12	σ_{x_e}/σ_n	3.28	0.29	0.047	16.46
13	n_{D2} [cm^{-3}]	$3.6 \cdot 10^{13}$	$5.3 \cdot 10^{13}$	$1.38 \cdot 10^{16}$	*
14	μ_{D2} [cm^2/V · s]	64870	−1840	−882	*
15	σ_{D2} [1/Ω cm]	0.37	−0.016	−1.95	*
16	σ_{D2}/σ_n	3.03	−0.00164	−0.014	*
17	Σn [cm^{-3}]	$4.6 \cdot 10^{13}$	$5.35 \cdot 10^{15}$	$1.233 \cdot 10^{17}$	$1.8 \cdot 10^{19}$
18	G_n [S]	$8.54 \cdot 10^{-5}$	$9.54 \cdot 10^{-4}$	$1.56 \cdot 10^{-2}$	$57.94 \cdot 10^{-2}$
19	G_{n_e} [S]	$4.02 \cdot 10^{-4}$	$1.845 \cdot 10^{-3}$	$1.31 \cdot 10^{-2}$	$5.85 \cdot 10^{-2}$
20	G_{x_e} [S]	$2.8 \cdot 10^{-4}$	$2.73 \cdot 10^{-4}$	$7.33 \cdot 10^{-4}$	9.538
21	G_{D2} [S]	$2.59 \cdot 10^{-4}$	$-1.6 \cdot 10^{-6}$	$-2.145 \cdot 10^{-4}$	*
22	$\Sigma G_{n_e+x_e+n_{D2}}$ [S]	$9.41 \cdot 10^{-4}$	$2.12 \cdot 10^{-3}$	$1.36 \cdot 10^{-2}$	9.59
23	$\Sigma G_{n_e+x_e+n_{D2}}/G_n$	11.02	2.22	0.872	16.55

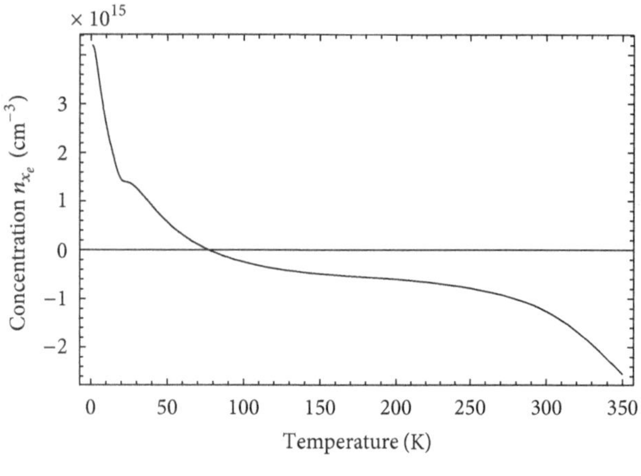

FIGURE 13: The x_e concentration versus temperature, using the same calculation parameters used in Figures 3–5; concentration values below 0 x_e indicate hole-like behaviour (see Section 1—Models).

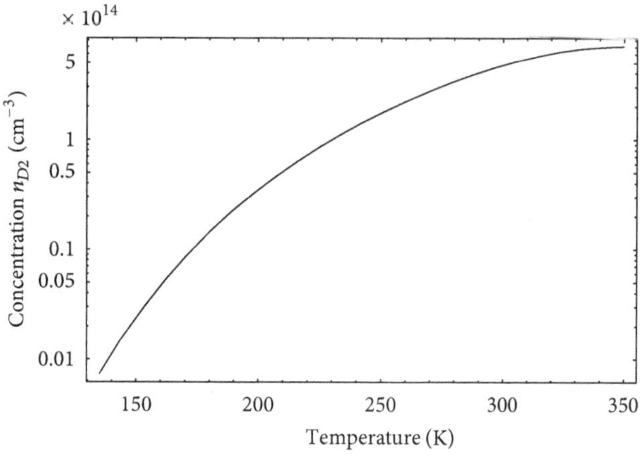

FIGURE 14: The n_{D2} concentration versus temperature, using the same calculation parameters used in Figures 3–5 (see Section 1—Models).

the bottom of the conduction band. This level is near $E_G/2$ for InAs (Table 2). This conductivity concentration (n) was used to calculate the resistivity-temperature (Figure 4) and the mobility-temperature curves (Figure 5). The calculated Hall concentration n_H below ~100 K lies distinctly under the experimental data.

We also verified the ratio of n to the density of states value in the conductivity band versus temperature curve. This ratio determines the degeneration probability in the investigated sample. Unfortunately, below ~15 K, its value is greater than 1; therefore, all calculations below that temperature that use Boltzmann statistics are problematic (Figure 6). It must be

TABLE 11: Conductance components for InGaAs samples at 300 K.

No.	Parameter	300 K			
		$In_{0.53}Ga_{0.47}As$ 7 μm	$In_{0.53}Ga_{0.47}As$ 1.0 μm	$In_{0.53}Ga_{0.47}As$ 1.1 μm	$In_{0.53}Ga_{0.47}As$ 1.0 μm
1	$n = n_e + x_e + n_{D2}$ [cm^{-3}]	$1.2 \cdot 10^{14}$	$6.75 \cdot 10^{15}$	$1.13 \cdot 10^{17}$	$1.796 \cdot 10^{19}$
2	μ_n [cm^2/V · s]	13000	8280	7310.5	1890
3	σ_n [1/Ω cm]	0.25	8.95	132.2	5431
4	σ_n/σ_n	1	1	1	1
5	n_e [cm^{-3}]	$1.6 \cdot 10^{14}$	$1.0 \cdot 10^{16}$	$9.76 \cdot 10^{16}$	$1.55 \cdot 10^{18}$
6	μ_{n_e} [cm^2/V · s]	12970	8400	7477	2970
7	σ_{n_e} [1/Ω cm]	0.33	13.4	116.8	732
8	σ_{n_e}/σ_n	1.32	1.5	0.88	0.13
9	x_e [cm^{-3}]	$-1.45 \cdot 10^{14}$	$-3.32 \cdot 10^{15}$	$1.59 \cdot 10^{16}$	$1.64 \cdot 10^{19}$
10	μ_{x_e} [cm^2/V · s]	−6500	−3230	875.6	25150
11	σ_{x_e} [1/Ω cm]	0.15	1.72	2.23	66076
12	σ_{x_e}/σ_n	0.6	0.19	0.017	12.17
13	n_{D2} [cm^{-3}]	$1.2 \cdot 10^{14}$	$6.9 \cdot 10^{14}$	$1.72 \cdot 10^{16}$	*
14	μ_{D2} [cm^2/V · s]	169360	25	443	*
15	σ_{D2} [1/Ω cm]	3.25	0.00276	1.22	*
16	σ_{D2}/σ_n	13.0	0.0003	0.0092	*
17	Σn [cm^{-3}]	$1.35 \cdot 10^{14}$	$7.37 \cdot 10^{15}$	$1.135 \cdot 10^{17}$	$1.795 \cdot 10^{19}$
18	G_n [S]	$1.75 \cdot 10^{-4}$	$8.95 \cdot 10^{-4}$	$1.45 \cdot 10^{-2}$	$54.31 \cdot 10^{-2}$
19	G_{n_e} [S]	$2.31 \cdot 10^{-4}$	$1.34 \cdot 10^{-3}$	$1.28 \cdot 10^{-2}$	$7.32 \cdot 10^{-2}$
20	G_{x_e} [S]	$1.05 \cdot 10^{-4}$	$1.72 \cdot 10^{-4}$	$2.45 \cdot 10^{-4}$	6.608
21	G_{D2} [S]	$2.275 \cdot 10^{-3}$	$2.76 \cdot 10^{-7}$	$1.34 \cdot 10^{-4}$	*
22	$\Sigma G_{n_e+x_e+n_{D2}}$ [S]	$2.61 \cdot 10^{-3}$	$1.35 \cdot 10^{-3}$	$1.34 \cdot 10^{-2}$	6.68
23	$\Sigma G_{n_e+x_e+n_{D2}}/G_n$	14.9	1.7	0.92	12.3

TABLE 12: Scattering mechanism involved in calculation of GaAs samples properties.

No.	Parameter	GaAs 4.25 μm	GaAs 1.7 μm	GaAs 2.6 μm	GaAs 2.4 μm
1	Mobility [cm^2/V · s]	$\tau_{ii}, \tau_{po}, \tau_{sc}$	$\tau_{ii}, \tau_{po}, \tau_{sc}$	$\tau_{ii}, \tau_{po}, \tau_{sc}$	$\tau_{ii}, \tau_{po}, \tau_{sc}$

explained here that our charge transport model was very accurate only for the GaAs samples (Table 13) and the calculated values using the numerical method were sufficiently reliable. With respect to the two other semiconductors, our model was less accurate. The neutrality equation was properly solved for all investigated samples; therefore, we believe that the concentration values are consistent with the experimental results. However the InAs and InGaAs samples presented problems with the wafers exhibiting very high concentrations at temperatures below 200 K (InAs (Table 5) and InGaAs (Table 9)).

The resistivity curves (conductivity parameters) show very satisfactory behaviour. The calculated curve approximately agrees with the experimental data at all temperatures with some discrepancies only at temperatures below 15 K. This disagreement is most marked below 100 K, when the concentration is near 10% of the number of states in the conduction band (Figure 6). The great influence of the space-charge scattering mechanism of the $N_S \cdot Q$ product (where N_S is the density of space charge regions and Q is their effective scattering cross-sectional area) must be underlined.

The Hall mobility measurements (as in Figure 5) lead to the conclusion that the theoretical r_H disagrees with the real value $r_{H,\text{experimental}} \approx 1$ at lower temperatures (approximately as low as ~100 K).

We now present the results of mobility calculations of each component obtained after solution of the neutrality equation (Figures 7–10). At lower temperatures, donors named *D*2 marginally influenced the total mobility; their numbers were so low that they could be omitted (they have important influence at temperatures 200–300 K). It must be noted that calculated Hall effect curves (Figures 3 and 5 curves in red) lightly disagree with experimental points. The conductivity effect curves show much better compliance with the experiment, especially at low temperatures. Using the measurements procedure as described for the GaAs sample, we obtained results that allowed us to determine the components of conductivity for these samples (Table 2). The scattering mechanism used for this calculation and the pertinent values are shown: for InAs in Tables 4, 5, 6, and 7, for $In_{0.53}Ga_{0.47}As$ in Tables 8, 9, 10, and 11, and for GaAs in Tables 12 and 13.

TABLE 13: Conductance components in GaAs.

No.	Parameter	10 K GaAs 1.7 μm	GaAs 4.25 μm	70 K GaAs 1.7 μm	GaAs 2.6 μm	GaAs 2.4 μm	GaAs 4.25 μm	300 K GaAs 1.7 μm	GaAs 2.6 μm	GaAs 2.4 μm
1	$n = n_e + x_e + n_{D2}$ [cm^{-3}]	$1.72 \cdot 10^{15}$	$4.3 \cdot 10^{14}$	$1.71 \cdot 10^{15}$	$3.47 \cdot 10^{15}$	$4.8 \cdot 10^{15}$	$4.6 \cdot 10^{14}$	$1.84 \cdot 10^{15}$	$5.24 \cdot 10^{15}$	$9.5 \cdot 10^{15}$
2	μ_n [$cm^2/V \cdot s$]	8000	62000	48200	20730	19940	6000	8000	4315.4	3117.9
3	σ_n [$1/\Omega$ cm]	2.2	4.27	13.5	11.52	15.33	0.45	2.36	3.62	4.74
4	σ_n/σ_n	1	1	1	1	1	1	1	1	1
5	n_e [cm^{-3}]	$1.35 \cdot 10^{15}$	$4.71 \cdot 10^{14}$	$1.85 \cdot 10^{15}$	$3.43 \cdot 10^{15}$	$4.83 \cdot 10^{15}$	$4.87 \cdot 10^{14}$	$2.02 \cdot 10^{15}$	$5.23 \cdot 10^{15}$	$1.0 \cdot 10^{16}$
6	μ_{n_e} [$cm^2/V \cdot s$]	7570	60793	48674	20730	19963	6141	7777	4323	3130
7	σ_{n_e} [$1/\Omega$ cm]	1.64	4.58	14.42	11.4	15.45	0.478	2.52	3.62	5.01
8	σ_{n_e}/σ_n	0.744	1.07	1.07	0.99	1.008	1.06	1.07	1.	1.056
9	x_e [cm^{-3}]	$3.4 \cdot 10^{14}$	$-4.31 \cdot 10^{13}$	$-1.41 \cdot 10^{14}$	$3.57 \cdot 10^{13}$	$-3.25 \cdot 10^{13}$	$-2.53 \cdot 10^{13}$	$-2.17 \cdot 10^{14}$	$1.2 \cdot 10^{13}$	$-5.8 \cdot 10^{14}$
10	μ_{x_e} [$cm^2/V \cdot s$]	1912	−5273	−3720	215.5	−134.6	−282.5	−837.6	10	−180
11	σ_{x_e} [$1/\Omega$ cm]	0.104	0.0364	0.084	0.00123	0.0007	0.00114	0.03	0.00002	0.017
12	σ_{x_e}/σ_n	0.05	0.0085	0.0062	0.000107	0.000046	0.0025	0.0123	0.0000053	0.0036
13	n_{D2} [cm^{-3}]	*	$4.37 \cdot 10^{8}$	$1.43 \cdot 10^{10}$	$3.6 \cdot 10^{9}$	$1.55 \cdot 10^{13}$	$4.98 \cdot 10^{13}$	$2.20 \cdot 10^{14}$	$1.0 \cdot 10^{15}$	$4.0 \cdot 10^{15}$
14	μ_{D2} [$cm^2/V \cdot s$]	*	0.062	0.4	0.02	65	741.5	1085.65	1017	2273.6
15	σ_{D2} [$1/\Omega$ cm]	*	*	*	*	0.000161	0.0059	0.04	0.163	1.46
16	σ_{D2}/σ_n	*	*	*	*	0.0000105	0.013	0.016	0.045	0.31
17	Σn [cm^{-3}]	$1.69 \cdot 10^{15}$	$4.28 \cdot 10^{14}$	$1.71 \cdot 10^{15}$	$3.46 \cdot 10^{15}$	$4.81 \cdot 10^{15}$	$5.11 \cdot 10^{14}$	$2.02 \cdot 10^{15}$	$6.24 \cdot 10^{15}$	$8.2 \cdot 10^{15}$
18	G_n [S]	$3.74 \cdot 10^{-4}$	$1.81 \cdot 10^{-3}$	$2.295 \cdot 10^{-3}$	$3.0 \cdot 10^{-3}$	$3.68 \cdot 10^{-3}$	$1.91 \cdot 10^{-4}$	$4.01 \cdot 10^{-4}$	$9.41 \cdot 10^{-4}$	$1.14 \cdot 10^{-3}$
19	G_{n_e} [S]	$2.79 \cdot 10^{-4}$	$1.95 \cdot 10^{-3}$	$2.45 \cdot 10^{-3}$	$2.96 \cdot 10^{-3}$	$3.71 \cdot 10^{-3}$	$2.03 \cdot 10^{-4}$	$4.28 \cdot 10^{-4}$	$9.41 \cdot 10^{-4}$	$1.20 \cdot 10^{-3}$
20	G_{x_e} [S]	$1.77 \cdot 10^{-5}$	$1.55 \cdot 10^{-5}$	$1.43 \cdot 10^{-5}$	$2.2 \cdot 10^{-7}$	$1.7 \cdot 10^{-7}$	$4.85 \cdot 10^{-7}$	$5.1 \cdot 10^{-7}$	$5.2 \cdot 10^{-9}$	$4.08 \cdot 10^{-7}$
21	G_{D2} [S]	*	*	*	*	$3.86 \cdot 10^{-8}$	$2.51 \cdot 10^{-6}$	$6.8 \cdot 10^{-7}$	$4.24 \cdot 10^{-5}$	$3.5 \cdot 10^{-4}$
22	$\Sigma G_{n_e+x_e+n_{D2}}$ [S]	$2.97 \cdot 10^{-4}$	$1.965 \cdot 10^{-3}$	$2.46 \cdot 10^{-3}$	$2.96 \cdot 10^{-3}$	$3.71 \cdot 10^{-3}$	$2.06 \cdot 10^{-4}$	$4.35 \cdot 10^{-4}$	$9.83 \cdot 10^{-4}$	$1.55 \cdot 10^{-3}$
23	$\Sigma G_{n_e+x_e+n_{D2}}/G_n$	0.794	1.086	1.072	0.987	1.008	1.08	1.085	1.57	1.36

*The values are so small that they can be omitted.

It may be concluded that with increased doping concentration of the InAs samples (Tables 5–7), fewer scattering mechanisms as determined by the calculation of the conductivity components apply (Table 4). The 23rd row of each table shows the values of the ratio of the calculated conductance $\sum G_{n_e+x_e+D2}$ used to measure G_n for each structure. This ratio must be approximately equal to 1. The actual values are presented in Figure 15. In Figures 16, 17, 18, and 19, the conductance components for a 4.7 μm InAs sample at temperatures of 10 K, 70 K, 200 K and 300 K are shown. This sample exhibits $\sum G_{n_e+x_e+D2}/G_n$ values of 0.71, 0.95, 0.98, and 0.88 at 10 K, 70 K, 200 K, and 300 K, respectively. Also, six scattering mechanisms are involved in the calculation of the transport parameters (Table 4).

Figures 16–19 show that the changes in the conductance components of InAs changes are independent of temperature. The main part of the conductance is attributed to the n_e part of the charge carriers. In Tables 5–7, some samples show a higher sum of the conductance components relative to the measured resultant conductance. We suppose that this is associated with the principles of the numerical calculation, which give approximate values. Also, the theoretical model that was used may be less accurate, due to the slightly higher doping of the semiconductor samples.

The scattering mechanism considered in for InGaAs calculation, the conductance components values, and the results of the calculated parameters are shown for 1.0–7 μm samples at temperatures of 70 K, 200 K, and 300 K in Tables 8–11 and in Figures 21, 22, and 23.

The actual the values given in line 23 of Tables 9–11 (InGaAs) are presented in Figure 20. The values of r_H were derived from experimental calculations and are equal to those for the samples with $n < 1.10^{16}$ cm^{-3}; however, for higher concentrations, the measured r_H is equal to 1 and is different from that calculated according to equation $r_H = \langle \tau_{n,h}^2 \rangle / \langle \tau_{n,h} \rangle^2$. The $\sum G_{n_e+x_e+D2}/G_n$ ratio for InGaAs samples (Figure 20) is approximately equal to 1 at concentrations between 10^{15} and 10^{18}, but for the samples with doping concentration of ~10^{13} and ~10^{19}, our procedure yields erroneous results (Tables 9–11).

The scattering mechanism used for the GaAs calculations, the conductance components values, and the results of the calculated parameters are shown for 1.7–4.25 μm samples at temperatures of 70 K, 200 K, and 300 K in Tables 12 and 13, and those for the 1.7 μm sample alone are shown in Figures 25, 26, and 27.

The actual values shown in line 23 of Table 13 (GaAs) are presented in Figure 24. In nearly all samples, the ratio

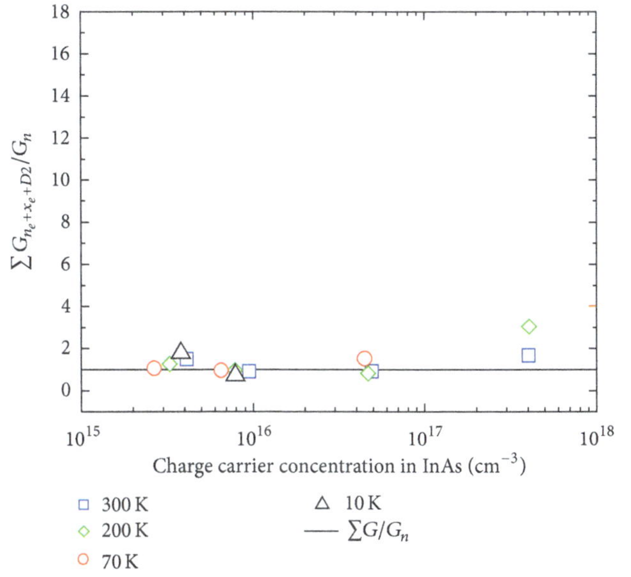

FIGURE 15: The actual values of the ratio $\sum G_{n_e+x_e+D2}/G_n$ versus charge carriers concentration in the investigated InAs samples.

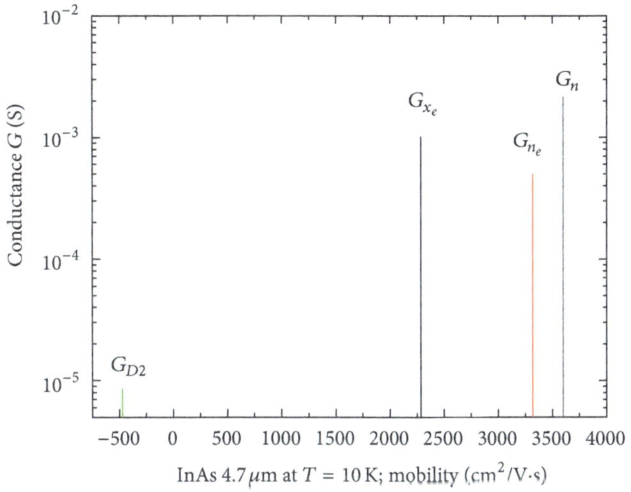

FIGURE 16: Conductance components for InAs 4.7 μm at 10 K.

$\sum G_{n_e+x_e+D2}/G_n$ (Figure 24) is approximately equal to 1. Only at 300 K is the concentration n_e slightly higher in GaAs 4.25 μm and GaAs 1.7 μm. From Figure 24, one can conclude that practically all GaAs samples present a $\sum G_{n_e+x_e+D2}/G_n$ value equal to ~1.

4. Conclusions

As a followup to our previous papers [3, 32], herein we present a potential application of the modelling and numerical simulation of material properties for three semiconductors InAs, $In_{0.53}Ga_{0.47}As$, and GaAs as MBE layers by our MCF method using Hall structures. We have found three sources of current carriers, n_e, n_{x_e}, and n_{D2}, whose sum should

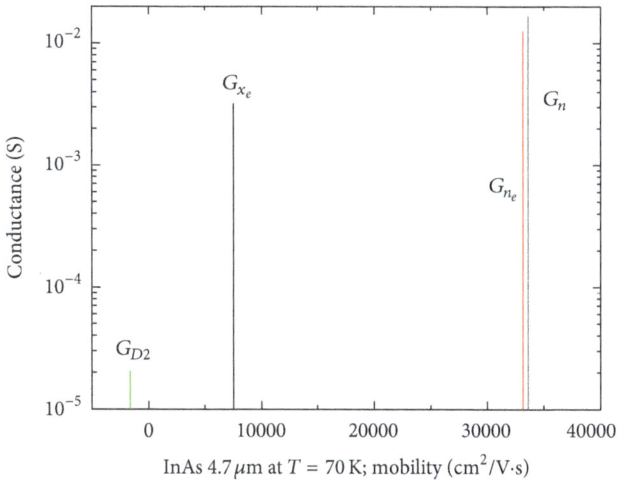

FIGURE 17: Conductance components for InAs 4.7 μm at 70 K.

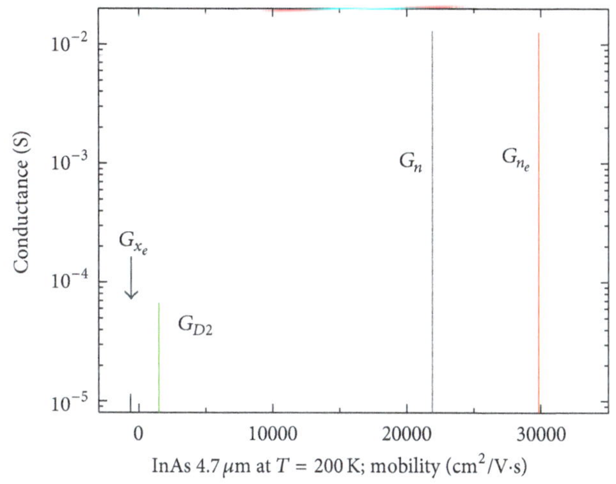

FIGURE 18: Conductance components for InAs 4.7 μm at 200 K.

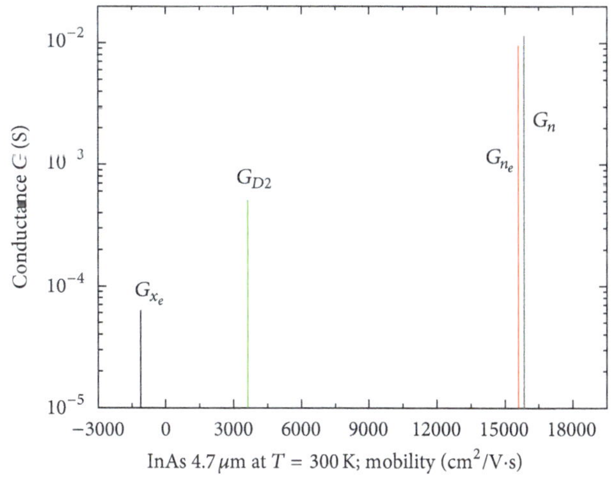

FIGURE 19: Conductance components for InAs 4.7 μm at 300 K.

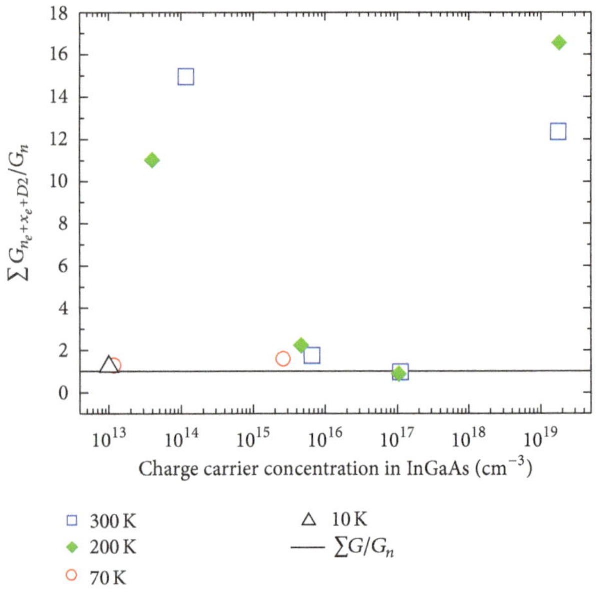

FIGURE 20: The actual values of the ratio $\sum G_{n_e+x_e+D2}/G_n$ versus charge carriers concentration in the investigated InGaAs samples.

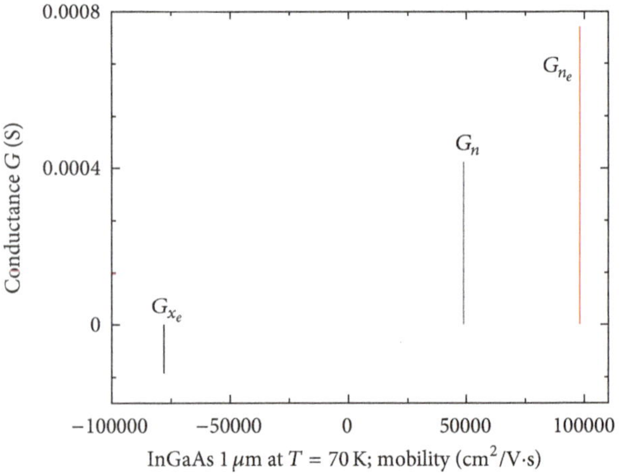

FIGURE 21: Conductance components for InGaAs 1.0 μm at 70 K.

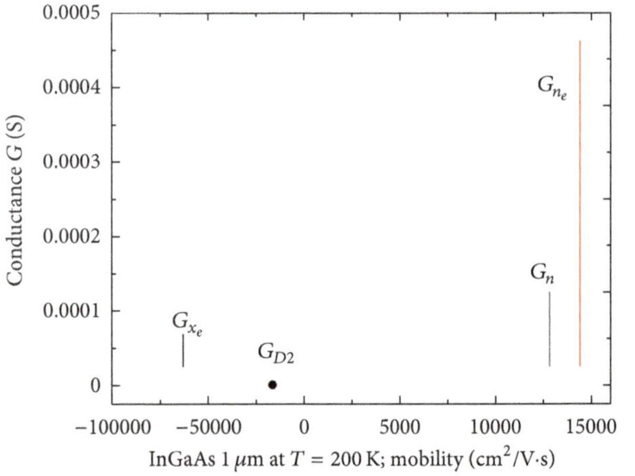

FIGURE 22: Conductance components for InGaAs 1.0 μm at 200 K.

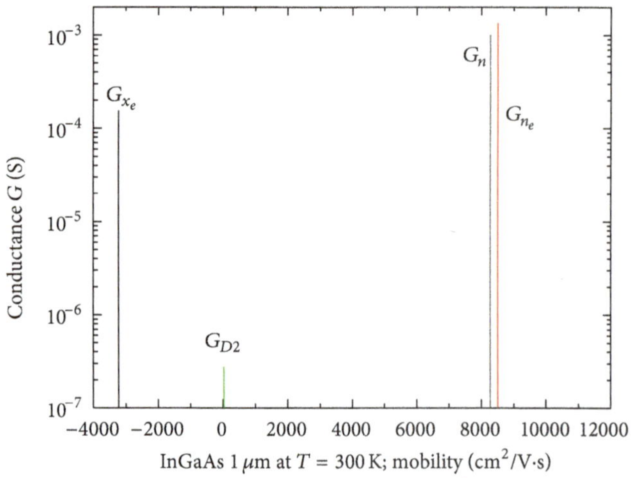

FIGURE 23: Conductance components for InGaAs 1.0 μm at 300 K.

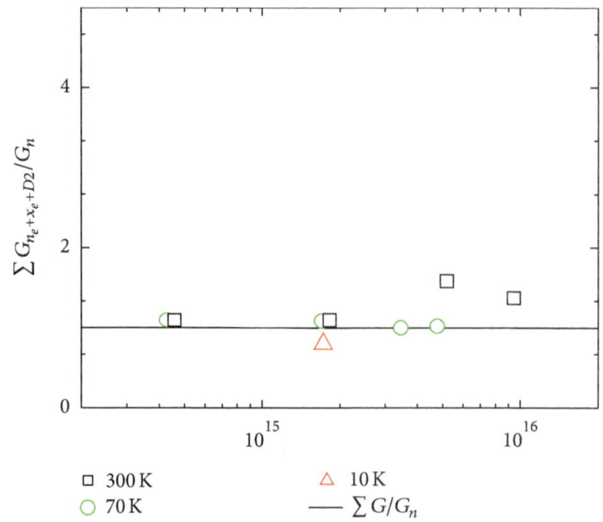

FIGURE 24: The actual values of the ratio $\sum G_{n_e+x_e+D2}/G_n$ versus charge carrier concentration in the investigated GaAs sample.

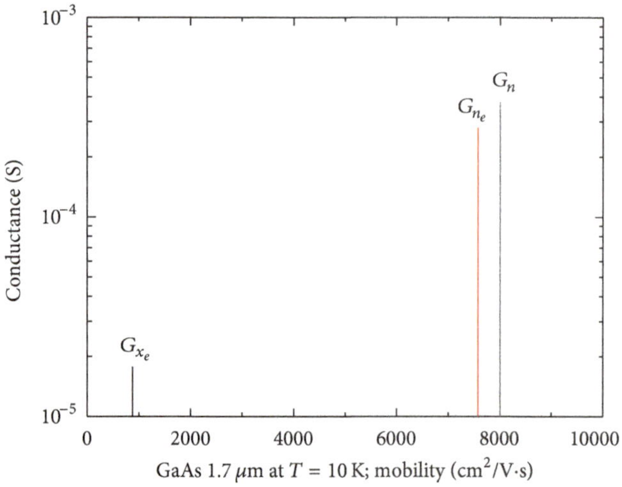

FIGURE 25: Conductance components for GaAs 1.7 μm at 10 K.

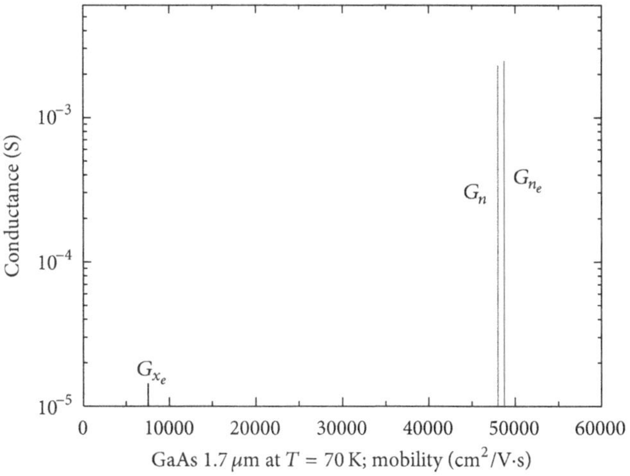

FIGURE 26: Conductance components for GaAs 1.7 μm at 70 K.

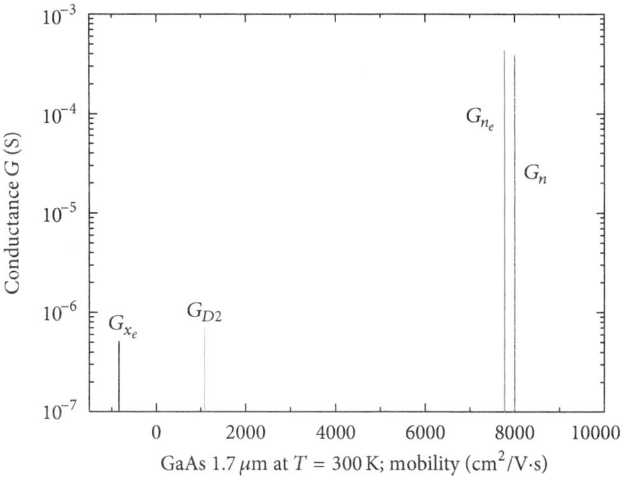

FIGURE 27: Conductance components for GaAs 1.7 μm at 300 K.

theoretically be n. We obtained the best results of numerical simulation for GaAs samples; for InAs and $In_{0.53}Ga_{0.47}As$, our method is limited. The use of different scattering mechanisms (Tables 4, 8, and 12) allows for a more exhaustive description of conduction in the investigated MBE layers. We believe, nevertheless, that, in this paper, we demonstrate a novel application of the solution of the neutrality equation and relaxation time approximation by numerical calculations. It is also clear to us that some improvement in the theory of conduction in thin semiconductor layers formed by MBE is required.

The scattering mechanisms used to calculate the charge transport properties (Tables 4, 8, and 12) indicate that the crystallographic quality of samples composed of the same material may vary. Only the GaAs samples (Table 13) present the same value for the scattering mechanisms (τ_{ii}, τ_{po}, τ_{sc}) in all of the investigated samples. The purest samples are those that have the largest thickness. Such behaviour agrees with our previous observations [3, 27–32].

The presented method for the numerical modelling and calculation of conductance components allows us to determine what parameters affect the total conductance. We calculated only the charge transport properties of carriers that originate from donors and acceptors using the solution of the neutrality equation. This solution is obtained such that the concentration versus temperature curve satisfactorily lies on the experimental data. The same applies to the resistivity and mobility. These three dependences underline the reliability of our calculations. In conclusion, we assume that our modelling method allows for the determination of the conductivity components strictly connected with experiments in relatively weak magnetic fields.

Acknowledgments

The author is indebted to Maciej Bugajski, Prof. and Janusz Kaniewski, DSc for their many helpful discussions. Special thanks are given to Tomasz Przesławski, PhD for the fruitful measurement results provided and Kazimierz Reginski, DSc for the samples preparation. The calculations are performed using math.8 at ICM Interdisciplinary Centre for Mathematical and Computational Modeling at University of Warsaw owing Calculation Grant G51-5.

References

[1] J. D. Grange, E. H. C. Parker, and R. M. King, "Relationship of MBE growth parameters with the electrical properties of thin (100) InAs epilayers," *Journal of Physics D*, vol. 12, no. 9, pp. 1601–1612, 1979.

[2] P. N. Fawcett, B. A. Joyce, X. Zhang, and D. W. Pashley, "Interface structure of InAs grown on GaAs(001) surfaces by molecular beam epitaxy," *Journal of Crystal Growth*, vol. 116, no. 1-2, pp. 81–86, 1992.

[3] A. Wolkenberg, T. Przeslawski, J. Kaniewski, E. Kowalczyk, and K. Reginski, "Conductivity analysis of n-GaAs molecular beam epitaxial layers using multicarrier fitting," *Journal of Applied Physics*, vol. 99, no. 9, Article ID 093708, 2006.

[4] M. C. Gold and D. A. Nelson, "Variable magnetic field Hall effect measurements and analysis of high purity, Hg vacancy (p-type)," *Journal of Vacuum Science & Technology*, vol. 4, pp. 2040–2046, 1986.

[5] J. R. Meyer, C. A. Hoffman, F. J. Bartoli, D. A. Arnold, S. Sivananthan, and J. P. Fauri, "Methods for magnetotransport characterization of IR detector materials," *Semiconductor Science and Technology*, vol. 8, pp. 805–823, 1993.

[6] W. A. Beck and J. R. Anderson, "Determination of electrical transport properties using a novel magnetic field dependent Hall technique," *Journal of Applied Physics*, vol. 62, no. 2, pp. 541–554, 1987.

[7] A. C. Beer, *Galvanomagnetic Effects in Semiconductors*, vol. 4 of *Solid state physics: Supplement*, 1963.

[8] Z. Dziuba, "Magnetotransport characterization of GaAs," *Acta Physica Polonica*, vol. 80, pp. 827–830, 1991.

[9] Z. Dziuba and M. Górska, "Analysis of the electrical conduction using an iterative method," *Journal de Physique III*, vol. 2, no. 1, pp. 99–110, 1992.

[10] J. Antoszewski, D. J. Seymour, L. Faraone, J. R. Meyer, and C. A. Hoffman, "Magneto-transport characterization using

quantitative mobility-spectrum analysis," *Journal of Electronic Materials*, vol. 24, no. 9, pp. 1255–1262, 1995.

[11] Z. Dziuba, J. Antoszewski, J. M. Dell et al., "Magnetic field dependent Hall data analysis of electron transport in modulation-doped AlGaN/GaN heterostructures," *Journal of Applied Physics*, vol. 82, no. 6, pp. 2996–3002, 1997.

[12] J. R. Meyer, C. A. Hoffman, J. Antoszewski, and L. Faraone, "Quantitative mobility spectrum analysis of multicarrier conduction in semiconductors," *Journal of Applied Physics*, vol. 81, no. 2, pp. 709–713, 1997.

[13] D. K. Schroder, *Semiconductor Material and Device Characterization*, Wiley Interscience, 2006.

[14] H. M. Lin, Y. L. Chen, J. Yang et al., "Synthesis and characterization of core-shell GaP@GaN and GaN@GaP nanowires," *Nano Letters*, vol. 3, no. 4, pp. 537–541, 2003.

[15] A. Amtout, S. Raghavan, P. Rotella, G. Von Winckel, A. Stintz, and S. Krishna, "Theoretical modeling and experimental characterization of InAs/InGaAs quantum dots in a well detector," *Journal of Applied Physics*, vol. 96, no. 7, pp. 3782–3786, 2004.

[16] G. Brammertz, H. C. Lin, K. Martens et al., "Capacitance-voltage characterization of GaAs-Oxide interfaces," *Journal of the Electrochemical Society*, vol. 155, no. 12, pp. H945–H950, 2008.

[17] H. F. Chiu, Y. S. Chang, J. Y. Wu et al., "Very high deposition rate of a-Si:H thin films by ECRCVD," *ECS Transactions*, vol. 34, pp. 1065–1070, 2011.

[18] F. A. Ali, P. Pattnaik, and S. K. Kamilla, "Characterisation of bulk GaSb using TCAD," in *Proceedings of the 2nd IEEE Students' Technology Symposium (TechSym '11)*, pp. 211–213, Kharagpur, India, January 2011.

[19] Q. Gao, H. H. Tan, H. E. Jackson et al., "Growth and properties of III-V compound semiconductor heterostructure nanowires," *Semiconductor Science and Technology*, vol. 26, no. 1, Article ID 014035, 2011.

[20] H. Arabshahi, "A new calculation method for thermal and electical characterization in CdTe and CdSe semiconductors," *International Journal of Physical Sciences*, vol. 6, no. 13, pp. 3155–3161, 2011.

[21] J. S. Kim, D. G. Seiler, and W. F. Tseng, "Multicarrier characterization method for extracting mobilities and carrier densities of semiconductors from variable magnetic field measurements," *Journal of Applied Physics*, vol. 73, no. 12, pp. 8324–8335, 1993.

[22] S. Kiatgamolchai, M. Myronov, O. A. Mironov, V. G. Kantser, E. H. C. Parker, and T. E. Whall, "Mobility spectrum computational analysis using a maximum entropy approach," *Physical Review E*, vol. 66, no. 3, Article ID 036705, 9 pages, 2002.

[23] D. Chrastina, J. P. Hague, and D. R. Leadley, "Application of Bryan's algorithm to the mobility spectrum analysis of semiconductor devices," *Journal of Applied Physics*, vol. 94, pp. 6583–6591, 2003.

[24] I. Vurgaftman, J. R. Meyer, C. A. Hoffman et al., "Improved quantitative mobility spectrum analysis for hall characterization," *Journal of Applied Physics*, vol. 84, no. 9, pp. 4966–4973, 1998.

[25] J. Antoszewski, L. Faraone, I. Vurgaftman, J. R. Meyer, and C. A. Huffman, "Application of quantitative mobility-spectrum analysis to multilayer HgCdTe structures," *Journal of Electronic Materials*, vol. 33, no. 6, pp. 673–683, 2004.

[26] A. Wolkenberg and T. Przesławski, "Application of galvanomagnetic measurements in temperature range 70-300K to MBE GaAs layers characterization," in *Metal/Nonmetal Microsystems: Physics, Technology, and Applications*, vol. 2780 of *Proceedings of SPIE*, pp. 170–182, 1995.

[27] A. Wolkenberg, T. Przesławski, J. Kaniewski, J. Bąk-Misiuk, and K. Regiński, "Thickness dependence of the structural and electrical properties of InAs layers epitaxially grown by MBE on GaAs (001)," *Materials Science and Engineering B*, vol. 77, pp. 250–254, 2000.

[28] T. Przesławski, A. Wolkenberg, K. Regiński, J. Kaniewski, and J. Bak–Misiuk, "Growth and transport properties of relaxed epilayers of InAs on GaAs," *Thin Solid Films*, vol. 367, no. 1-2, pp. 232–234, 2000.

[29] A. Wolkenberg, T. Przesławski, K. Regiński, and J. Kaniewski, "Electrical charge transport of n-InAs epitaxial films on GaAs," *Materials Science and Engineering B*, vol. 90, no. 1-2, pp. 176–179, 2002.

[30] A. Wolkenberg, T. Przesławski, J. Kaniewski, and K. Regiński, "Experimental confirmation by galvanomagnetic methods of a complex transport model in $In_{0.53}Ga_{0.47}As$ layers deposited by MBE on SI-InP," *Journal of Physics and Chemistry of Solids*, vol. 64, no. 1, pp. 7–14, 2003.

[31] A. Wolkenberg, T. Przesławski, J. Kaniewski, and K. Reginski, "A method for calculating the conductivity mobility spectrum using multi-carrier fitting," *Materials Science and Engineering B*, vol. 110, no. 1, pp. 79–86, 2004.

[32] A. Wolkenberg and T. Przesławski, "Parametry galwanomagnetyczne i przewodnościowe warstw epitaksjalnych MBE z InAs/GaAs, $In_{0,53}Ga_{0,47}$ As/InP i GaAs/GaAs," *Elektronika*, vol. 48, pp. 26–38, 2007.

11

Evidence for Stable High-Temperature Ferromagnetism in Fluorine-Treated C_{60}

Rajen Patel[1] and Frank J. Owens[2]

[1] *Armament Research, Development and Engineering Center, Picatinny, NJ 07806, USA*
[2] *Department of Physics, Hunter College and Graduate Center, City University of New York, 696 Park Avenue, New York, NY 10065, USA*

Correspondence should be addressed to Frank J. Owens; owensfj@gmail.com

Academic Editor: Zhongfang Chen

It is shown by magnetic field dependent ac susceptibility, magnetic force microscopy, and ferromagnetic resonance that exposure of C_{60} to fluorine at 160°C produces a stable ferromagnetic material with a Curie temperature well above room temperature. The exposure to fluorine is accomplished by decomposing a fluorine-rich polymer, trifluorochloroethylene $[F_2C\text{–}CFCl]_n$, which has C_{60} imbedded in it. Based on previous experimental observations and molecular orbital calculations, it is suggested that the ferromagnetism is arising from crystals of C_{60}–F.

1. Introduction

There has been much interest in the material science community in synthesizing molecular-based ferromagnetic materials because of the potential to chemically engineer their properties and the possible ease of production. The C_{60} molecule has played an important role in this possibility because of a number of reports of ferromagnetism in it. In 1991 a complex of C_{60} and the strong electron donor molecule $C_2N_4(CH_3)_8$ (TDAE) were found to be ferromagnetic at 16.1 K [1]. Subsequently ferromagnetism having a Curie temperature of 500 K was reported in a two-dimensional polymeric form of C_{60} produced by high pressure (6 GPa) and high temperature in the vicinity of 1000 K [2–4]. It is known that C_{60} when subjected to UV light forms oligomers most of which are dimers [5]. It has been reported that when the photolysis is done in the presence of oxygen the material is ferromagnetic well above room temperature [6, 7]. There are a number of reports of ferromagnetism in halogenated C_{60}. C_{60} subjected to a heat treatment in the presence of iodine is shown to be ferromagnetic having a Curie temperature of 60 K [8, 9]. Doping with a mix of iodine and bromine produced a material which was ferromagnetic below 30 K [10, 11].

In this work it is shown by magnetic field dependent AC susceptibility, magnetic force microscopy, and ferromagnetic resonance that C_{60} treated with fluorine is ferromagnetic well above room temperature. The C_{60} was exposed to fluorine by embedding it in a fluorine-rich polymer, polytrifluorochloroethylene, $[F_2C\text{–}CFCl]_n$, (PTFCE) and decomposing the polymer-C_{60} mix at high temperature to produce fluorine.

2. Experimental

The paramagnetic and ferromagnetic resonance measurements were made using a Varian E-9 spectrometer operating at 9.2 GHz with 100 KHz modulation. The temperature of the sample was controlled by flowing heated or cold nitrogen gas through a double-walled quartz tube, which is part of an ADP Heli-Tran system. This system is inserted through the center of the microwave cavity. The magnetization was obtained by measuring the magnetic field dependence of the ac susceptibility at 350 kHz using a method similar to that described by Clover and Wolf [12]. The system consists of an HP 204C LC oscillator modified to have an external coil. The sample is contained in the coil, which is in a cryogenic dewar between the poles of a magnet. The change in the frequency of the oscillator, which is proportional to the change in susceptibility, is measured as a function of dc magnetic field strength using a HP 5314 frequency counter. The relative susceptibility as a function of dc magnetic field is measured by taking the difference in frequency between zero field and a given applied field. This method of measuring the susceptibility is quite sensitive but not widely used. Magnetic force microscope images were obtained using a Veeco Nano scope IV equipped with a magnetic tip. Raman measurements were made using a J. Y. Horiba confocal micro-Raman system employing a 25 mW He–Ne laser having a wavelength of 632.8 nm and focused to a spot of a 15-micron radius.

Crystalline C_{60}, having 99.99% purity, was obtained from the Aldrich Chemical Company. Polytrifluorochloroethylene (PTFCE) was obtained from the Halocarbon Products Corporation. In this paper we will report clear evidence for ferromagnetism in a mixture of C_{60} and PTFCE subjected to heat treatment. In any report of ferromagnetism in an organic material, the purity of the starting materials is a critical issue, and we have paid considerable attention to sample analysis. It is particularly important to make sure no magnetic impurities are present in the starting materials. The starting C_{60} was analyzed by induction coil plasma mass spectrometry (ICP-MS). The results indicated that signals from all metals were less than 1 part per billion (PPB). Electron paramagnetic resonance (EPR) of the C_{60} showed no evidence of the presence of magnetic elements or complexes. EPR is sensitive to 1 part in 10^{10}. A very narrow line at $g = 2.000$ was detected which has previously been identified as the C_{60} anion [13]. The PTFCE was also subjected to detailed analysis. Energy dispersive X-ray spectroscopy of PTFCE showed only the presence of carbon, fluorine, and chlorine, the constituents of the polymer. No other elements were detected. No magnetic species were detected by EPR. ICP-MS showed no magnetic metals above PPB in the PTFCE.

In a typical synthesis, 0.084 grams of PTFCE were dissolved in acetone and 0.040 grams of C_{60} added to the solution. The solution, while subjected to sonication, was allowed to slowly evaporate. The resulting residue was dried for some hours at 50°C to remove any entrapped acetone. The composite was heated to 160°C for two minutes and then rapidly quenched to room temperature.

3. Results

3.1. PTFCE as a Source of Fluorine. PTFCE was heated to 160°C for one hour and the Raman spectra recorded before and after the heating. No difference was observed between the frequencies of the Raman spectra indicating the material is not decomposing in the condensed phase.

However, the material sublimes at this temperature. In a second experiment the polymer was heated to 160°C for one hour in a small beaker which was on a hot plate. The beaker had a chilled slide on the top which enabled collection of the condensed vapor. The Raman spectra of the material condensed on the slide is shown at the bottom of Figure 1. On the top are the spectra in the same frequency region obtained from the polymer before heating. The two lines at

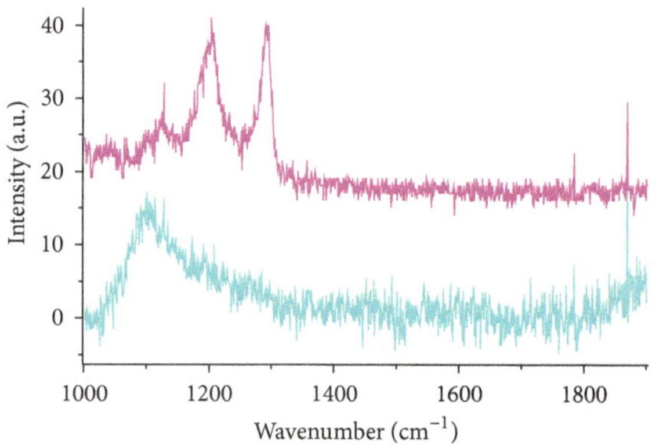

FIGURE 1: Raman spectra before decomposition of PTFCE showing the C–F vibrations at 1198 cm^{-1} and 1298 cm^{-1} and decomposed vapor of PTFCE condensed on a cold slide (bottom) after heating to 160°C showing absence of C–F vibrations.

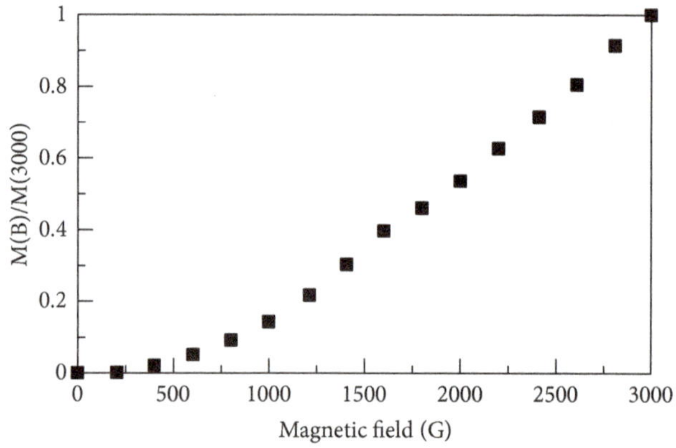

FIGURE 2: Magnetic field dependence of the ac susceptibility measured at 300 K and normalized to its value at 3000 Gauss.

1198 cm^{-1} and 1298 cm^{-1}, which are due to C–F vibrations, are completely gone in the spectra of the vapor indicating single fluorines are removed from the carbon atoms of the polymer in the vapor phase providing a source of atomic fluorine rather than diatomic fluorine.

3.2. Magnetization Measurements. Figure 2 is a plot of the ac susceptibility versus dc magnetic field at 300 K for the heat-treated C_{60}-TFPCE. The susceptibility has been normalized to the measured value at 3 Kilo Gauss. By comparing with a measurement on a sample of known magnetization, it is estimated the magnetization at 3000 Gauss is 0.07 emu/gm. No field-dependent magnetization is observed in the separated starting materials subjected to the same treatment. Figure 3 shows the details at lower fields for increasing and decreasing dc magnetic field to and from 3000 Gauss showing a small hysteresis. Figure 4 presents the temperature dependence of the magnetization above room temperature measured in

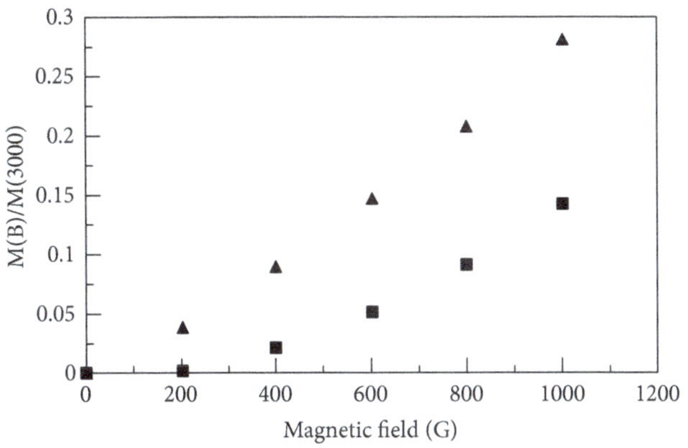

FIGURE 3: Magnetic field dependence of ac susceptibility at low fields for increasing and decreasing magnetic field to and from 3000 Gauss.

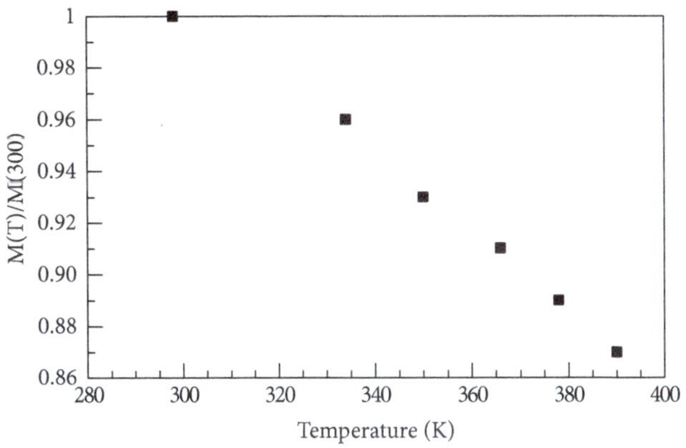

FIGURE 4: Temperature dependence of the magnetization above room temperature measured in a 3000-Gauss magnetic field and normalized to its value at 300 K.

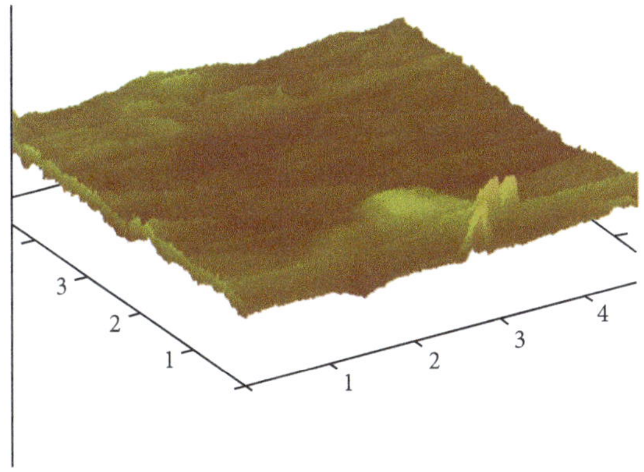

FIGURE 5: Magnetic Force microscope image of the flattened surface of the heated treated polymer containing C_{60}. The units on the x- and y-axis are microns.

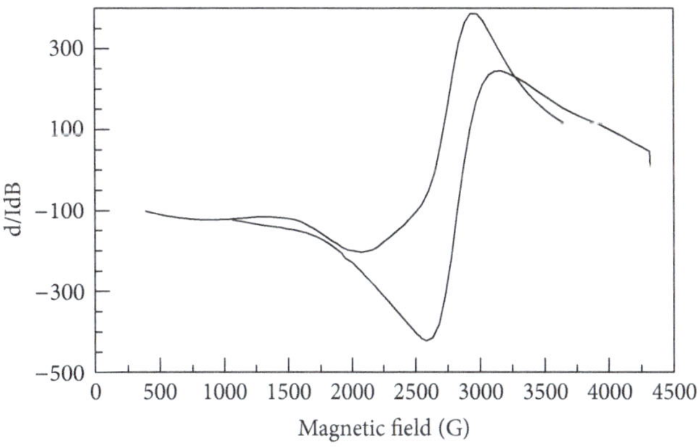

FIGURE 6: Ferromagnetic resonance spectra of the material melted and field cooled, as described in text, for the magnetic field parallel and perpendicular to the cooling field.

a 3000 Gauss field and normalized to its value at 300 K. A fit of this data to the Bloch equation,

$$M(T) = Ms(0)\left[1 - AT^{3/2}\right], \quad (1)$$

yields a value of A of 4.1246×10^{-5}. The temperature at which $M(T)$ is zero, the Curie temperature, is estimated to be 837 K, from this fit. This is likely an overestimate because of deviations from the Bloch law near T_c due to critical fluctuations. However, it is close to estimates of the Curie temperature obtained in C_{60} photolyzed in the presence of oxygen.

3.3. Magnetic Force Microscope Measurement. The PTFCE-C_{60} material which was heat treated was pressed flat on a slide in order to have a smooth surface. Figure 5 shows a magnetic force microscope image of the material. The brighter regions indicate areas of ferromagnetism.

3.4. Ferromagnetic Resonance Measurements. The fluorine-treated material was heated above its melting point and a 4000 G magnetic field applied. The material was then cooled below the melting temperature in the magnetic field. This aligns and locks in the direction of maximum magnetization parallel to the direction of the applied magnetic field. Figure 6 shows the ferromagnetic resonance spectra for the sample oriented perpendicular and parallel to the direction of the cooling field. For a particle having axial symmetry, the dependence of the field position of the FMR signal is given by [14],

$$H_r = H_0 - H_A\left(\frac{1}{2}\right)\left(3\cos^2\theta - 1\right), \quad (2)$$

where $H_A = 4K/M$ is the anisotropy field, K is the magnitude of the anisotropy constant, and M is the magnetization. The angle, θ, is between the direction of maximum magnetization and the applied dc field, H_r is the magnetic field at the center of the FMR signal, and H_0 determines the g value. Because H_r is lower than the parallel orientation compared to the perpendicular orientation, it can be concluded that K is positive. Fitting the data in Figure 6 to (2) allows determination of H_0 and H_A. The values for H_0 and H_A are 2730 G and 93 G, respectively, at 300 K. The g value is 2.287. One of the characteristics of an FMR signal as opposed to an

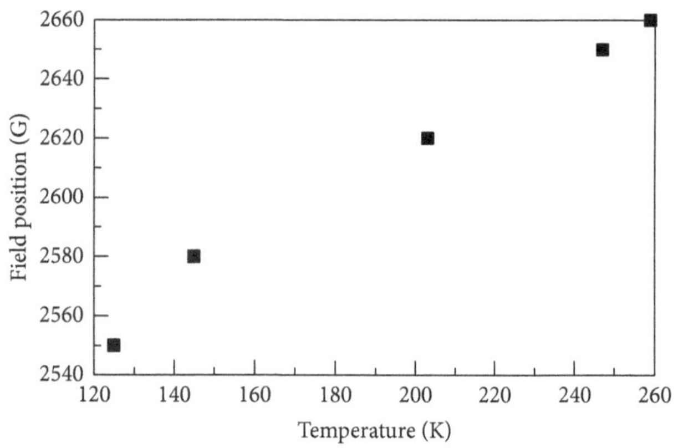

Figure 7: Temperature dependence of the field position of the FMR signal for the magnetic field perpendicular to the direction of the cooling field.

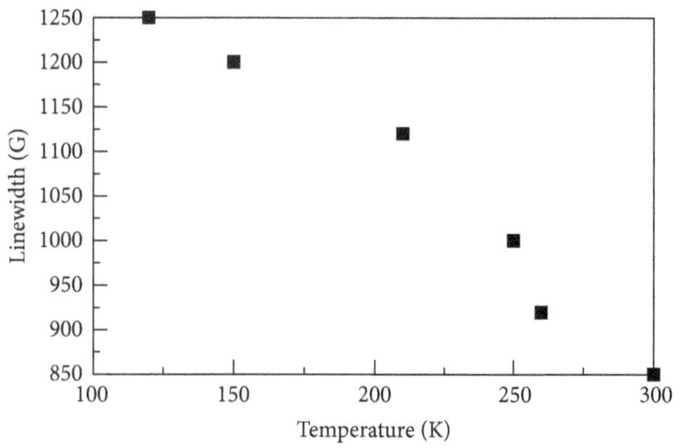

Figure 8: Temperature dependence of the line width of the FMR signal for the same orientation as in Figure 7.

EPR signal is a strong temperature dependence of the line width and field position of the resonance. Figure 7 is a plot of the temperature dependence of the field position for the sample oriented perpendicular to the direction of the cooling field showing a pronounced decrease in the field position with decreasing temperature. Figure 8 plots the line width as a function of decreasing temperature showing a marked broadening as the temperature is lowered. The data in Figures 7 and 8 confirm that the signal is an FMR signal.

3.5. A Possible Model for the Origin of Ferromagnetism. There has been a previous report of ferromagnetism in C_{60} subjected to a different fluorine treatment than that used here [15]. It was observed in C_{60} ultrasonically dispersed in dimethylformamide (DMF) solution of polyvinyl difluoride (PVDF), $[H_2C\text{-}CF_2]_n$. However, the observed magnetism was not stable having a half-life of 30 hours at room temperature. Further the results were only reproducible 1 out of 15 times. In our observation the results are stable over years and reproducible once the correct synthesis conditions are determined. However, an interesting observation in the PVDF solution made material is that ionization time of flight mass spectrometry showed the formation of C_{60} oligomers having fluorine atoms bonded to each C_{60}.

Previous density functional molecular orbital calculations of the minimum energy structure of the $F\text{–}C_{60}\text{–}C_{60}\text{–}F$ dimer indicated that the triplet state has a lower energy than the singlet state by 0.55 eV [16, 17]. This suggests that the fluorinated dimer could be a possible source of the unpaired spin necessary to form a ferromagnetic state. However a calculation of the bond dissociation energy (BDE) to dissociate the dimer into $2(F\text{–}C_{60})$ indicates the dimer would not be stable above 400 K in disagreement with experimental observations.

As discussed in the introduction there have been a number of reports of ferromagnetism at lower temperatures in C_{60} subjected to heat treatment in the presence of I, IBr, and H [8–11]. X-ray diffraction measurements of these materials indicated the ferromagnetism was arising from a cubic phase of C_{60} where the C_{60} was functionalized with halogens or hydrogens. This is a likely possibility for the structure of the ferromagnetic phase of the fluorinated C_{60} observed here. A calculation of the BDE to remove F from C_{60} gives a value of 4.36 eV indicating that $C_{60}\text{–}F$ is stable above 400 K further supporting this possibility [17].

There has been considerable work done on the fluorination of C_{60} which has been discussed in a number of reviews [18, 19]. Generally the fluorination is achieved by exposing C_{60} to F_2 gas at high temperature, which of course results in $C_{60}F_N$ where N is even. An even N would have no net spin. The method of synthesis used here is likely exposing C_{60} to atomic fluorine produced by the removal of F from the TFPCE polymer. This would result in N being odd giving the entity a net unpaired spin.

4. Conclusion

Field dependent ac susceptibility, ferromagnetic resonance, and magnetic force microscopy clearly show that exposing C_{60} to fluorine at high temperature produces a ferromagnetic phase having a Curie temperature well above room temperature. The results cannot be explained by the presence of magnetic impurities such as Fe, Ni, or Co or compounds of them because ICP-MS and EPR of the starting materials indicate that such materials are present at less than one part per billion. While the structure of the ferromagnetic phase is not determined, some possibilities can be considered. Dimers of C_{60} have been suggested to be the origin of the unpaired spin. DFT calculations of the BDE of $F\text{–}C_{60} = C_{60}\text{–}F$ indicate it unlikely to be stable above 400 K. On the other hand $C_{60}\text{–}F$ is predicted to be stable at this temperature. This suggests the structure of the ferromagnetic phase may be cubic lattice of $C_{60}\text{–}F$, as observed for ferromagnetism in $C_{60}\text{–}I$, $C_{60}\text{–}IBr$, and $C_{60}\text{–}H$. However further work is needed to confirm this possibility.

Conflict of Interests

There is no conflict of interests.

References

[1] P. M. Allemand, K. C. Khemani, A. Koch et al., "Organic molecular soft ferromagnetism in a fullerene C_{60}," *Science*, vol. 253, no. 5017, pp. 301–303, 1991.

[2] T. L. Makaravoa et al., "Magnetic carbon," *Nature*, vol. 413, pp. 716–718, 2001.

[3] R. A. Wood, M. H. Lewis, M. R. Lees et al., "Ferromagnetic fullerene," *Journal of Physics*, vol. 14, article L385, 2002.

[4] T. L. Makarova, B. Sundqvist, and R. Höhne, "Retraction: magnetic carbon," *Nature*, vol. 440, article 707, 2006.

[5] P. Zhou, Z. H. Dung, A. M. Rao, and P. C. Eklund, "Reaction mechanism for the photopolymerization of solid fullerene C_{60}," *Chemical Physics Letters*, vol. 211, pp. 337–340, 1993.

[6] Y. Murakami and H. Suematsu, "Magnetism of C_{60} induced by photo-assisted oxidation," *Pure and Applied Chemistry*, vol. 68, no. 7, pp. 1463–1467, 1996.

[7] F. J. Owens, Z. Iqbal, L. Belova, and K. V. Rao, "Evidence for high-temperature ferromagnetism in photolyzed C_{60}," *Physical Review B*, vol. 69, Article ID 033403, 4 pages, 2004.

[8] V. Buntar, H. W. Weber, and M. Ricco, "Magnetic irreversibility of C_{60}, C70 and I-C_{60} fullerenes at low temperature: transition to a frozen magnetic glass state?" *Solid State Communications*, vol. 98, pp. 175–178, 1995.

[9] L. S. Grigoryan and M. Tokumoto, "New iodine containing polymers of C_{60}: synthesis, characterization and magnetic properties," *Solid State Communications*, vol. 96, pp. 523–528, 1995.

[10] H. Sekine, "Magnetic behavior and structure of the halogen-doped fullerene C_{60}," *Journal of Applied Physics*, vol. 72, no. 11, article 5448, 3 pages, 1992.

[11] M. Tokumoto, N. Kinoshita, Y. Tanaka et al., "Structure and magnetic properties of halogen-doped C_{60}," *Synthetic Metals*, vol. 70, no. 1, pp. 1387–1388, 1995.

[12] R. B. Clover and W. P. Wolf, "Magnetic susceptibility measurements with a tunnel diode oscillator," *Review of Scientific Instruments*, vol. 41, no. 5, article 617, 5 pages, 1970.

[13] S. G. Kukolich and D. R. Huffman, "EPR spectra of C_{60} anion and cation radicals," *Chemical Physics Letters*, vol. 182, pp. 263–265, 1991.

[14] R. S. De Biasi and T. C. Devezas, "Anisotropy field of small magnetic particles as measured by resonance," *Journal of Applied Physics*, vol. 49, no. 4, pp. 2466–2469, 1978.

[15] M. Ata, M. Machida, H. Watanabe, and J. Seto, "Polymer-C_{60} composite with ferromagnetism," *Japanese Journal of Applied Physics*, vol. 33, pp. 1865–1871, 1994.

[16] O. E. Kvyatkovskii, I. B. Zakharova, A. L. Shelankov, and T. L. Makarova, "Spin-transfer mechanism of ferromagnetism in polymerized fullerenes: Ab initio calculations," *Physical Review B*, vol. 72, no. 21, Article ID 214426, 8 pages, 2005.

[17] M. Miller and F. J. Owens, "Stability of functionalized C_{60} paramagnetic dimers and monomers," *Chemical Physics*, vol. 394, pp. 52–55, 2012.

[18] O. V. Boltalina, "Fluorination of fullerenes and their derivatives," *Journal of Fluorine Chemistry*, vol. 101, no. 2, pp. 273–278, 2000.

[19] R. Taylor, "Why fluorinate fullerenes?" *Journal of Fluorine Chemistry*, vol. 125, no. 3, pp. 359–368, 2004.

12

Spectral and Thermal Degradation of Melamine Cyanurate

V. Sangeetha,[1] N. Kanagathara,[2] R. Sumathi,[3] N. Sivakumar,[4] and G. Anbalagan[4]

[1] *Department of Physics, D.G.Vaishnav College, Arumbakkam, Chennai 106, India*
[2] *Department of Physics, Vel Tech Multi Tech Dr. Rangarajan Dr. Sakunthala Engineering College, Avadi, Chennai 62, India*
[3] *Department of Physics, Vel Tech Polytechnic College, Avadi, Chennai 62, India*
[4] *Department of Physics, Presidency College, Chennai 5, India*

Correspondence should be addressed to G. Anbalagan; anbu24663@yahoo.co.in

Academic Editor: Carmen Alvarez-Lorenzo

Melamine cyanurate, an organic crystalline complex was, synthesized by evaporation of an aqueous solution containing equimolar quantities of melamine and cyanuric acid. The synthesized compound has been subjected to various characterizations like Powder XRD, FT-IR, TG-DTG, SEM, and SHG. The presence of sharp diffraction peaks in the XRD confirms that the products are highly crystalline. The average particle size was calculated using the Debye-Scherrer formula, and it was found to be 3.067 μm. Thermal behavior of the grown crystal has been studied by TG-DTG analysis. From TG-DTG, it is found that the title crystal possesses good thermal stability. The activation energy was calculated using the Broido, Coats-Redfern, and Horowitz-Metzger methods. A sharp peak exothermic peak at 405.40°C was assigned as the melting point of the title material. SEM reveals the morphology of the synthesized salt. No detectable signal was observed during the Kurtz-Perry technique.

1. Introduction

The rapid growth in material and processing technology makes it essential for its use in various industrial applications. Cyanuric acid which is also known as triazinetriol or trihydroxy cyaniding is a polymer of cyanic acid and arises by dry heating of urea. The organic derivatives of cyanuric acid find wide industrial applications today. The composition of cyanurates includes the s-triazine ring formed as a result of trimerization of the cyanato groups. The cyanuric acid derivatives containing the s-triazine ring (C_3N_3) are considered to be promising compounds for the synthesis of complexes [1]. The strong interaction of the π electrons of the cyanuric acid with the unshared electron pair of the amine nitrogen atom imparts basic properties to the melamine amino groups. Two forms of hydrogen bonding between melamine and cyanuric acid are suggested with a planar structure [2, 3]. Flame retardancy is one of the important properties in industrial application. Melamine cyanurate offers better thermal stability (up to 377°C) than pure melamine. This helps in polymer processing steps. Both melamine and cyanuric acid have got several industrial applications [4]. Melamine cyanurate is particularly effective in improving fire safety of nitrogen-based polymers, such as polyamides (nylons) and thermoplastics (polyurethane). It can be used in epoxy polymers and in a variety of other substrates. These complexes are very promising light-emitting materials because of their good thermal stability [5]. The thermochemical properties of cyanuric acid were characterized using mass spectrometry measurements along with computational studies was already reported by Mukherjee and Ren [6]. Cyanuric acid is a triazine derivative and may also be described as two structural isomers, the enol-like triazine-triol and the keto-like tautomer. Cyanuric acid is commonly used as a component of bleaches in swimming pools [7, 8]. Cyanuric acid can form hydrogen bonded self-assemblies with specific surface structures, and these assemblies have been used as surface templates in supramolecular chemistry [9]. Cyanuric acid is used in the preparation of high performance magnet-wire enamels and in electrical varnishes, plastics with enhanced properties, flame retardant resins and solid lubricants, and cross-linking and curing agents in the manufacture plastics and coatings. Cyanuric acid is also used to reduce nitrogen oxides (NOX) in stationary diesel engine exhaust gases. In the present work, melamine cyanurate has been synthesized and characterized by subjecting it to various studies like XRD, FT-IR, and TG-DTG, and results are discussed in detail.

Cyanuric acid is mainly used as a precursor to N-chlorinated cyanurates, which are used to disinfect water. Cyanuric acid is also used in the manufacture of specialty intermediates used in the production of plastics and coatings. This paper discusses the structure, properties, analysis, chemistry, manufacture, and uses of cyanuric acid.

$$(HNCO)_3 + (H_2NCN)_3 \longrightarrow (HNCO)_3 \cdot (H_2NCN)_3$$
$$\text{cyanuric acid} + \text{melamine} \longrightarrow \text{melamine cyanurate} \quad (1)$$

Cyanuric acid (1,3,5-triazine-2,4,6-triol) is the cyclic trimer of cyanic acid [10]. It is an organic compound with the chemical formula $C_3H_3N_3O_3$. It has wide industrial applications. This triazine derivative is a stable, white, odorless, and hygroscopic solid at room temperature. It is used as a stabilizer in recreational water treatment to minimize the decomposition of hypochlorous acid by light in outdoor swimming pools and hot tubs. Chlorinated derivatives of cyanuric acid, such as trichloro-*s*-triazinetrione and sodium dichloro-*s*-triazinetrione, are used as algacides or microbiocides in swimming pool water and large-scale water systems in industry [11]. The ring in its molecule has an aromatic character, and the hydroxyl groups in the triol form of the molecule take on a phenolic character, becoming somewhat more acidic than hydroxyls in an alcohol. The hydrogen of any one of cyanuric acid hydroxyls can be neutralized to form a cyanurate salt [12]. Melamine cyanurate is a halogen-free flame retardant that can be used in thermoplastic urethanes (TPUs) for electrical wire coatings [13–15].

2. Materials and Methods

AR grade samples of melamine and cyanuric acid were taken in 1 : 1 ratio, and the hot solution of cyanuric acid was added dropwise to the hot solution of melamine. The mixed solution was stirred well for nearly 4 hours and then allowed to cool at room temperature. During a few days, the entire solvent was evaporated and the white colour melamine cyanurate was obtained which are insoluble in water and common organic solvents like acetone, chloroform, benzene, alcohol, and so forth but soluble in strong acids or basic solutions [16]. The obtained compound was then washed with distilled water to remove the presence of impurities and dried at a constant temperature of 80°C.

3. Characterization

The synthesized compound has been subjected to various characterization studies like Powder XRD, FT-IR, TG-DTG, SEM, and SHG. The synthesized compound has been characterized by X-ray powder diffraction technique using Rich Seifert X-ray powder diffractometer with CuKα radiation of $\lambda = 1.5406$ Å. The 2θ range was analyzed from 10° to 70° by employing the reflection mode for scanning. The detector used was a scintillation counter. A Perkin Elmer

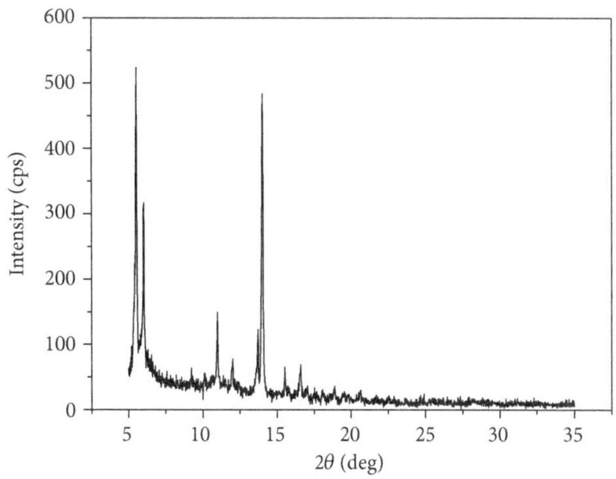

FIGURE 1: Powder X-ray diffraction pattern of melamine cyanurate complex.

Spectrum ONE FT-IR spectrometer was employed to record the IR spectrum to analyze the functional groups present in the synthesized melamine cyanurate compound. The sample for this measurement was finely grounded and mixed with KBr. TG-DTG spectra were recorded for the synthesized melamine cyanurate compound using SDT Q 600 V 8.0 build 95 thermal analyzer with a heating rate of 20°C/min in the temperature up to 1000°C in nitrogen atmosphere. The Quanta 200 FEG scanning electron microscope (SEM) was used to know the surface morphology of the synthesized salt with a resolution of 1.2 nm gold particle separation on a carbon substrate with a magnification from a minimum of 12X to greater than 1,00,000X. The nonlinear optical activity of melamine cyanurate was identified by illuminating it with an Nd:YAG laser. In this technique, sample packed in a cell was sandwiched between two glass plates and was subjected to the output of a Q switched Nd:YAG laser emitting a functional λ of 1064 nm with 10 ns pulse width with a power output 20 mW.

4. Results and Discussion

4.1. Powder X-Ray Diffraction Analysis. Figure 1 shows the powder XRD pattern of melamine cyanurate. The presence of sharp diffraction peaks in the XRD confirms that the products are highly crystalline. The average grain size (D) of the synthesized was calculated by the Debye-Scherrer formula, $D = 0.9\lambda/\beta\cos\theta$, where λ is the wavelength of CuKα line ($\lambda = 1.5418$ Å), β is the full width at half maximum height, and θ is the diffraction angle. From the XRD analysis, the average particle size of the title material was estimated as 3.067 μm.

4.2. FT-IR Studies. Figure 2 shows the FT-IR spectrum of melamine cyanurate, and the vibration band assignment is listed in the Table 1. Internal vibrations of melamine molecule were recently published [17, 18]. According to crystallographic data, melaminium residues often form hydrogen

Table 1: Vibration band assignment of melamine cyanurate.

Observed wavenumber (cm^{-1})	Assignment	References
3396	NH_2 symmetric stretching type of vibrations of 3 triazine NH_2 groups	[3]
3234	N–H symmetric stretching	[4]
3041	N–H···N stretching	[8]
2825	N–H deformation (cyanuric acid)	[12]
2698	NH_2 asymmetric stretching	[10]
1782	Stretching vibration of the carbonyl group C=O	[2]
1743	NH_2 scissoring	[9]
1667	NH_2 bending vibration	[4]
1528	C=N symmetric stretching	[2]
1449	C–N symmetric stretching	[2]
1036	Ring breathing type of vibration	[11]
924	Ring breathing type of vibration	[13]
808	Ring-sextant out-of-plane bending type of vibration	[9]
771	Ring-sextant out-of-plane bending type of vibration	[9]
596	Ring bending vibration	[11]
529	Side chain in plane C–N bending vibration	[12]

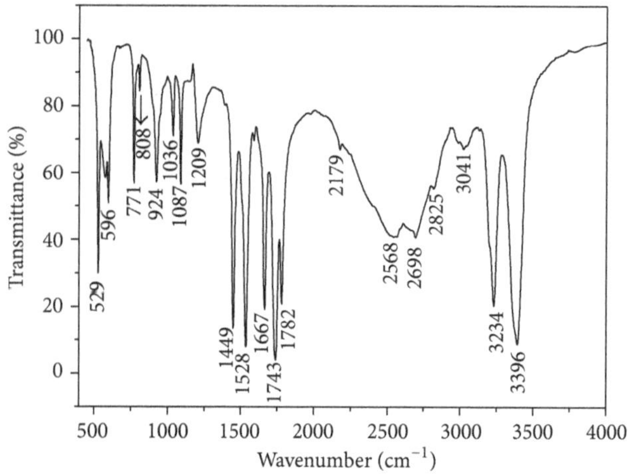

Figure 2: FT-IR spectrum of melamine cyanurate.

bonds of N–H···N and N–H···O type. Melamine cyanurate complex is held together by an extensive two-dimensional network of hydrogen bonds between the two compounds. The bands observed in the measured region 4000–350 cm^{-1} arise from internal vibrations of melaminium cations, cyanuric acid anions and water molecules, the vibrations of both N–H···O and O–H···O types of hydrogen bond, and from the vibrations of lattice. The three amino groups in the melamine molecule have different types of hydrogen bonds. The melamine molecules are not linked by hydrogen bonds to each other. Due to this system of hydrogen bond network, the visible modes of NH_2, OH and the six-membered ring of melamine are expected to give unusual frequencies and intensities for the IR bands [19]. A strong peak at 3396 cm^{-1} is due to the NH_2 symmetric stretching type of vibrations of triazine groups [3], and this band usually occurs at 3328 cm^{-1} in melamine crystal [20]. The difference of 68 cm^{-1} is known as blue shift. Similar shift is observed in Debrus et al. [21]. The strong intense peak at 3234 cm^{-1} and medium peak at 3041 cm^{-1} are due to the formations of hydrogen bonding between NH_2/NH groups. N–H stretching vibration occurs at 3205 cm^{-1} and 3079 cm^{-1} which shifted to lower wave numbers compared with those of melamine [5, 12]. The medium peak at 2825 cm^{-1} is assigned to N–H deformation. Usually this band appears in the range of 2828 cm^{-1}–2907 cm^{-1} in the cyanuric acid spectrum [22]. A medium peak at 2698 cm^{-1} can be assigned to NH_2 asymmetric stretching of vibration [19]. The stretching vibration of the carbonyl group C=O at 1782 cm^{-1} is due to cyanate anion [4]. The strong peak at 1743 cm^{-1} is assigned to NH_2 scissoring [18]. Several pronounced differences have been observed between the IR spectrum of melamine crystal and that for melamine cyanurate in region of NH_2 bending type of vibration. A weak band appears at 1667 cm^{-1}, and this occurs usually at higher frequencies (1666 cm^{-1}). This may be due to intermolecular interaction through the NH_2 groups of melamine molecule. This causes the rising of their frequencies for bending type of motion [22]. The triple bond C–N stretching frequency in the IR spectra of cyanurates should be assigned to a strong polarization of the cyanurate anion by melamine cation [4]. The benzene ring has two intense absorption bands corresponding to the vibrations of C–N and C=N bonds. This appears in the region of 1450 cm^{-1}–1500 cm^{-1} and 1530 cm^{-1}–1600 cm^{-1}, respectively which indicates the presence of cyanuric acid. Similar bands are shifted to lower frequency of 1449 cm^{-1} and 1528 cm^{-1} in our spectra [4]. The band corresponds to C=N stretching frequency vibration occurs in the range of 1590 cm^{-1}–1600 cm^{-1}. Which may either be appearing weak or disappear. The medium band at 1036 cm^{-1} [20] and 924 cm^{-1} originates from ring breathing type of vibration. And the peaks at

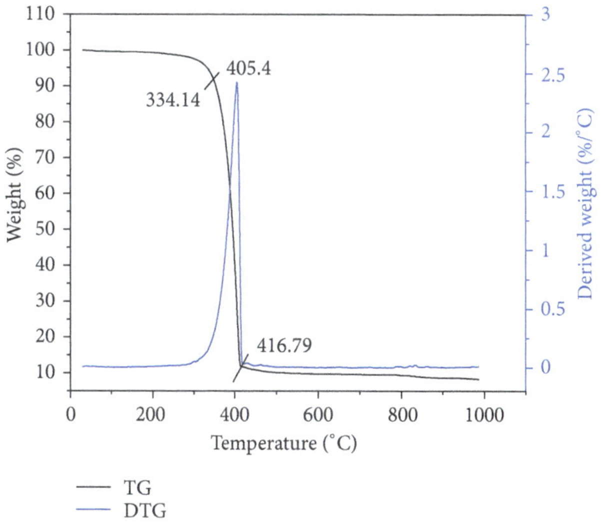

Figure 3: TG-DTG spectrum of melamine cyanurate.

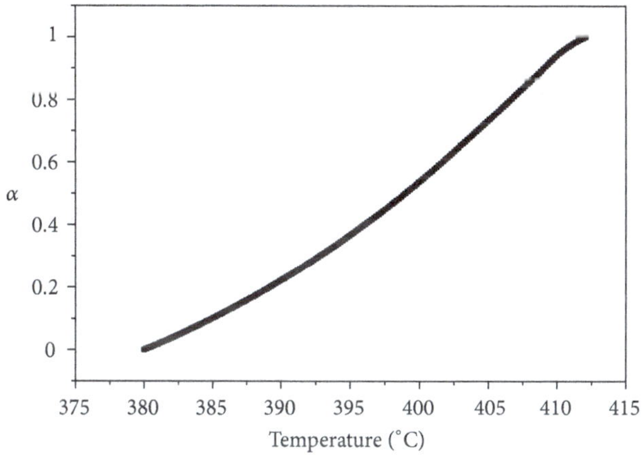

Figure 4: Fraction reacted α versus temperature.

808 cm^{-1} and 771 cm^{-1} are attributed to ring-sextant out-of-plane bending type of vibration [18], and the medium peak at 596 cm^{-1} is due to ring bending [20], and the peak at 529 cm^{-1} is due to the side chain in plane C–N bending vibration [22].

4.3. Thermogravimetric Analysis. Figure 3 shows the TG-DTG spectra recorded for the synthesized melamine cyanurate. The TG curve exhibits mass losses in a single stage which indicates that decomposition of the grown crystals takes place sharply. Initial mass is taken as 3.0210 mg. The thermal decomposition of melamine cyanurate is accompanied by the detachment of ammonia and gradual linking of cyanuric rings through imide bridges [4]. The initial weight loss starts at 334.14°C and ends at 416.79°C with about 88.25% of weight loss which corresponds to liberation of ammonia and water molecules followed by the decomposition of cyanuric acid to volatile cyanates. There is a sharp weight loss which is indicated by the derivative curve as peak. This suggest that there is no other processes of decomposition except transition from solid to liquid state which is designated as melting point,

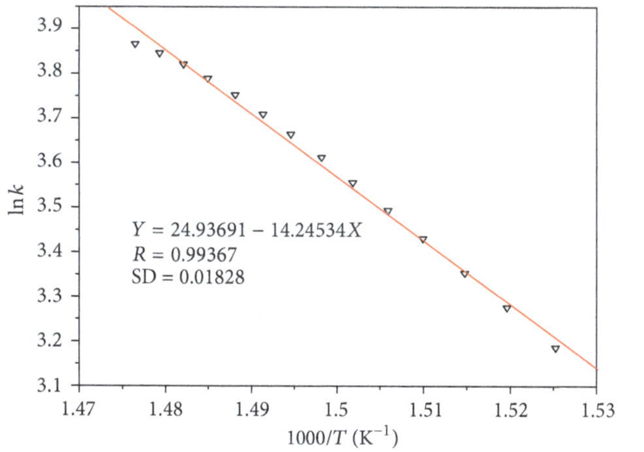

Figure 5: The Arrhenius plot.

and this occurs at 405.4°C as indicated in Figure 3. For pure melamine this occurs at 354°C [4]. Thus it can be understood that melamine cyanurate offers better thermal stability than pure melamine (up to 320°C). This enables its use in polymers for processing. The decomposition at 407°C may be due to decay of cyanates. The thermal decomposition of melamine cyanurate may be decomposition of melamine and cyanuric acid. Above 250°C, carbonyl groups are replaced with amine groups in cyanuric acid. The decomposition ends with a final residue of 0.2513 mg.

Various researchers put forward integral method, which can be applied to TG data assuming order of reaction and from which the activation energy is calculated. Generally, mass conversion can be calculates as

$$\alpha = \frac{m_o - m}{m_o - m_f}, \tag{2}$$

where m_o is the initial mass, m is the corresponding mass, and m_f is the final mass.

Kinetic studies assumed that the isothermal rate of conversion, $d\alpha/dt$, is a linear function of reactant concentration loss and of temperature-independent function of the conversion α:

$$\frac{d\alpha}{dt} = kf(\alpha), \tag{3}$$

where $f(\alpha)$ is reaction model that depends on the mechanism of degradation. The function k is described by the Arrhenius expression

$$k = A\exp\left(-\frac{E_a}{RT}\right), \tag{4}$$

$$\ln k = \ln A - \frac{E_a}{RT}, \tag{5}$$

where A is preexponential factor independent of temperature, E_a is activation energy, and R is the universal gas constant.

The linear Arrhenius plot of $\ln k$ versus $1/T$ is plotted from the TG data. From the slope $(-E_a/R)$, the activation energy (E_a) was calculated.

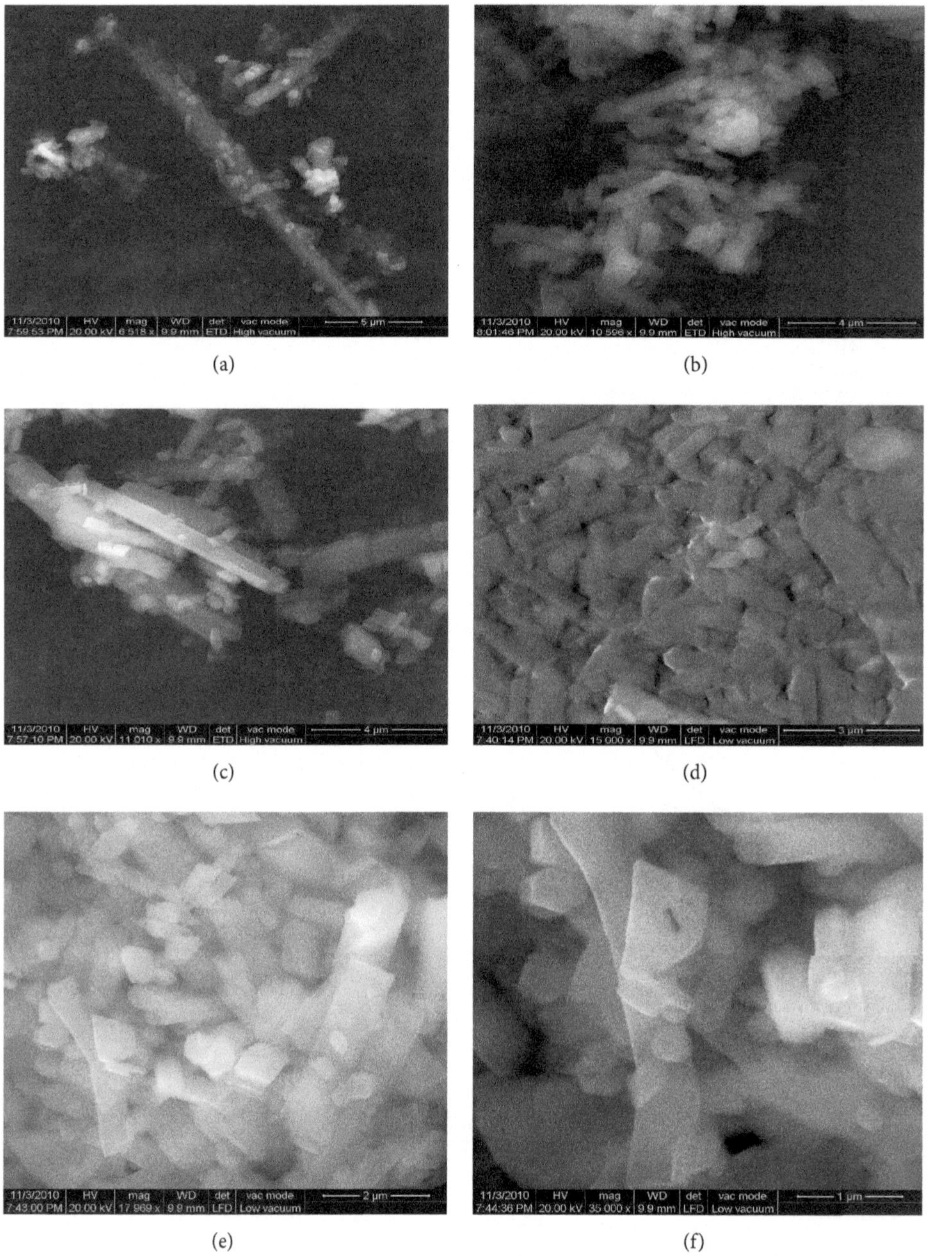

FIGURE 6: (a) SEM photograph with 6518X, (b) SEM photograph with 10500X, (c) SEM photograph with 11010X, (d) SEM photograph with 15000X, (e) SEM photograph with 17900X, and (f) SEM photograph with 35000X.

Equation (3) can be writtens as

$$\frac{d\alpha}{dt} = \beta\left(\frac{d\alpha}{dT}\right) = Af(\alpha)\exp\left(-\frac{E_a}{RT}\right), \quad (6)$$

where β is the heating rate (20°C/min) and $f(\alpha)$ is the reaction model.

Figure 4 shows the fraction reacted α versus temperature. The Arrhenius plot is shown in Figure 5.

4.3.1. Broido's Method. Consider,

$$\ln\left[\ln\left(\frac{1}{\alpha}\right)\right] = \left(-\frac{E_a}{RT}\right) + \left[\frac{RAT^2}{E_a\beta}\right]. \quad (7)$$

In Broido's approximation [23], the thermal degradation is considered as first-order and the calculations are done accordingly. The value of the activation energy (E_a) can be calculated from the slope of the plot of graph between $\ln[\ln(1/y)]$ and $1/T$. The calculated activation energy is 325.21 KJ mol^{-1}.

Table 2: Kinetic parameters obtained by different methods.

Method	Activation energy (kJ mol^{-1})	R	S.D
Arrhenius	118.44	0.99367	0.01828
Broido	325.21	0.9964	0.03063
Coats-Redfern	393.33	0.99939	0.00598
Horowitz-Metzger	336.57	0.98575	0.03021

4.3.2. The Coats-Redfern Method. Consider,

$$\ln\left[-\frac{\ln(1-\alpha)}{T^2}\right] = \ln\left(\frac{AR}{\beta E_a}\right)\left[1-\frac{2RT}{E_a}\right] - \frac{E_a}{RT} \quad \text{for } n = 1. \tag{8}$$

By using the Coats-Redfern expression [24], the value of the activation energy (E_a) can be calculated from the slope of the plot of graph between $\ln[-\ln(1-\alpha)/T^2]$ and $1/T$. The calculated activation energy is 393.33 KJ mol^{-1}.

4.3.3. The Horowitz-Metzger Method [25]. Consider,

$$\ln(1-\alpha) = \frac{E_a(T-T_p)}{RT_p} \quad \text{for } n = 1. \tag{9}$$

By using the expression above (9), the activation energy (E_a) can be calculated from the slope of the plot of graph between $\ln(1-\alpha)$ and $1/T$. The calculated activation energy is 336.57 KJ mol^{-1}. The activation energy (E_a) and pre-exponential factor (A) calculated from all the above methods were listed in Table 2.

4.4. Scanning Electron Microscope Analysis. Scanning electron microscopy was used to study the surface features of the synthesized salt. Figures 6(a)–6(f) shows different types of morphologies exhibited by synthesized salt at different magnifications. The various types of morphologies include spherulites, platelets, cuboids, and coalesced and rod-shaped crystals. It is seen from SEM observation that the decomposition takes place starting from a relevant number of nuclei which grow rapidly enough. Melamine cyanurate forms spoke-like crystals in aqueous solutions. Figures 6(a)–6(f), shows the SEM photograph of melamine cyanurate under different magnifications with 1–5 μm in size.

4.5. Kurtz Perry Technique. The sample was found to be NLO inactive. There is no emission of green (λ = 520 nm) radiation which confirms that there is no production of second harmonic generation.

5. Conclusion

Melamine cyanurates were obtained by employing slow evaporation solution growth technique. The grown crystals are of 1 μm to 5 μm size. The Powder XRD pattern of the complex ensures its crystallinity. The FTIR spectrum clearly indicates the presence of functional groups in the complex. The TG-DTG studies were used to formulate the decomposition pattern of the complex compound, and it is found that thermal stability for melamine cyanurate is more than that of pure melamine and hence finds applications in polymers for processing at high temperatures. From the Arrhenius expression, a linear straight line equation is obtained. Activation energy was calculated by four different methods the Arrhenius, Broido, Coats-Redfern, and Horowitz-Metzger methods. Among four different methods, activation energy obtained by Coats-Redfern is higher. And activation energy obtained by Arrhenius method E_a is 118.4 kJ/mol which is much lower than that obtained from other methods. This is so because the data are taken at a single heating rate (20°C/min) in all the cases which may not be applicable. This is more so if mechanism of decomposition is going to change, and this will be confirmed at different heating rates like 5, 10 and 15°C/min. Also these complexes are very promising light-emitting materials due to their good thermal stability. There is no emission of green signal through Kurtz-Perry technique which confirms that the synthesized material is not having second harmonic generation efficiency. Further research has to be carried out and can be extended to several industrial and research applications.

References

[1] H. Zhu, Z. Yu, X. You, H. Hu, and X. Huang, "The crystal and molecular structure of bis(melamine)silver(I) perchlorate, $Ag(C_3H_6N_6)_2ClO_4$," *Journal of Chemical Crystallography*, vol. 29, no. 2, pp. 239–242, 1999.

[2] Y. Wang, B. Wei, and Q. Wang, "Crystal structure of melamine cyanuric acid complex (1:1) trihydrochloride, MCA·3HCl," *Journal of Crystallographic and Spectroscopic Research*, vol. 20, no. 1, pp. 79–84, 1990.

[3] F. H. Herbstein, "Purported "melamine cyanuric acid trihydrochloride" $C_3H_6N_6{\cdot}C_3H_3N_3O_3{\cdot}3HCl$ is actually "diprotonated-melamine cyanuric acid dichloride dihydrate" $(C_3H_8N_6)2+ {\cdot}C_3H_3N_3O_3{\cdot}2Cl{-}{\cdot}2H_2O$," *Journal of Chemical Crystallography*, vol. 33, no. 7, pp. 527–529, 2003.

[4] G. R. Seifer, "Cyanuric acid and cyanurates," *Russian Journal of Coordination Chemistry*, vol. 28, no. 5, pp. 301–324, 2002.

[5] Y. Qiu and L. Gao, "Blue-emitting cyanuric acid-melamine complexes from urea thermolysis," *Materials Research Bulletin*, vol. 40, no. 5, pp. 794–799, 2005.

[6] S. Mukherjee and J. Ren, "Gas-Phase acid-base properties of melamine and cyanuric acid," *Journal of the American Society for Mass Spectrometry*, vol. 21, no. 10, pp. 1720–1729, 2010.

[7] A. Zaknich, "Swimming pool and Spa water testing process," Granted innovation Pat. (Australia). Application, AU 2009100474, 8, 2009.

[8] V. I. Teichberg, "Methods and compositions and Devices for maintaining chemical balance of chlorinated water," Yeda Research and Development co.Ltd., Israel, In PCT Int. Appl; Application, WO 2007107981, 58, 2007.

[9] K. Damodaran, G. J. Sanjayan, P. R. Rajamohanan, S. Ganapathy, and K. N. Ganesh, "Solid state NMR of a molecular self-assembly: multinuclear approach to the cyanuric acid-melamine system," *Organic Letters*, vol. 3, no. 12, pp. 1921–1924, 2001.

[10] N. Kriebizsch, N. V. Degussa-Antwerpen, H. Klenk, and W. H. Degussa, in *Ullmann's Encyclopedia of Industrial Chemistry*,

W. Gerhartz, Y. S. Yamamoto, L. Kaudy, R. Pfefferkorn, and F. Rounsavaille, Eds., vol. 8, p. 191, Wiley, New York, NY, USA, 5th edition, 1987.

[11] G. Herzberg and C. I. Reid, "Infra-red spectrum and structure of the HNCO molecule," *Discussions of the Faraday Society*, vol. 9, pp. 92–99, 1950.

[12] P. G. Maiella and T. B. Brill, "Spectroscopy of hydrothermal reactions," *Applied Spectroscopy*, vol. 50, pp. 829–835, 1996.

[13] J. Zhang, M. Lewin, E. Pearce, M. Zammarano, and J. W. Gilman, "Flame retarding polyamide 6 with melamine cyanurate and layered silicates," *Polymers for Advanced Technologies*, vol. 19, no. 7, pp. 928–936, 2008.

[14] P. Gijsman, R. Steenbakkers, C. Fürst, and J. Kersjes, "Differences in the flame retardant mechanism of melamine cyanurate in polyamide 6 and polyamide 66," *Polymer Degradation and Stability*, vol. 78, no. 2, pp. 219–224, 2002.

[15] Z. Y. Wu, W. Xu, Y. C. Liu, J. K. Xia, Q. X. Wu, and W. J. Xu, "Preparation and characterization of flame-retardant melamine cyanurate/polyamide 6 nanocomposites by in situ polymerization," *Journal of Applied Polymer Science*, vol. 113, no. 4, pp. 2109–2116, 2009.

[16] R. Henricus and M. Kierkels, "Patent application title: Melamine cyanurate in crystalline form Inventors," JoAnn Villiamizar, Ciba Corporation/Patent Department, EP1799655A1, 2007.

[17] R. J. Meier, J. R. Maple, M. J. Hwang, and A. T. Hagler, "Molecular modeling urea- and melamine-formaldehyde resins. 1. A force field for urea and melamine," *Journal of Physical Chemistry*, vol. 99, no. 15, pp. 5445–5456, 1995.

[18] P. J. Larkin, M. P. Makowski, N. B. Colthup, and L. A. Flood, "Vibrational analysis of some important group frequencies of melamine derivatives containing methoxymethyl, and carbamate substituents: mechanical coupling of substituent vibrations with triazine ring modes," *Vibrational Spectroscopy*, vol. 17, no. 1–3, pp. 53–72, 1998.

[19] C. Y. Panicker, H. T. Varghese, A. John, D. Philip, and H. I. S. Nogueira, "Vibrational spectra of melamine diborate, $C_3N_6H_62H_3BO_3$," *Spectrochimica Acta A*, vol. 58, no. 8, pp. 1545–1551, 2002.

[20] W. J. Jones and W. J. Orville-Thomas, "The infra-red spectrum and structure of dicyandiamide," *Transactions of the Faraday Society*, vol. 55, pp. 193–202, 1959.

[21] S. Debrus, M. K. Marchewka, M. Drozd, and H. Ratajczak, "Vibrational, calorimetric and nonlinear optical studies of melaminium-bis(trichloroacetate) monohydrate molecular-ionic crystal," *Optical Materials*, vol. 29, no. 8, pp. 1058–1062, 2007.

[22] E. García-López, G. Marci, N. Serpone, and H. Hidaka, "Photoassisted oxidation of the recalcitrant cyanuric acid substrate in aqueous ZnO suspensions," *Journal of Physical Chemistry C*, vol. 111, no. 49, pp. 18025–18032, 2007.

[23] A. Broido, "A simple, sensitive graphical method of treating thermogravimetric analysis data: part A-2," *Journal of Polymer Science*, vol. 7, no. 10, pp. 1761–1773, 1969.

[24] A. W. Coats and J. P. Redfern, "Kinetic parameters from thermogravimetric data," *Nature*, vol. 201, no. 4914, pp. 68–69, 1964.

[25] H. H. Horowitz and G. Metzger, "A new analysis of thermogravimetric traces," *Analytical Chemistry*, vol. 35, no. 10, pp. 1464–1468, 1963.

Morphological and Cell Growth Assessment in Near Dense Hydroxyapatite Scaffold

Florencia Edith Wiria,[1] Bee Yen Tay,[1] and Elaheh Ghassemieh[2]

[1] *Singapore Institute of Manufacturing Technology, 71 Nanyang Drive, Singapore 638075*
[2] *Department of Mechanical Engineering, The University of Sheffield, Sir Frederick Mappin Building, Sheffield S1 3JD, UK*

Correspondence should be addressed to Florencia Edith Wiria; florencia@simtech.a-star.edu.sg

Academic Editor: Amit Bandyopadhyay

This paper reports the preliminary results on the morphology of low porosity hydroxyapatite scaffold and its compatibility as a substrate for osteoblast cells. Although having low porosity, the hydroxyapatite scaffold was found to be capable of sustaining cell growth and thus assisting bone ingrowth. Due to the low porosity nature, the scaffold provides higher strength and therefore more suitable for applications with load-bearing requirements such as spinal spacer. The hydroxyapatite scaffolds are prepared via powder processing techniques, using a combination of wet mixing, powder compaction, and sintering processes. The scaffold porosity is estimated via image analysis and micro-CT, which detect porosity level of approximately 16% and pore size of 13 μm. Cell culture investigation demonstrates that the hydroxyapatite substrate is able to provide favourable cell attachment and collagen matrix production, as compared to the commonly used cell culture control substrates. These results indicate that despite the low porosity in the hydroxyapatite scaffolds, they do not hinder being a preferred substrate to provide conducive environment osteoblast cell growth.

1. Introduction

Over the years, many healthcare products related to the improvement of a person's wellbeing as one ages have seen an increase in demand [1]. One of the examples is the use of an artificial implant to replace dysfunctional body parts or to assist the healing process of a diseased body part. Artificial bone implantation has an advantage over the use of autograft (implant sourcing from the same individual) or allograft (implant sourcing from different individual), mainly due to the limited source of bony parts that can be removed and reimplanted. Secondly, autograft and allograft require secondary operations to perform organ removal from the donor site and implantation into the targeted site. Therefore, artificial materials have been researched for bony implant purposes.

In the context of bone-related diseases, hydroxyapatite (HA) has been widely used for bone replacement material due to its similarity in composition with respect to bone. The composition of bone mineral is comparable to sintered HA. The mineral component of bone is a biological apatite, where carbonate substitutes phosphate ions by about 3–5 wt.% [2].

HA has been reported to have bioactivity nature and thus assists integration and interaction of osteoblast with the implants [3]. Apart from its compatibility with osteoblast, HA surface is biocompatible with several other types of cells, such as macrophages, fibroblasts, osteoclasts, and periodontal ligament cells [1]. This makes the potential applications of HA even wider.

There are typically two types of HA being researched, namely, dense and porous HA. Both dense and porous types of HA have been applied in many biomedical applications. Dense HA is usually referred to as HA implants with porosity up to 5% and pore size of approximately 1 μm [4]. It is the most commonly used material in the case of bone substitution for the ear, such as for ossicle (small bone in the middle ear) connection. The advantage of dense HA is its excellent biocompatibility with minimal risk of resorption due to its slow dissolution rate, even among other calcium-phosphate ceramics [5]. Its precipitation rate is also amongst the slowest [6]. These properties give dense HA an edge over highly porous HA for implant or bone replacement applications that are needed for longer period of time. Dense HA also provides sufficient mechanical strength, as compared to porous HA which is much more susceptible to breaking *in vivo* [7, 8]. Applications of dense HA include a spacer in a disc space, percutaneous access assist devices for continuous ambulatory peritoneal dialysis, and optical examination of internal tissues and maxillofacial reconstruction [4, 9, 10]. However, no osteoconduction occurs in dense HA as the component is lacked in porosity.

The current research focuses on the use of near dense HA structures. By having porosity, however minimal, the HA structures will be able to maintain their advantage of assisting cell ingrowth and osteoconductive properties, but at the same time, the structure is mechanically strong enough to withstand breakage. A dense HA spacer for cervical spine is one of the several applications that would benefit from a near dense HA structures with porosity to a certain degree. A clinical outcome of a dense HA spacer was previously reported as equal to autologous iliac crest interbody fusion and that patients who underwent the HA fusion had less need for secondary operation due to plug slippage [7]. With the introduction of porosity, HA fusion in an implant could be further improved without sacrificing its mechanical strength.

HA scaffold is typically prepared from axial powder compaction of HA powder, followed by binder removal process and further densification of the HA material [4, 11, 12]. To produce straightforward components, axial powder compaction produces little material waste, and production can be carried out at a high rate [13]. Densification processes can be carried out by pressureless sintering, hot isostatic pressing, and microwave sintering [14, 15].

In this research study, low porosity HA scaffolds are prepared via wet powder mixing process, followed by pressureless sintering for the pore former removal. The samples are then characterized and tested for their morphology using microcomputed tomography and image processing. The potential cell growth on the manufactured scaffolds is also tested using MTT assays and collagen staining quantification.

2. Materials and Methodology

Hydroxyapatite (HA) powder was supplied by Himed (powder size distribution of D10 = 95.95 μm, D50 = 145.6 μm, and D90 = 227.6 μm). Corn starch was supplied by Yiak Say Hang Food Industries. The starch particle size was estimated using scanning electron microscope (SEM), and the average particle size was 10 μm.

2.1. Hydroxyapatite Scaffold Fabrication. The scaffold was processed via powder processing route. HA powder was first mixed with corn starch, as a binder-cum-pore forming agent, with the percentage of 80 weight (wt.)% of HA and 20 wt.% of starch. The starch was first dissolved in warm water (70°C) while continuously stirring the solution. The HA powder was added little by little into the dissolved starch solution. Deionized water was added to the solution. The subsequent mixing process was carried out in a jar mill (Gardco 764 AVM) for 3 hours at 120 rpm. Alumina balls were used in the mixing process, with the powder-to-ball mass ratio of 1 : 4.

The HA/starch solution was then air dried for 48 hours and pulverized in room temperature by crushing and grinding the dried HA/starch cake using a mortar and pestle. The HA/starch powder was then compacted with the force of 49.03 kN. The pellets were subsequently subjected to thermal debinding-cum-sintering process (CM Rapid Temp Furnace) in atmospheric environment. The starch was burnt out from the green part during the debinding stage. The debinding process was conducted at holding temperatures of 250°C, 300°C, and 600°C for 1 hour at each holding temperature to ensure complete removal of the binder. The subsequent sintering was performed at 1250°C for 2 hours. The final manufactured scaffolds were 24.4 mm in diameter and 4.8 mm in height.

2.2. Morphological Study

2.2.1. Surface Porosity Evaluation. Surface porosity of the scaffolds was evaluated using SEM (Leica S360) and image analysis (Adobe Photoshop Element v.3). The samples were first gold coated (sputter coated SC7640) prior to SEM inspection using EHT supply of 15 kV. Seven distinct areas ($n = 7$) were taken to obtain porosity using image analysis technique. A grey scale threshold with the limit of 115 was performed, and the resultant image was binarized. The porosity was obtained from the histograms of grey scale. The reported final porosity was obtained by averaging the porosity values of each image analysis.

2.2.2. Internal Morphological Test. Internal morphological test was conducted using microcomputed tomography (micro-CT, SkyScan 1172) scans carried out with sample sized 2 × 2 × 2 mm. The pixel size used was 4.3 μm. Reconstruction of the scan to image was carried out with conversion values from 0 (minimum) to 0.12 (maximum). Analysis was carried out using CTAn software with region of interest volume (ROI) in the form of a circular area of 1 mm diameter with 1 mm thickness. Six ROIs were taken to obtain the porosity measurement ($n = 6$), with histogram greyscale limit set from 75 to 125. These ROIs extend through the data set slice by slice to create 3D volume of interest (VOI), which can then be analysed. The values of interest produced for the processed VOIs are pore size, closed porosity, open porosity, and total porosity.

The porosity of the scaffold was indirectly calculated from the sample's true volume, measured using a pycnometer (Micromeritics, AccuPyc 1330). Ten samples were used for the measurement ($n = 10$).

2.3. Hardness Measurement. Hardness measurement of the HA scaffolds was conducted using a Vickers microhardness measurement tool (MMT-X3, Matsuzawa Co., Ltd.). A diamond indenter with 490.3 N load was used to indent eight distinctly different locations on the scaffold ($n = 8$). These readings were calculated to obtain the average and standard deviations of the final hardness.

2.4. Cell Culture Experiment. The scaffolds were sterilized using autoclave process. Subsequent to the sterilization, the scaffolds were washed with phosphate buffer saline (PBS) and were conditioned with the cell media (Invitrogen, MEM Alpha Medium, with Ribonucleosides and Deoxyribonucleosides, cat. 22571).

The scaffolds were divided into two groups, each having three scaffolds ($n = 3$). The cells used were MLO-A5 osteoblastic cells, which is a postosteoblast and preosteocytes type of cells. Group 1 was seeded with cell density of 50,000 cells per scaffold. Group 2 was seeded with cell density of 500,000 cells per scaffold. Each scaffold well was given 4 mL of media, sufficient to cover the whole scaffold. Three controls were used for each of the group. Each control well was given 2 mL of media, which was typically sufficient for cells seeded on 6-well culture plates. Media for cells cultured on both scaffolds and control wells were changed every 3 days. The cell-seeded scaffolds and controls were kept in a static incubator for 7 days. Microscope observations were carried out to ensure that there were no contaminations and that the cells were growing well. The examination of the cell culture result was carried out by two methods, namely, MTT assay and collagen staining observations.

MTT ((3-(4,5-dimethylthiazol-2-yl)-2,5-diphenyltetrazolium bromide) concentration of 0.5 mg powder (Sigma) per 1 mL of PBS was used. The culture plates were incubated at 37°C for 1 hr. Subsequently, the MTT solution was removed off the plate. Acidified isopropanol (prepared by mixing 25 μL of concentrated HCl to 20 mL isopropanol) was given to the scaffold to elute the stain and form the purple formazan. Two samples were taken from each of the purple formazan resulting from the scaffolds and control wells. The optical density was then obtained using a spectrophotometer (Bio-Tek ELx800), using 540 nm wavelength.

Collagen stain was prepared by mixing a solution of sirius red (direct red 80, Fluka) 1 mg/mL in saturated picric acid (Fluka). The samples were washed with distilled water prior to addition of the collagen stain. The immersed samples were then shaken at room temperature overnight. Following that, the samples were washed with distilled water until the red coloring stopped eluting. Subsequently, the samples were destained with a mixture of 0.2 M sodium hydroxide (Sigma) and methanol in a ratio of 1 : 1 and were mildly shaken for 15 minutes at room temperature. Two liquid samples were taken from each solution of the scaffolds and control wells, such that the density could be read. The optical density was then measured using a spectrophotometer (Bio-Tek ELx800), using 490 nm wavelength.

3. Observation, Results, and Discussion

This section discusses the observations gained from the surface porosity, micro-CT, and *in vitro* cell culture examinations.

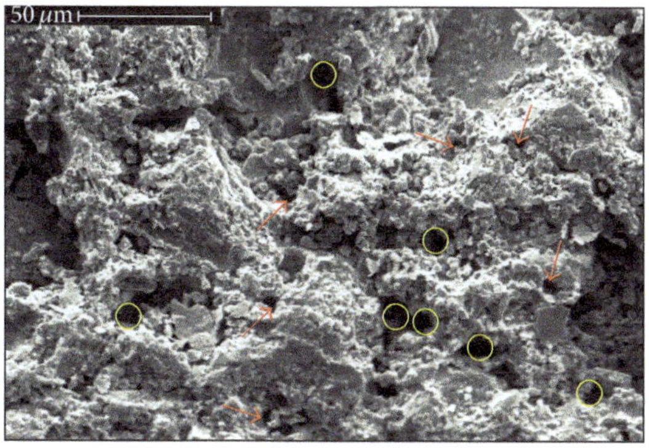

FIGURE 1: Cross-section image of the HA samples.

TABLE 1: Micro-CT measurement result.

	Measurement
Pore size	13.21 μm ± 0.35 μm
Closed porosity	8.35% ± 0.78%
Open porosity	8.07% ± 1.24%
Total porosity	15.76% ± 0.53%

3.1. Porosity Analysis. SEM micrographs were used to estimate the surface pore size and porosity level of the scaffold. Figure 1 shows the cross-section area of the sintered HA samples. The image also shows voids that are approximately 10 μm in diameter, as indicated by the circles. These voids are presumably the pores created by the pore forming agents during the debinding process. There are also voids with irregular shape and size smaller than 10 μm, as shown by the arrows. These voids were caused by the sintering and densification process of the HA powder. As the pore-forming agents were burnt, they leave behind empty space. These voids were then taken up by the HA powders that are being sintered, such that connections among the HA particles were formed irregularly. These observations show that the debinding-cum-sintering processes is sufficient to remove the pore-forming agents, as well as to sinter the HA skeleton.

Images of the sample surface were used in the image analysis to estimate the porosity of the scaffolds. Figure 2(a) shows a typical SEM image used of the sample's surface area for surface porosity measurement using image analysis, with the inset showing typical pores found on the scaffold surface. Figure 2(b) shows the analyzed image, with the black contrast colour highlighting pore locations on the scaffold's surface. The image analysis approximated the HA surface scaffold porosity to be 15.08% (±0.90%).

Further analysis with the micro-CT confirms the porosity estimation of the HA scaffold within its volume. Micro-CT estimates the morphological parameters of the scaffolds, such as the porosity and average pore size. The micro-CT measurement result is reported in Table 1, and a section of the HA sample used for the micro-CT porosity analysis is shown in Figure 3.

Micro-CT shows that the total porosity of the sample is 15.8% (±0.5%). This porosity is approximated to be a combination of open porosity and closed porosity. Open porosity

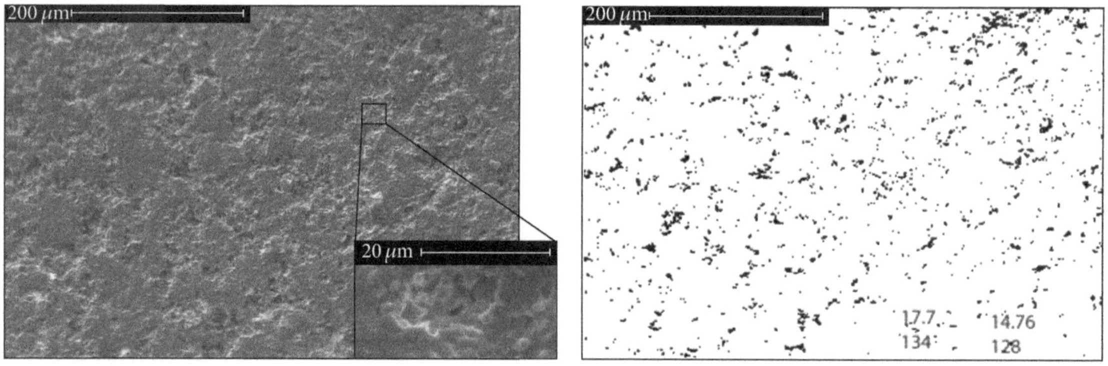

(a) (b)

FIGURE 2: Typical image analysis of HA scaffold. (a) Micrograph of surface structure (magnified 200 times); inset: typical pores on scaffold surface (magnified 1000 times). (b) Surface porosity of the scaffold, highlighted by the black contrast colour.

FIGURE 3: Micro-CT image of the HA sample.

refers to the pores that are interconnected between one another, and closed porosity refers to isolated pores within the scaffold.

The total porosity is shown to be almost evenly divided between the open and closed porosities. It indicates that there are similar amount of both interconnected and isolated pores. Assuming that the pore-forming agents were evenly distributed in the scaffold during the processing, the isolated pores would be predominant in the scaffold. However, the micro-CT finding showed that approximately half of the porosity was open and interconnected pores. There were two reasons of this finding. Firstly, there were pore formers that were not homogenously mixed, presumably during the wet mixing process. Secondly, densification of HA typically starts at temperature range from 700 to 800°C [16]. To ensure complete removal of the pore-forming agents, the debinding process was held at temperature 600°C. Hence, there are voids left by the pore-forming agents, which allow the HA particles to rearrange themselves in temperature range from 600 to 700°C. During the particle rearrangement, there could be movements of the HA particles such that interconnected pores are formed. However, as the range of temperature allowing the particle rearrangement was relatively short, there were limited interconnected pores formed, resulting in the remaining isolated pores.

To further verify the porosity analysis, the porosity of the HA samples ($n = 3$) are calculated indirectly with the help of a pycnometer. The porosity measurement was based on the following formulation:

$$\text{porosity\%} = \left(1 - \frac{\text{true volume}}{\text{bulk volume}}\right) \times 100\%. \tag{1}$$

The pycnometer measures the sample's true volume, whereas the bulk volume is calculated from the measured height and diameter of each sample. The average porosity obtained from the measurement was 22.7% (±2.96%).

It is noted that there is a difference between the porosity calculated using the pycnometer and the porosity estimated by both image analysis and micro-CT. One of the main sources of differences is that there could be due to the loss of data resolution and incorrect identification of the pore edges during the CTAn analysis of the micro-CT volume of interests. Hence, the total pore areas and volume could be underestimated.

From the micro-CT, the average pore size of the HA scaffolds is analyzed to be 13 μm. This pore size estimation is expected as the pore-forming agent used is approximately 10 μm in diameter. Therefore, assuming that the pore-forming agents were removed completely and the HA particle rearrangement during sintering was minimal, the pore formers would leave behind void volumes similar to their original size.

The porosity and pore size obtained from the process are well below the so-called definition of highly porous structures that are commonly reported, with minimum 40% porosity and pore size of 100 μm and above [17–20]. This clearly indicates that the selection of pore formers and ratio between the pore former and main material, as well as sintering process parameter, are able to construct HA samples with lower porosity.

3.2. Hardness of the Scaffolds. The HA samples were found to have microhardness property of 5.28 (±1.08) GPa. This

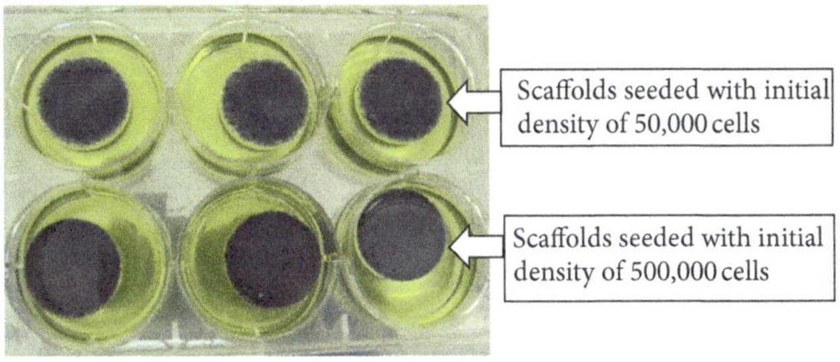

FIGURE 4: HA scaffolds after MTT staining.

hardness property is distinctively higher compared to similar HA scaffold with porosity of 20%, as reported by Ma et al. [21]. It was noted, however, that Ma et al. applied a different processing technique to prepare the HA scaffold green part. Multiple deposition of HA colloid was carried out using electrophoretic deposition technique in order to obtain bulk HA scaffold. The green part was sintered at 1200°C, which was slightly lower as compared to the sintering temperature used in the current study. However, the sintering duration is the same as the one used in the current study. The electrophoretic disposition technique employed in the previous report could have produced green part with lower strength as compared to the powder-processing techniques utilized to fabricate the HA samples in this paper, as the the deposition was carried out by repeating single deposition several times and as such allowing the possibility of delamination to occur. Microhardness is an indicator of local structural integrity [22] and is related to the yield strength of a material [23]. As such, this suggests that the HA scaffold processed by the powder compaction has more superior structural integrity as compared to HA samples with similar porosity level prepared by other methods. This is attributed to the formation of a dense strong matrix inside the scaffold.

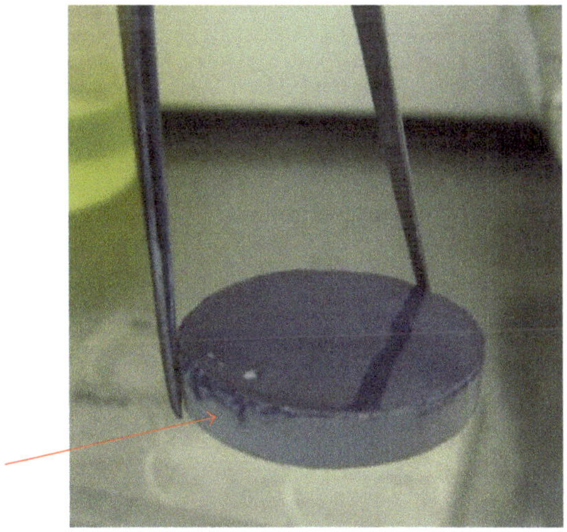

FIGURE 5: Cells indicated at the scaffold circumference.

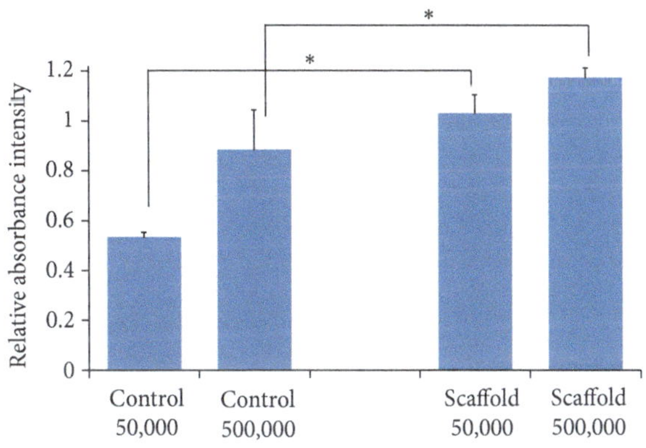

FIGURE 6: MTT quantification of relative numbers of cells after 7 days of seeding (*indicates significant difference, $P < 0.05$).

3.3. Cell Culture Experiments. MTT is a colorimetric assay that is used to qualitatively identify the cell locations in a substrate. This assay measures the reduction of yellow MTT solution by mitochondrial succinate dehydrogenase that is found in live cells. The MTT enters the cells and passes into their mitochondria where it is reduced to an insoluble, dark, purple formazan product. The cells are then solubilized with an organic solvent, and the released dark purple formazan reagent is measured spectrophotometrically. Quantitatively, for a 2D substrate, the colour intensity of the purple liquid indicates the relative number of cells that were alive in the substrate.

After the addition of MTT solution, both the HA scaffolds and control plates showed signs of live cells occupying the space. This was indicated by the purple colour that appeared on the scaffolds and the control plates as a reaction between the MTT solution and live cells, as shown in Figure 4.

As the reduction of MTT into dark purple formazan can only occur in metabolically active cells, the appearance of the dark purple colour is a sign of the cell viability in the scaffolds and on the control plates.

It is noted the cells are able to slightly penetrate to the circumference of the scaffolds. It is observed in Figure 5 that the dark purple colour appears at the scaffold's circumference, as pointed by the arrow. This implies that even though the scaffolds has relatively low amount of porosity of 15 to 22%, limited amount of cells could still penetrate into the scaffold's interior. The cellular activity of infiltrating into the scaffold's circumference regardless of the scaffold's low porosity could be due the sintering of the HA that is adopted to produce the samples. Rodrigues et al. reviewed that sintering synthetic HA aids in the proliferation and differentiation of osteogenic cells [24]. In another work, the use of sintered HA suggested that the sintering causes the material's surface to be reactive and hence interact favourably with the seeded osteoblast cells [25]. As such, the sintered HA obtains added chemical attraction to the cellular activity, which attracts cells to infiltrate into the scaffold.

Upon complete elution of the dark purple formazan, the colour intensity from the formazan is read by a spectrophotometer. Figure 6 shows the relative number of cells, quantified by MTT assay, after 7 days of continuous seeding.

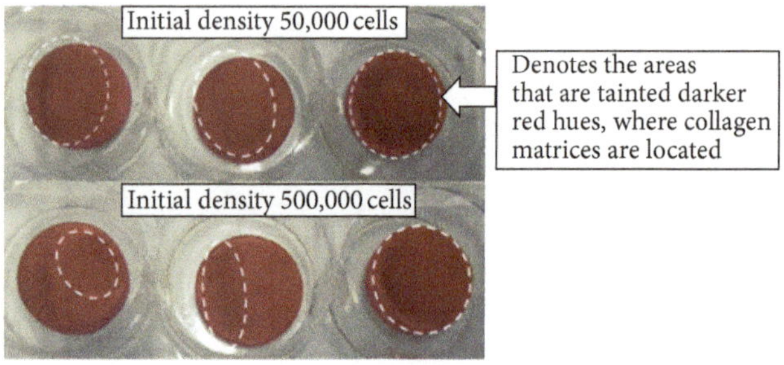

FIGURE 7: Collagen staining of the HA samples.

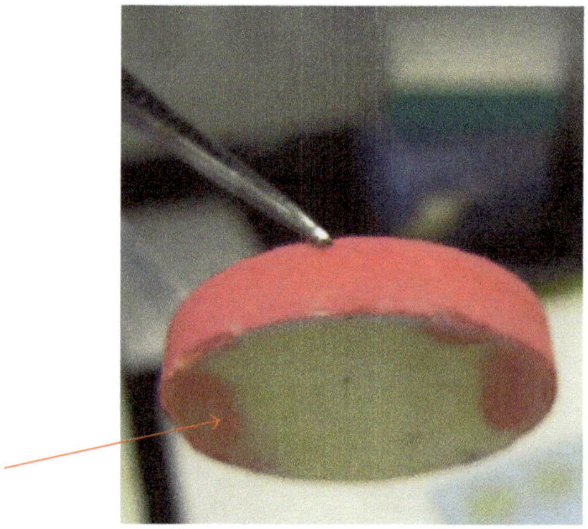

FIGURE 8: Collagen matrices indicated at the scaffold circumference.

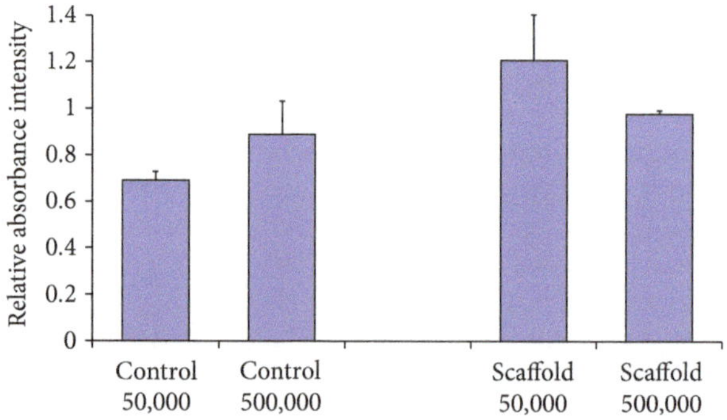

FIGURE 9: Quantification of collagen matrix production after 7 days of culture.

The numbers 50,000 and 500,000 refer to the initial cell seeding density of 50,000 or 500,000 cells per scaffold or per control plate, respectively.

The MTT reading shows that there is a significant difference between the numbers of cells found on the scaffolds as compared to those found on the control plates. Both readings with 50,000 and 500,000 cell density have $P < 0.05$ confidence level. This result is expected as HA, the scaffold material, is an inorganic mineral with high chemical similarity to that of bone mineral [2, 4]. This result shows that HA is a better substrate to facilitate bone cells growth, as compared to the polymeric control plates, despite of the low porosity of the current HA scaffolds. The HA scaffolds are able to provide more cell growth and proliferation compared to the control plates.

It is noted that the initial amount of cells seeded in the two groups of scaffolds and the controls have a 10-fold difference. However, this is not reflected in the growth and proliferation outcomes after 7 days of culture. This could be explained by the fact that in general cells do not grow in a linear relationship. There are various conditions that affect cell growth. Firstly, overcrowded culture could result in cell death as there is lack of growth space and nutrients from the media. Cells could also fall off the substrate and get washed away in an overcrowded culture. Secondly, there is a limit to the number of cells that a substrate can support for a period of time. Again, this is because the substrate is a static structure which does not have any changes in the dimensions. After a certain period of time, the culture will become overcrowded. Normal cells stop dividing after they reach a certain density. It is understood that after a cell makes contact with other cells on the substrate, specific signal is relayed to inform the cell if all cellular contact points have been satisfied. If all contact points have been satisfied, the genes that signal proliferation are turned off and as such, the proliferation stops [26]. This shows that there is a cell growth saturation point that cannot be surpassed for a specified culture substrate. Therefore, it is not a guarantee that a 10-fold difference in the initial seeding will result in the same fold of cell growth and increase after a certain period of culture time.

As MLO-A5 is an osteoblastic cell type, it would normally secrete collagen when it is proliferating healthily. The collagen staining test is used to measure the amount of collagen matrix formed by the MLO-A5 cells. The collagen staining visual observation is shown in Figure 7. The location where collagen matrices were profoundly found on the scaffold was tainted darker red hues, which is denoted by the dotted lines.

The collagen staining showed that the cells were able to penetrate into the circumference of the scaffolds, as can be seen from Figure 8. However, it was observed that the cell penetration was not in-depth, with mostly limited to the outer circumference of the scaffold. It was seen that the outer circumference of the scaffold was tainted red, and this showed the area that has collagen matrices. However, due to the small pore size and low porosity level possessed by the scaffold, there is a limitation in the spread of collagen matrices into the scaffold. After the baseline reading subtraction, the intensity reading of the collagen matrices amount after 7 days of culture is reported in Figure 9.

In this observation, it is seen that the collagen matrix production in the HA scaffolds was more than that produced on the control substrates. This supports the argument that a 3D HA scaffold, in this case HA scaffold, is a more suitable substrate material, preferred by the bone cells to grow, proliferate, and perform their collagen matrix production function, as compared to the polymeric 2D cell culture plate. This observation was also seen in other work, in which osteocalcin secretion and alkaline phosphatase activity were detected at a significantly higher level on 3D scaffold than on 2D cell culture plate [27]. This is presumably due to the bioactivity nature of HA, which indicates that the material provides substrates that are capable of attracting cells for bone tissue formation. The bioactive surface is thus capable of significantly facilitating differentiation of osteogenic cells the subsequently and the apposition of tissue matrix *in vitro* [3].

It is also observed that collagen matrices produced by the scaffold with initial seeding density of 500,000 cells were less than that seeded ones with 50,000 cells. This was an interesting finding, as the earlier MTT readings showed that by the end of the 7-day culture, the cells on scaffolds seeded with initial 500,000 cells were able to proliferate more than that of scaffolds seeded with 50,000 cells. Several postulations are given. Firstly, there is a limited area in the scaffold for the osteoblast cells to contain both their proliferation and collagen matrix production activities. When there are too many cells on the substrate, there will not be enough areas for the produced collagen matrices to attach, and the layer of collagen matrices may easily fall off the substrate or get washed away by the PBS. Hence, as the cell quantity in the 50,000-cell scaffolds was less than that of the 500,000-cell scaffold, as was estimated using the MTT test, the cells in the prior scaffolds had more areas to contain the produced collagen matrix. Secondly, when the cells in the 500,000-cell scaffold were occupied in the proliferation activity, it was harder for these cells to work as much to produce the collagen matrices. In the molecular level, there is a certain driving force and energy required for any molecular activities [28]. Therefore, the collagen production could be slowed down if there is less driving energy in the 500,000-cell scaffold due to the continuous proliferation activity.

The cell culture results have confirmed that the HA scaffolds are able to demonstrate its suitability to support anchorage and maintain viability of live cells. Future study would be to continue with the *in vivo* experimental work. It is expected that the low porosity HA will have interactions with its surrounding tissues as normally occurs in implantation of an active bioceramic, as was exhibited elsewhere. A 5-year-HA dental implant retrieved from a human mandible showed a thick bone addition formed on the implant, and chemical bonding existed between the HA implant and the bone [29]. The chemical bonding introduces bioactive fixation between the surface-reactive HA with the bone [30].

4. Conclusions

The current research reported here has shown that low porosity HA scaffold could be well produced using compaction and sintering methods. The microporosity of the HA scaffold has been conducted and has shown that the pore size and porosity level of the scaffold are within the range for a low porosity scaffold. The mechanical hardness strength of the scaffold was also found to be comparable, if not higher than other HA scaffolds with similar porosity value.

The cell culture experiment has shown that HA scaffold is a more preferred substrate material as compared to the polymeric cell culture plate for culturing osteoblast cells. This is despite the standard condition of the cell culture control plates having been treated by vacuum-gas plasma to provide conducive cell attachment environment. The osteoblast cells seeded onto the HA scaffolds are found to yield favourable results in both the numbers of cells and in cellular activity after a certain period of time. This results demonstrate that the HA scaffolds are suitable to support anchorage and maintain viability of live cells. Initial cell seeding of 50,000 MLO-A5 cells is found to be sufficient for culture on the HA scaffolds. This amount of initial cell seeding density is adequate to demonstrate favourable cell growth and proliferation. This initial cell seeding density is also sufficient in order to avoid cell overcrowding for effective collagen matrix production.

It is noted that this scaffold with lower porosity or near dense properties are still capable of providing cell attachment into the pores of the scaffolds, while it simultaneously provides sufficient handling strength to potentially avoid breakage during the implantation. It is hence suggested that this type of HA scaffold could be used as effective bioactive scaffolds for applications that required high strength, such as bone or spinal implant grafts. The future works that could be carried out based on the current study is to evaluate the degradation and resorption mechanisms of the HA scaffold *in vitro*.

Acknowledgments

The authors would like to thank Dr. Gwendolen Reilly and Dr. Anuphan Sittichokechaiwut (Department of Engineering Materials, The Kroto Research Institute, North Campus, University of Sheffield) for the valuable advice and discussion, provision of the osteoblast cells and cell culture consumables, and assistance in the cell culture process. Dr. F. E. Wiria would like to acknowledge the funding from the British Council Researcher Exchange Programme that provides the collaboration opportunity and the use of micro-CT equipment.

References

[1] M. J. Lysaght and A. L. Hazlehurst, "Tissue engineering: the end of the beginning," *Tissue Engineering*, vol. 10, no. 1-2, pp. 309–320, 2004.

[2] D. Tadic and M. Epple, "Mechanically stable implants of synthetic bone mineral by cold isostatic pressing," *Biomaterials*, vol. 24, no. 25, pp. 4565–4571, 2003.

[3] K. Y. Lee, M. Park, H. M. Kim et al., "Ceramic bioactivity: progresses, challenges and perspectives," *Biomedical Materials*, vol. 1, no. 2, pp. R31–R37, 2006.

[4] R. Z. LeGeros and J. P. LeGeros, "Dense hydroxyapatite," in *An Introduction to Bioceramics*, L. L. Hench and J. Wilson, Eds., vol. 1, pp. 139–180, World Scientific, Singapore, 1993.

[5] O. Malard, F. Espitalier, P. Bordure, G. Daculsi, P. Weiss, and P. Corre, "Biomaterials for tissue reconstruction and bone substitution of the ear, nose and throat, face and neck," *Expert Review of Medical Devices*, vol. 4, no. 5, pp. 729–739, 2007.

[6] P. Ducheyne and Q. Qiu, "Bioactive ceramics: the effect of surface reactivity on bone formation and bone cell function," *Biomaterials*, vol. 20, no. 23-24, pp. 2287–2303, 1999.

[7] P. Ylinen, M. Raekallio, R. Taurio et al., "Coralline hydroxyapatite reinforced with polylactide fibres in lumbar interbody implantation," *Journal of Materials Science*, vol. 16, no. 4, pp. 325–331, 2005.

[8] T. Iguchi, A. Kanemura, A. Kurihara et al., "Cervical laminoplasty: evaluation of bone bonding of a high porosity hydroxyapatite spacer," *Journal of Neurosurgery*, vol. 98, no. 2, pp. 137–142, 2003.

[9] S. Barinov and V. Komlev, *Calcium Phosphate Based Bioceramics for Bone Tissue Engineering*, Trans Tech Publications, Enfield, NH, USA, 2008.

[10] L. L. Hench, "Bioceramics: from concept to clinic," *Journal of the American Ceramic Society*, vol. 74, no. 7, pp. 1487–1510, 1991.

[11] G. With, H. J. A. Dijk, N. Hattu, and K. Prijs, "Preparation, microstructure and mechanical properties of dense polycrystalline hydroxy apatite," *Journal of Materials Science*, vol. 16, no. 6, pp. 1592–1598, 1981.

[12] M. H. Nazarpak, M. Solati-Hashjin, and F. Moztarzadeh, "Preparation of hydroxyapatite ceramics for biomedical applications," *Journal of Ceramic Processing Research*, vol. 10, no. 1, pp. 54–57, 2009.

[13] R. H. Todd, D. K. Allen, and L. Alting, *Manufacturing Processes Reference Guide*, Industrial Press, New York, NY, USA, 1st edition, 1994.

[14] D. S. Seo and J. K. Lee, "AFM analysis of anisotropic dissolution in dense hydroxyapatite," *Ultramicroscopy*, vol. 108, no. 10, pp. 1157–1162, 2008.

[15] J. W. Nicholson, *The Chemistry of Medical and Dental Materials*, Royal Society of Chemistry, Cambridge, UK, 2002.

[16] E. Landi, A. Tampieri, G. Celotti, and S. Sprio, "Densification behaviour and mechanisms of synthetic hydroxyapatites," *Journal of the European Ceramic Society*, vol. 20, no. 14-15, pp. 2377–2387, 2000.

[17] F. Chevalier, D. Chulia, C. Pouget, and M. Viana, "Fabrication of porous substrates: a review of processes using pore forming agents in the biomaterial field," *Journal of Pharmaceutical Sciences*, vol. 97, no. 3, pp. 1135–1154, 2008.

[18] K. A. Hing, S. M. Best, K. E. Tanner, W. Bonfield, and P. A. Revell, "Quantification of bone ingrowth within bone-derived porous hydroxyapatite implants of varying density," *Journal of Materials Science*, vol. 10, no. 10-11, pp. 663–670, 1999.

[19] H.-W. Kim, J. C. Knowles, and H.-E. Kim, "Hydroxyapatite/poly(ε-caprolactone) composite coatings on hydroxyapatite porous bone scaffold for drug delivery," *Biomaterials*, vol. 25, no. 7-8, pp. 1279–1287, 2004.

[20] H. Omae, Y. Mochizuki, S. Yokoya, N. Adachi, and M. Ochi, "Effects of interconnecting porous structure of hydroxyapatite ceramics on interface between grafted tendon and ceramics," *Journal of Biomedical Materials Research A*, vol. 79, no. 2, pp. 329–337, 2006.

[21] J. Ma, C. Wang, and K. W. Peng, "Electrophoretic deposition of porous hydroxyapatite scaffold," *Biomaterials*, vol. 24, no. 20, pp. 3505–3510, 2003.

[22] R. A. Ayers, S. J. Simske, C. R. Nunes, and L. M. Wolford, "Long-term bone ingrowth and residual microhardness of porous block hydroxyapatite implants in humans," *Journal of Oral and Maxillofacial Surgery*, vol. 56, no. 11, pp. 1297–1301, 1998.

[23] G. P. Evans, J. C. Behiri, J. D. Currey, and W. Bonfield, "Microhardness and Young's modulus in cortical bone exhibiting a wide range of mineral volume fractions, and in a bone analogue," *Journal of Materials Science*, vol. 1, no. 1, pp. 38–43, 1990.

[24] C. V. M. Rodrigues, P. Serricella, A. B. R. Linhares et al., "Characterization of a bovine collagen-hydroxyapatite composite scaffold for bone tissue engineering," *Biomaterials*, vol. 24, no. 27, pp. 4987–4997, 2003.

[25] F. B. Bagambisa and U. Joos, "Preliminary studies on the phenomenological behaviour of osteoblasts cultured on hydroxyapatite ceramics," *Biomaterials*, vol. 11, no. 1, pp. 50–56, 1990.

[26] G. M. Fuller and D. Shields, *Molecular Basis of Medical Cell Biology*, Prentice Hall, New Jersey, NJ, USA, 1998.

[27] W. Liu, M. K. Bergenstock, W. Lau, W. Sun, and Q. Liu, "Comparison of osteogenic cell differentiation within 2D and 3D culture systems," in *Proceedings of the 2nd World Congress on Tissue Engineering and Regenerative Medicine*, Seoul, Korea, August-September 2009.

[28] H. F. Noller, "The driving force for molecular evolution of translation," *RNA*, vol. 10, no. 12, pp. 1833–1837, 2004.

[29] M. Ogiso, T. Tabata, T. Ichijo, and D. Borgese, "Examination of human bone surrounded by a dense hydroxyapatite dental implant after long-term use," *Journal of Long-Term Effects of Medical Implants*, vol. 2, no. 4, pp. 235–247, 1992.

[30] L. L. Hench, "Bioceramics," *Journal of the American Ceramic Society*, vol. 81, no. 7, pp. 1705–1728, 1998.

14

Flocculation Efficiency of Poly(Acrylamide-*Co*-Acrylic Acid) Obtained by Electron Beam Irradiation

Gabriela Craciun,[1] Elena Manaila,[1] and Maria-Daniela Stelescu[2]

[1] *National Institute for Lasers, Plasma and Radiation Physics, Accelerators Laboratory, 409 Atomistilor Street, 077125 Magurele, Romania*
[2] *National Research and Development Institute for Textile, 93 Ion Minulescu Street, 031215 Bucharest, Romania*

Correspondence should be addressed to Gabriela Craciun; gabriela.craciun@inflpr.ro

Academic Editor: Maurizio Ferrari

A correlation between physicochemical characteristics of flocculants obtained by electron beam irradiation and their efficiency for wastewater treatment is presented. For real wastewater treatment, our interest was focused upon total suspended solids, fatty matter, and chemical oxygen demand. Flocculation studies were carried out using a standard jar test. A treatment option based on poly(acrylamide-*co*-acrylic acid) for wastewater taken from a slaughterhouse plant is presented.

1. Introduction

Wastewater treatment is a major problem around the world. Growing along with the population growth, industries create environmental problems and health hazards for the population. Hence, environmental concerns behoove scientists and engineers to develop materials and methods to lower the extent of pollution of the environment [1].

Coagulation and flocculation still play a dominant role in many water and Wastewater treatment schemes. Because of the complex interdependence of numerous factors inherent in the coagulation and flocculation processes, a well understanding of the phenomena involved is essential. Coagulation is the process whereby destabilization of given suspension or solution is affected; that is, the function of coagulation is to overcome the factors that promote the stability of a given system. The manifestation of destabilization depends on the particular system; for example, in the case of disperse suspensions or solutions, visible floc or precipitate formation occurs; in the case of highly concentrated suspensions, dewatering of the sludge mass occurs, and so on. Flocculation is the process whereby destabilized particles, or particles formed as a result of destabilization, are induced to come together, make contact and thereby form large(r) agglomerates. In this case, the manifestation of destabilization is realized in practicable terms: in effect, flocculation accelerates floc formation, influences the physical characteristics of flocs formed (e.g., their strength, size, and density), and governs the final concentration of destabilized particles. In water and Wastewater treatment, coagulation and flocculation phenomena are extremely important [2].

In Wastewater treatment, flocculation is essential. Flocculating agents, or flocculants, are important components which cause flocculation, the process of bringing together small particles to form larger particles by adding small quantities of chemicals in water and Wastewater treatment [3, 4]. Flocculants are classified into inorganic and organic categories. Though the inorganic flocculants (also called coagulants) with multivalent metals like aluminum and iron are widely employed, organic flocculants based on acrylamide-based polymers, like polyacrylamide and its derivatives, are generally more effective than their inorganic counterparts as they possess the advantages, such as low dose, ease in handling, no interference with pH of the suspensions, and larger floc-forming capability [3, 4].

Flocculants are applied in potable water, industrial raw and process water, municipal sewage treatment, mineral processing and metallurgy, oil drilling and recovery, and so forth

[5–11]. These materials have the ability to induce an advanced coagulation process, after which a large amount of bacteria and viruses from the water are precipitated together with the suspended solids of this. Especially flocculants based on poly(acrylamide-*co*-acrylic acid) obtained by electron beam irradiation are very efficacious for Wastewater treatment. But a strict condition of use is that residual acrylamide monomer content is to be less than 0.05%. The major advantages of radiation induced polymerization processes are (1) very easy to manipulate the molecular weight, from low to very high, by simply changing the feed composition as well as the composition of the product by incorporating different monomers; (2) precise control of charge density as the monomer feed composition is controlled at the initial stages only; (3) precise control of molecular weight distribution; (4) no flammable and toxic solvents used; (5) no production of waste matter or evolution of noxious gases; (6) no production of hazardous effluents; (7) very low monomer contents; (8) very clean process [12].

Physical and chemical properties of flocculants based on poly(acrylamide-*co*-acrylic acid) obtained by electron beam irradiation are strictly related to the efficiency of Wastewater treatment expressed by the level of treated water quality indicators. There are many situations in which organic flocculants should be used together with classic coagulation aids (inorganic flocculants) such as $Al_2(SO_4)_3$, $FeSO_4$, or $Ca(OH)_2$ because in the case of very charged Wastewaters, treatments based only on organic flocculants or inorganic flocculants (classical treatment) are less efficient than in the case of their combined use. Moreover, the so-called flocs are larger and more strongly bound than the aggregates obtained by coagulation [1].

It is well known that each Wastewater type, depending on its origin, has its particularities and more than that the same type of Wastewater shows significant variation during the same day. For this reason treatments based on polyelectrolyte should be seen as treatment schemes for each type of Wastewater. These treatment schemes should also be easily adaptable to the significant variations of the same type of Wastewater characteristics. There is no polyelectrolyte with universal destination regardless of the method of production. This is one of the reasons why the field of flocculants obtaining by radiation technologies is still open.

In the present work, we present a treatment option based on poly(acrylamide-*co*-acrylic acid) for Wastewater taken from a slaughterhouse plant.

2. Methods

2.1. Materials and Sample Preparation. For flocculants obtaining, the following materials (from Sigma-Aldrich) were used: acrylamide, AMD (molar mass 71.08 g mol^{-1}, density 1.13 g/cm^3, solubility in water 2.04 kg/L at 25°C); acrylic acid, AA (molar mass 72.06 g mol^{-1}, density 1.051 g/mL, solubility in water: miscible); sodium chloride, NaCl (molar mass 58.44 g mol^{-1}, density 2.165 g/cm^3, solubility in water: 359 g/dm^3 at 20°C); sodium hydroxide, NaOH (molar mass 39.99 g mol^{-1}, density 2.13 g/cm^3, solubility in water: 1110 g/dm^3 at 20°C); sodium formate, HCOON (molar mass 68.01 g mol^{-1}, density 1.92 g/cm^3, solubility in water: 97 g/100 mL at 20°C, serves as chain transfer agent in the copolymerization process); potassium persulfate, $K_2S_2O_8$ (molar mass 270.322 g mol^{-1}, density 2.477 g/cm^3, solubility in water: 1.75 g/100 mL at 0°C, serves as initiator in the copolymerization process).

Table 1: The typical chemical composition of the monomer aqueous solutions.

Chemical composition	Characteristics
Total monomer concentration	40%
Monomer ration in terms of percentage, R = AMD/AA	9/1
NaCl concentration	8%
Initiator concentration, potassium persulfate	0%–0.015%
Chain transfer agent concentration, sodium formate	0.0075%

For the preparation of acrylamide-acrylic acid monomer solution, 40% AMD : AA (mass ratio 9 : 1) and natrium chloride 8% were dissolved in 1000 mL distilled water. The optimum pH monomer solution, which is in the range of 6 to 6.5 in the case of electron beam polymerization, was obtained by gradually adding sodium hydroxide to the monomer solution. From the obtained monomer solution, samples of 100 mL were distributed in quartz vessels. Different amounts of initiator (potassium persulfate) and chain transfer agent (sodium formate) were added and then each quartz vessel was irradiated with electron beam in atmospheric conditions and at room temperature of 25°C. The thickness of monomer aqueous solution in each quartz vessel was 23 mm. The typical chemical composition of the monomer aqueous solutions irradiated in order to produce acrylamide-acrylic acid copolymers is presented in Table 1. Electron beam dose rate was fixed at 2.4 kGy/min, in order to accumulate the working dose.

2.2. Experimental Installation. The facility used for poly(acrylamide-*co*-acrylic acid) flocculants was a linear accelerator called ALIN 10, which is a travelling-wave type, operating at a wavelength of 10 cm and having 164 W maximum output power. The accelerating structure is a disk-loaded tube operating in the $\pi/2$ mode. The optimum values of the EB peak current I_{EB} and EB energy E_{EB} to produce maximum output power P_{EB} for a fixed pulse duration τ_{EB} and repetition frequency f_{EB} are as follows: $E_{EB} = 6.23$ MeV, $I_{EB} = 75$ mA, and $P_{EB} = 164$ W ($f_{EB} = 100$ Hz, $\tau_{EB} = 3.5\,\mu$s) [13–15].

2.3. Poly(Acrylamide-Co-Acrylic Acid) Characterization. EB irradiation is extremely efficacious in polymerization processes. EB induced polymerization involves the Coulomb interaction of the accelerated electrons with atoms or molecules in the irradiated medium. These interactions generate reactive ions, thermalized electrons, excited states, and radicals to drive the chemical reactions. Polymerization is subsequently induced by ionizing radiation in much the same way as free radical addition polymerization. The main function of the ionizing radiation in a radiation-initiated polymerization is limited to the primary events, the initiation step which leads to the production of free radicals and to a few specific secondary effects. The subsequent steps of propagation, termination, and chain transfer proceed as in a chemically catalyzed process [15].

Preparation of poly(acrylamide-*co*-acrylic acid) flocculants was based on copolymerization by EB irradiation of the aqueous solutions containing appropriate mixtures of acrylamide and acrylic acid monomers and certain agents, such as chain transfer agents to mitigate cross-linking of the polymer structure and initiators to optimize the monomer conversion process. Our interest was focused on the physicochemical properties involved in Wastewater treatment, such as conversion coefficient (CC), residual monomer concentration (M_r), intrinsic viscosity (η_{intr}) or average M_w (molecular weight), and a linearity coefficient given by the Huggins' constant (k_H). These characteristics of poly(acrylamide-*co*-acrylic acid) obtained by electron beam irradiation technology are influenced by the following factors: chemical composition of the solutions to be irradiated, absorbed dose level, and absorbed dose rate level [15]. Many years of investigations of polymeric material properties [16–19] demonstrated that EB induced polymerization, under proper irradiation conditions and for fixed chemical composition of the monomer mixtures to be irradiated, leads to higher conversion efficiency (near 100%) and lower residual monomer concentration (under 0.01%) than classical polymerization of the acrylamide-acrylic acid aqueous solutions.

Two grams of poly(acrylamide-*co*-acrylic acid) flocculants obtained as above were placed in 200 mL distilled water for 24 hours and then every sample was stirred for 1 hour at 400 rpm to ensure a very well mixing. The conversion coefficient (CC) and the residual monomer concentration (M_r) were determined based on the bromation reaction of the double-bond method [11, 20]. After complete dissolution in water, the copolymer AMD/AA was treated excessively with a bromide-bromate solution and the bromine excess was determined by means of a iodometric method in presence of sodium thiosulfate solution (1 M).

The intrinsic viscosity (η_{intr}) and the Huggins' constant (k_H) were determined by the viscometry method, using the Hoppler type BH-2 [21]. Hoppler viscometer is a falling ball one. The measured parameter is the time of fall of the ball in a cylindrical tube inclined by 10 deg with respect to the vertical plane and filled with the liquid subject to investigation. The principle of falling-ball viscometers is based on the fact that the viscosity of liquid modifies the speed of ball falling through this liquid. The time of falling through the copolymer solution was measured for five different concentrations of it. As a solvent sodium nitrate 1 N ($NaNO_3$) was used and the working temperature was 30°C. From the time of falling through the copolymer solution (t) and the time of falling through the solvent (t_0), relative viscosity ($\eta_{\text{rel.}} = t/t_0$) was obtained. Specific viscosity was calculated from the relation, $\eta_{\text{sp.}} = \eta_{\text{rel.}} - 1$. Reduced viscosity was determined using the relation $\eta_{\text{red.}} = \eta_{\text{sp.}}/c$, where c is the copolymer concentration (g/dL). From the graphical representation of the $\eta_{\text{red.}}$ as a function of the copolymer concentration (c), by extrapolation, the intrinsic viscosity $\eta_{\text{intr.}}$ and $tg\alpha$ was obtained.

The linearity constant was determined from the following relation: $k_H = tg\alpha/[\eta_{\text{intr.}}]^2$. Intrinsic viscosity is a measure of the hydrodynamic volume occupied by the individual polymer molecules in isolation [22]. In dilute solutions, the polymer chains are separate and the $[\eta_{\text{intr.}}]$ of a polymer in solution depends only on the dimensions of the polymer chain. Because $[\eta_{\text{intr.}}]$ indicates the hydrodynamic volume of the polymer molecule and is related to the molecular weight, it provides deep insights on the molecular characteristics [23].

From the Mark-Houwink equation, the relationship among the molecular weight and viscosity is given below: $[\eta] = K * [M_w{}^a]$, where $[\eta]$ is the intrinsic viscosity (dL/g), M_w is molecular weight (Dalton), and K and a are constants for a particular polymer-solvent system (K and a are $3.73*10^{-4}$ and 0.66 in 1 N aqueous sodium nitrate at 30°C) [21, 24].

Radius of gyrations is a parameter which characterizes the size of a particle of any shape. The higher the radius of gyration, the more the molecular dimension of the polymer and that will affect the flocculation characteristics synergistically [25]. Radius of gyration was determined by the following relation: $R_g = 0.0147 * M_w{}^{0.59}$ [26, 27]. Flocculation studies were carried out on real Wastewater taken from a slaughterhouse plant, at room temperature (20–25°C) using the standard jar test. The jar-test procedure is commonly used to simulate the water treatment process in the laboratory in order to determine the proper coagulant dosage and accumulate data for process control. Beakers of 1000 mL capacity each equipped with a variable speed (0–100 rpm) agitator were used. In each beaker, 500 mL of Wastewater was taken and requisite amount of polymer was added into it [28] by means of a pipette. The agitator was first adjusted to 75 rpm for 2 minutes and then for another 5 minutes at 25 rpm. The flocs were then allowed to settle down for 2 minutes [3–19, 28]. A measured volume of samples from each beaker (with and without polymeric flocculant) was taken to determine the suspended solids [29]. It is important that the process using a flocculator involves significant mechanical agitation in order to increase the collision of reagents coagulant and flocculant and the colloidal particles in suspension. However, the cutting action must not be too strong; otherwise, it will destroy the flocs that are formed [30].

2.4. Determination of Wastewater Quality Indicators. The primary purpose of a Wastewater treatment is to remove solids from water. *Total suspended solids* (TSS) represent the solids which are suspended in Wastewater and would be caught by a filter. Suspended solids are measured by passing Wastewater sample through a filter. The solids caught by the filter, once dried, are the total suspended solids. To measure TSS, the water sample was filtered through a preweighed filter. The residue retained on the filter was dried in an oven at 103 to 105°C until the weight of the filter no longer changes. The increase in weight of the filter represents the total suspended solids. To calculate the concentration of total suspended solids in the sample the following formula was used:

$$\text{TSS (mg/L)} = \frac{(A - B) \times 1000}{V}, \quad (1)$$

where: A = sample and filter weight, (mg), B = sample and filter weight, (mg), V = volume of water sample, (L).

Determination of extractable compounds with solvents (fatty matter—FM) is a gravimetric method which is used to determine the content of substances that can be extracted with organic solvents from Wastewater and surface water. Solvents used are petroleum ether or hexane or carbon-tetrachloride. The extractable substances are animal fat and vegetable oil (mineral oil and heavy oil), compounds with functional groups like hydroxyl, carbonyl, carboxyl, organic nitrogen compounds, insecticides, soaps, waxes, resins, and tars that are extracted with solvents. The final result represents the mean value of three determinations made in the same time. The procedure consists of sample acidification to pH less than or equal to 2 and FM extraction with organic solvent three or four times. The organic extract is evaporated to dryness and the amount of extracted substances is determined by weighing. To calculate the extractable compounds (FM) in the sample, the following formula was used:

$$\text{FM (mg/L)} = \frac{m_2 - m_1}{V} - \frac{m_4 - m_3}{V}, \quad (2)$$

where m_1 = capsule mass with residue from sample, (mg); m_2 = capsule mass without residue, (mg); m_3 = capsule mass without residue, blank sample, (mg); m_4 = capsule mass with residue from blank sample, (mg); and V = volume of water sample, (L).

Determination of organic matter (OM) is an expression of CCOMn consumption. The method is based on permanganate ability to oxidize organic compounds in acid medium if chlorides content is above 300 mg/L. The working procedure is as follows: over the Wastewater sample add 5 mL of sulfuric acid (diluted 1 : 2) and 10 mL potassium permanganate solution (0.01 N) and then boil for 10 minutes; after that add 10 mL of oxalic acid (0.01 N) and titrate with potassium permanganate (0.01 N) to pale pink lasting 1 minute. The temperature during the titration should by between 50 and 60°C. To calculate the organic matter in the sample, the following formula was used:

$$\text{CCO} - \text{Mn (mg KMnO}_4\text{/L)} = \frac{1000 \times [(V_1 + V_2) \times F - V_3] \times 0.316}{V}, \quad (3)$$

where V_1 = volume of potassium permanganate solution 0.01 N initially added in the sample (10 mL); V_2 = volume of potassium permanganate solution 0.01 N used for sample titration (mL); V_3 = volume of oxalic acid solution 0.01 N added in the sample (10 mL); F = factor of potassium permanganate solution 0.01 N; V = volume of water sample, (mL).

3. Results and Discussion

Wastewater contains solid particles with a wide variety of shapes, sizes, densities, and composition. Specific properties of these particles affect their behaviour in liquid phases and thus the removal capabilities. Many chemical and microbiological contaminants found in Wastewater are adsorbed on or incorporated in the solid particles. Thus, the removal of solid particles and organic matter is essential for purification and recycling of Wastewater [1].

EB irradiation is extremely efficacious in polymerization processes. Polyelectrolytes obtained by electron beam irradiation are branched and exhibit high CC, high η_{intr}, and low M_r, but sometimes low water solubility, especially when η_{intr} is higher than 10 dL/g. In our flocculation studies, we used flocculants (F) having the following physicochemical properties: $CC > 95\%$, a wide variety of η_{intr} between 2 and more than 9 dL/g, small k_H values which ensure good water solubility, and $M_r < 0.05\%$. Intrinsic viscosity of a polymer is indicative of its hydrodynamic volume in solution, which depends on its molecular weight, structure, and nature of the solvent as well as the temperature of the medium. For two polymers with approximately similar molecular weight, the branched polymer has lower hydrodynamic volume compared to its linear counterpart and thus has lower intrinsic viscosity value. Furthermore, long branches determine higher intrinsic viscosity and vice versa [3]. Intrinsic viscosity of polymers is dependent on their molecular weight and chain dimension [31, 32]. In dilute solutions, the polymer chains are separate, so intrinsic viscosity of a polymer in solution depends only on the dimensions of the polymer chain and and such the molecular weight [23, 31].

In Table 2 the characteristics of five flocculants chosen to be used for Wastewater treatment experiments are presented.

In order to provide measurable improvements in water quality, we have studied the effects of different classical treatments with electrolytes and the effects of various combined treatments with electrolytes and flocculants. Classical treatments (CT) were based on the use of $Al_2(SO_4)_3$ and $CaCO_3$, products that are traditionally used in the coagulation-flocculation treatment of wastewaters. For the Wastewater treatment, our interest was focused upon the following quality indicators established by the Romanian Standards concerning the conditions for Wastewater evacuation in the urban sewerage system: total suspended solids (TSS), extractable substances with petroleum ether (fatty matter—FM), and chemical oxygen demand by potassium permanganate method (organic matter—OM). Table 3 presents the results of different classical treatments applied on mentioned quality indicators.

For Wastewater treatment, the poly(acrylamide-*co*-acrylic acid) flocculants are used in very small amounts and their application reduces the amount of flocculants necessary in combined treatments in the range of 20%–40%. We present below the results of flocculation experiments in which we used the poly(acrylamide-*co*-acrylic acid) amount of 4 mg/L together with classic treatment number 2 (CT_2). Although classical treatments lead to good results in terms of quality indicators reducing, we decided to use CT_2 because

TABLE 2: The physicochemical properties of flocculants obtained by EB irradiation.

Sample	Properties of flocculants obtained by EB irradiation					
	CC (%)	η_{intr} (dl/g)	k_H	M_r (%)	M_w	R_g (nm)
Flocculant 1 (F_1)	100	2.77	0.31	0	0.73×10^6	43
Flocculant 2 (F_2)	97.63	4.83	0.17	0.00426	17.01×10^6	69
Flocculant 3 (F_3)	98.91	6.01	0.18	0.00284	23.69×10^6	85
Flocculant 4 (F_4)	98.66	7.94	0.24	0.00284	36.12×10^6	109
Flocculant 5 (F_5)	96.09	9.18	0.17	0.0426	45.01×10^6	124

TABLE 3: The quality indicators values after classic treatment applying.

	Classic treatment type (CT)		Quality indicators		
	$Al_2(SO_4)_3$, mg/dm^3	$CaCO_3$, mg/dm^3	TTS, mg/dm^3	OM, mg $KMnO_4$/dm^3	FM, mg/dm^3
CT_1	100	100	56	224.83	156
CT_2	200	200	74	179.03	154
CT_3	300	300	60	137.4	84
CT_4	400	400	58	124.91	60
CT_5	500	500	60	170.71	66
CT_6	600	600	54	133.23	44
CT_7	700	700	74	195.76	58
	Raw water		676	566.2	1280

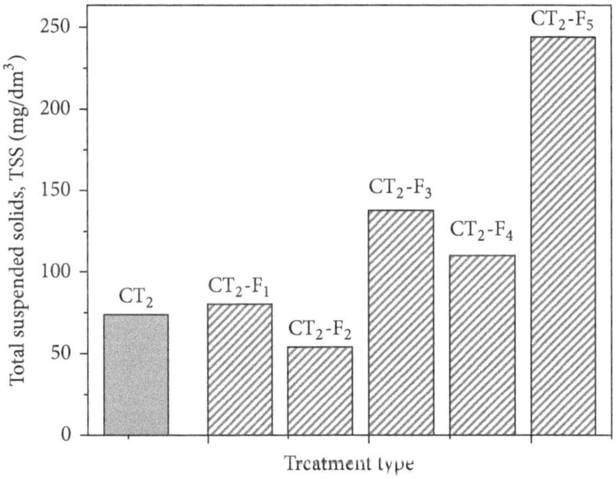

FIGURE 1: The effect of classic treatment CT_2 versus combined treatments $CT_2 + F_{1-5}$ on TSS.

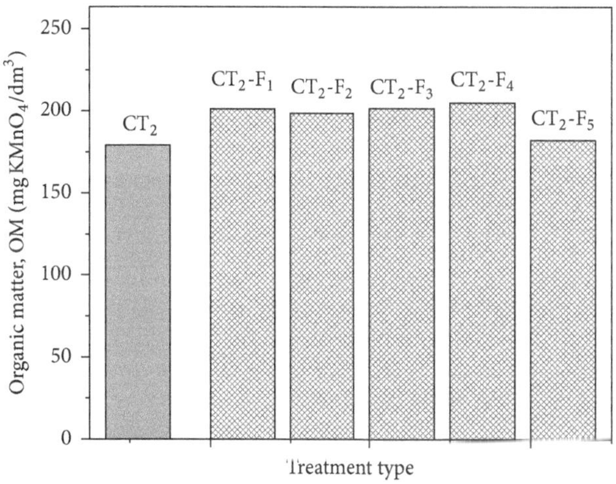

FIGURE 2: The effect of classic treatment CT_2 versus combined treatments $CT_2 + F_{1-5}$ on OM.

it gave relatively good results and the amount of added aluminum sulphate does not generate what other authors call a secondary pollution [3]. In Figures 1–3 the variations of treated water quality indicators TSS, OM and FM as a function of the applied treatment (CT + F) are represented.

Figure 1 shows that combined treatment $CT_2 + F_2$ reduce TSS below the level obtained by applying classical treatment CT_2 only. Flocculant number 2 is a modest one from an intrinsic viscosity point of view, and the result of its use in the combined treatment suggests that to reduce TSS, a high intrinsic viscosity is not required.

In the case of relatively charged Wastewaters, it is difficult to reduce OM quality indicator. This can be seen from the results presented in Figure 2 which show that both classical treatment and all combined treatments gave modest results. However, the combined treatments $CT_2 + F_5$ caused an OM reduction slightly below the one obtained by applying classical treatment CT_2 only. Flocculant number 5 has the highest intrinsic viscosity of all those used in these flocculation experiments. The result obtained after its use as a flocculation adjuvant leads to the idea that flocculant should have a high intrinsic viscosity.

Quality indicator FM is the most difficult to be reduced from all Wastewater quality indicators. All flocculants used in combination with CT_2 greatly reduced this quality indicator, but the best results were obtained from the use of the

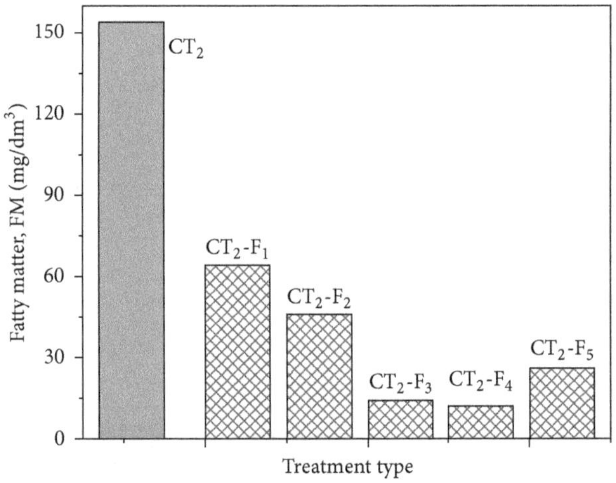

FIGURE 3: The effect of classic treatment CT_2 versus combined treatments $CT_2 + F_{1-5}$ on FM.

combined treatment CT_2-F_3 and CT_2-F_4. The reduction was almost 90% (87% and 89%, resp.). Flocculants F_3 and F_4 have high intrinsic viscosities but not the highest. If the intrinsic viscosity increases, we observe a downward trend in terms of FM reduction efficiency.

4. Conclusions

Poly(acrylamide-*co*-acrylic acid) flocculants obtained by EB irradiation have the capability to provide measurable improvements in Wastewater quality, especially leading to a significant reduction of the FM quality indicator (the most difficult to be reduced), compared with the classical treatments. Are used in very small amounts and their application reduces the necessary amount of classical electrolytes. So, these products exhibit good performance in Wastewater treatment and eliminate shortcomings such as secondary pollution of classical treatments.

References

[1] W. Brostow, S. Pal, and R. P. Singh, "A model of flocculation," *Materials Letters*, vol. 61, no. 22, pp. 4381–4384, 2007.

[2] J. Bratby, "Introduction," in *Coagulation and Flocculation in Water and WasteWater Treatment*, chapter 1, pp. 1–8, IWA, London, UK, 2006.

[3] C. Xie, Y. Feng, W. Cao, H. Teng, J. Li, and Z. Lu, "Novel biodegradable flocculating agents prepared by grafting polyacrylamide to konjac," *Journal of Applied Polymer Science*, vol. 111, no. 5, pp. 2527–2536, 2009.

[4] J. Bratby, "Testing and control of coagulation and flocculation," in *Coagulation and Flocculation in Water and Wastewater Treatment*, chapter 8, pp. 279–290, IWA, London, UK, 2006.

[5] J. K. Edzwald, "Coagulation in drinking water treatment: particles, organics and coagulants," *Water Science and Technology*, vol. 27, no. 11, pp. 21–35, 1993.

[6] J. Fettig, H. Ratnaweera, and H. Odegaard, "Synthetic organic polymers as primary coagulants in wastewater treatment," *Water Supply*, vol. 9, no. 1, pp. 19–26, 1991.

[7] C. L. McCormick, R. D. Hester, S. E. Morgan, and A. M. Safieddine, "Water-soluble copolymers. 31. Effects of molecular parameters, solvation, and polymer associations on drag reduction performance," *Macromolecules*, vol. 23, no. 8, pp. 2132–2139, 1990.

[8] C. L. McCormick, R. D. Hester, S. E. Morgan, and A. M. Safieddine, "Water-soluble copolymers. 30. Effects of molecular structure on drag reduction efficiency," *Macromolecules*, vol. 23, no. 8, pp. 2124–2131, 1990.

[9] C. L. McCormick, S. E. Morgan, and R. D. Hester, "The roles of molecular structure and solvation on drag reduction in aqueous solutions," in *Water Soluble Polymers: Synthesis, Solutions Properties and Applications*, vol. 467 of *ACS Symposium*, chapter 21, pp. 320–327, American Chemical Society, Washington, DC, USA, 1991.

[10] N. Narkis, B. Ghattas, M. Rebhun, and A. J. Rubin, "The mechanism of flocculation with aluminium sALts in combination with polymeric flocculants as flocculant aids," *Water Supply*, vol. 9, no. 1, pp. 37–44, 1991.

[11] P. Selvapathy and M. J. Reddy, "Effect of polyelectrolytes on turbidity removal," *Water Supply*, vol. 10, no. 4, pp. 175–178, 1992.

[12] D. Martin, G. Craciun, E. Manaila et al., "Waste treatment by microwave and electron," in *Proceedings of the 2nd Environmental Physics Conference*, pp. 91–100, 2006.

[13] D. I. Martin, D. I. Ighigeanu, E. N. Mateescu et al., "Combined Microwave and Accelerated Electron Beam Irradiation Facilities for Applied Physics and Chemistry," *IEEE Transactions on Industry Applications*, vol. 40, no. 1, pp. 41–52, 2004.

[14] D. I. Martin, E. Mateescu, G. Craciun, D. Ighigeanu, and A. Ighigeanu, "Polymeric flocculants processing by accelerated electron beams and microwave heating," *Radiation Physics and Chemistry*, vol. 64, no. 5-6, pp. 423–428, 2002.

[15] E. Manaila, D. Martin, G. Craciun, D. Ighigeanu, C. Oproiu, and N. Iacob, "IAEA-CN-115-49 Electron beam processed polyelectrolytes," in *Proceedings of International Symposium on Utilization of Accelerators (CD-ROM)*, pp. 1–8, Dubrovnik, Croatia, 2005.

[16] D. Martin, M. Fiti, L. A. Radu et al., "Low power-high energy linacs for irradiation in polymeric systems," *Radiation Physics and Chemistry*, vol. 45, no. 4, pp. 615–621, 1995.

[17] D. Martin, M. Dragusin, A. Radu et al., "IAP linacs in applied research," *Nuclear Instruments and Methods in Physics Research B*, vol. 113, no. 1–4, pp. 106–109, 1996.

[18] D. Martin, M. Dragusin, M. Radoiu et al., "Polymers for waste water treatment," *Progress in Colloid and Polymer Science*, vol. 102, pp. 147–151, 1996.

[19] D. Martin, M. Radoiu, I. Calinescu et al., "Combined electron beam and microwave treatment for flue gas purification," *Materials and Manufacturing Processes*, vol. 14, no. 3, pp. 365–382, 1999.

[20] M. J. Caulfield, X. Hao, G. G. Qiao, and D. H. Solomon, "Degradation on polyacrylamides. Part II. Polyacrylamide gels," *Polymer*, vol. 44, no. 14, pp. 3817–3826, 2003.

[21] M. Dimonie, C. Boghina, C. Cincu, M. Marinescu, and N. Marinescu, *Poliacrilamida*, Editura Tehnica, Bucuresti, Romania, 1986.

[22] R. K. Richardson and S. Kasapis, "Rheological methods in the characterisation of food biopolymers," *Developments in Food Science C*, vol. 39, pp. 1–48, 1998.

[23] M. A. Rao, "Introduction," in *Rheology of Fluid and Semisolid Foods*, pp. 1–24, Aspen Publishers, Gaithersburg, Md, USA, 1999.

[24] M. E. Zeynali and A. Rabbii, "Alkaline hydrolysis of polyacrylamide and study on poly(acrylamide-*co*-sodium acrylate) properties," *Iranian Polymer Journal*, vol. 11, no. 4, pp. 269–275, 2002.

[25] S. Pal, D. Mal, and R. P. Singh, "Synthesis and characterization of cationic guar gum: a high performance flocculating agent," *Journal of Applied Polymer Science*, vol. 105, no. 6, pp. 3240–3245, 2007.

[26] R. L. Shogren, "Flocculation of kaolin by waxy maize starch phosphates," *Carbohydrate Polymers*, vol. 76, no. 4, pp. 639–644, 2009.

[27] P. M. Patterson and A. M. Jamieson, "Molecular weight scaling of the transport properties of polyacrylamide in water," *Macromolecules*, vol. 18, no. 2, pp. 266–272, 1985.

[28] S. Pal, D. Mal, and R. P. Singh, "Cationic starch: an effective flocculating agent," *Carbohydrate Polymers*, vol. 59, no. 4, pp. 417–423, 2005.

[29] A. Jha, S. Agrawal, A. Mishra, and J. P. Rai, "Synthesis, characterization and flocculation efficiency of poly(acrylamide-*co*-acrylic acid) in tannery waste-water," *Iranian Polymer Journal*, vol. 10, no. 2, pp. 85–90, 2001.

[30] Y.-H. Bae, H.-J. Kim, E.-J. Lee, N.-C. Sung, S.-S. Lee, and Y.-H. Kim, "Potable water treatment by polyacrylamide base flocculants coupled with an inorganic coagulant," *Environmental Engineering Research*, vol. 12, no. 1, pp. 21–29, 2007.

[31] H. A. Khouryieh, T. J. Herald, F. Aramouni, and S. Alavi, "Intrinsic viscosity and viscoelastic properties of xanthan/guar mixtures in dilute solutions: effect of salt concentration on the polymer interactions," *Food Research International*, vol. 40, no. 7, pp. 883–893, 2007.

[32] P. J. Flory, "The structure of vinyl polymers," in *Principles of Polymer Chemistry*, chapter 6, pp. 308–314, Cornell University Press, New York, NY, USA, 1953.

15

Processing and Properties of Natural Fiber-Reinforced Polymer Composite

Jyoti Prakash Dhal[1] and S. C. Mishra[2]

[1] *Department of Chemistry, National Institute of Technology, Rourkela, Odisha 769008, India*
[2] *Department of Metallurgical & Materials Engineering, National Institute of Technology, Rourkela, Odisha 769008, India*

Correspondence should be addressed to Jyoti Prakash Dhal; jyoti84.chem@gmail.com

Academic Editor: Víctor M. Castaño

A novel low cost polymer composite using brown grass flower broom reinforcement is prepared. The prepared polymer composite has the lowest porosity, homogeneous surface structure, and the greatest interface bonding. From the physico-mechanical characterization such as: hardness measurement, density measurement, void fraction or porosity measurement, and flexural strength measurement, it is found that the prepared composite is of light weight and high strength. Again, from dielectric behaviour of this polymer composite, it is found that this material has an efficiency that is considered as a high valued marketable product. As the composite is made using bio-materials from local resources, its cost is less compared to other polymer composites available today.

1. Introduction

Natural fiber-reinforced polymer composites have raised great attention and interest among materials scientists and engineers in recent years due to that the composites give a combination of superior mechanical property, dielectric property, and environmental advantages such as renewability and biodegradability. Due to various disadvantages such as: high progressing technologies, rising prices of finite resources, and ecounfriendly, the conventional petroleum-based plastic, glass or carbon fiber materials are compensated by natural/bi-based fibers. These fiber composites are well suited as wood substitutes in the housing and construction sector. Using such natural/biofibers with polymers based on renewable resources will allow many environmental issues to be solved. The various natural fibers such as jute, coir, sisal, pineapple, ramie, bamboo, and banana are used as reinforcement of polymer composite, nowadays [1–12]. In recent years, a number of investigations have been made which prove that the worth of natural fibers against their synthetic counterparts such as glass and/or carbon fiber-reinforced polymer composites [13–16].

The potential of fiber-reinforced polymer composites was recognized more than 50 years ago, now they can find their applications in almost every industry including construction, aerospace, automotive, and electronics. Composite materials are increasingly used for dielectric applications, that is, applications that make use of electrically insulating or nearly insulating behaviour. This is because of the need of the electronic industry for dielectric materials in electrical insulation, encapsulation, multilayer ceramic chip, and capacitors and for piezoelectric, ferroelectric, and pyroelectric devices that provide sensing, actuation, and so forth. Development of dielectric material with low dielectric constant k and low dielectric loss is considered to be one of the main issues in high-speed microelectronics. The dielectric constant of any material depends upon the polarizability of its molecules and is determined by different contributions: interfacial, dipole, atomic, and electronic polarizations. Interfacial polarization in the composite influences the dielectric properties at very low frequencies and usually decreases with increasing frequency [17–19].

Developing an efficient and light weight dielectric material with high strength from sustainable resources, such as

FIGURE 1: Brown grass flower broom.

FIGURE 2: Polymer composites of brown grass flower broom and epoxy.

brown grasses broom fibre, is quite attractive from both application and environmental point of view. This paper aims at development of a biofibre-reinforced polymer composite using natural fibre, that is, brown grass flower broom which is lightweight, commonly available, biodegradable, and low cost. The raw material is commonly available in agricultural sector.

2. Materials and Methods

2.1. Materials Used. Epoxy LY 556 (common name: bisphenol A diglycidyl ether) is used as matrix material in the composite fabrication. The hardener used here is HY-951(IUPAC name: NN′-bis (2-aminoethylethane-1, 2-diamin). Epoxy Resin and hardeners are mixed in a ratio of 10 : 1 by weight. Brown grass flower broom (shown in Figure 1) is commonly used in house. Short fiber of the broom that is nearly 10 mm length was prepared and used as the reinforcing agent in the composite preparation.

2.2. Preparation of Test Samples. The biofiber (brown grasses fiber) was mixed with the epoxy by stirring at room temperature, in a glass beaker with the help of suitable glass rod. Hardener was added into the beaker containing mixture at the time of stirring. With proper stirring for 10 minutes, uniform mixing of the reinforcing agent and the polymer matrix were possible. Proper stirring was required for uniform mixing of the reinforcing agent and the polymer matrix and they were poured into suitable moulds to obtain disc-shaped samples of 12 mm diameter and 2.5 mm thickness. Three different broom fiber-reinforced epoxy composites were fabricated varying the amount of reinforcement. Test specimens of suitable dimensions are cut from the composite (shown in Figure 2). The different composites prepared are described in Table 1.

The surface morphology of the sample (Epoxy + 10% brown grass flower broom) was examined with JEOL T-330 Scanning Electron Microscope. Samples were coated with 60 Å thick platinum in JEOL sputter ion coater for surface conductivity and then observed under SEM, operated at an acceleration voltage of 20.0 kV. Hardness test was conducted in a Vickers hardness tester, Leitz, Germany. The Vickers hardness numbers (VHN) of the hybrid composite were measured under a load of $F = 0.3$ Kgf and Vickers hardness number was calculated by using the formula:

$$\mathrm{HV} = 0.1889\frac{F}{L^2}, \qquad L = \frac{X+Y}{2}, \tag{1}$$

where F is the applied load, L is the diagonal of square impression (mm), X is the horizontal length (mm), and Y is the vertical length (mm). The density of neat epoxy and the composites were measured by measuring its mass and volume. The bulk density and void fraction of the composite materials will be obtained from following equations by using Archimedes' principle:

$$\text{density} = \frac{\text{dry weight}}{\text{soaked weight} - \text{suspended weight}},$$

$$\text{void fraction or porocity} = \frac{\text{soaked weight} - \text{dry weight}}{\text{soaked weight} - \text{suspended weight}} \times 100, \tag{2}$$

where dry weight is the weight of the sample at completely dried condition, soaked weight is the weight of the sample that soaked in kerosene oil, and suspended weight is the weight of the sample suspended in the oil through a string. To evaluate the value of flexural strength (FS), the short-beam shear (SBS) tests (3-point bend test) were performed on the samples at room temperature. The SBS test was conducted as per ASTM D2344-84 using the Instron 1195 UTM. The span length was 40 mm, and the cross head speed was 5 mm/min. The FS of any composite can be calculated by using the following formula:

$$\mathrm{FS} = \frac{3FL}{2bt^2}, \tag{3}$$

where F is the applied load, L is span length, and t and b are the thickness and width of the specimen, respectively.

Dielectric measurements were carried out with the help of a Solartron 1296 Impedance Analyser. For that, the samples

Table 1: Test samples prepared.

Samples	Compositions
Sample A	Pure epoxy
Sample B	Epoxy + 10% brown grass flower broom
Sample C	Epoxy + 20% brown grass flower broom

Table 2: Variation of density with different weight percent of fiber reinforced in epoxy matrix.

Sample	Fiber (%)	Density (gm/cc)
Sample A	Nil	1.92
Sample B	10	1.852
Sample C	20	1.614

of the composites had to be cut into thin circular shape, and their surfaces were polished. Then graphite coating was given on their surfaces to make surface conducting and for allowing measurements over frequency interval from 100 Hz to 1 MHz. The dielectric constant and dielectric loss were determined as follows.

$$\text{Dielectric constant}\,(k) = \frac{C'}{C}, \tag{4}$$

where C' (pF) is the measured capacitance, and C (pF) is calculated using the equation

$$C = \varepsilon_0 \left(\frac{A}{d}\right), \tag{5}$$

where A is area of the electrode (mm^2) and d is the thickness of the sample (mm).

The dielectric loss is given by

$$\tan\delta = \frac{G\,(S)}{wC'\,(F)}, \tag{6}$$

where $w = 2\pi f$, f is the measuring frequency, and $G = G_0(R - R_0)$ [20].

3. Results and Discussion

The surface morphology of the prepared test sample with 10% broom fiber is shown in Figure 3. The interface bonding of the broom fiber with epoxy resin is clearly visible. This composite has lowest porosity and homogeneous surface structure.

Hardness values of the samples are given in Figure 4. From the figure, it can be seen that the hardness of broom fiber-reinforced polymer composite is more than that of the pure epoxy and also increases with the increase in amount of reinforcement. The increase in hardness of the composite may be due to stronger interface bonding of the broom fiber with epoxy resin.

The actual density of the composite is determined experimentally by simple water immersion technique by using Archimedes principle. The densities of the samples are given in Table 2. It is observed that the presence of broom fibres

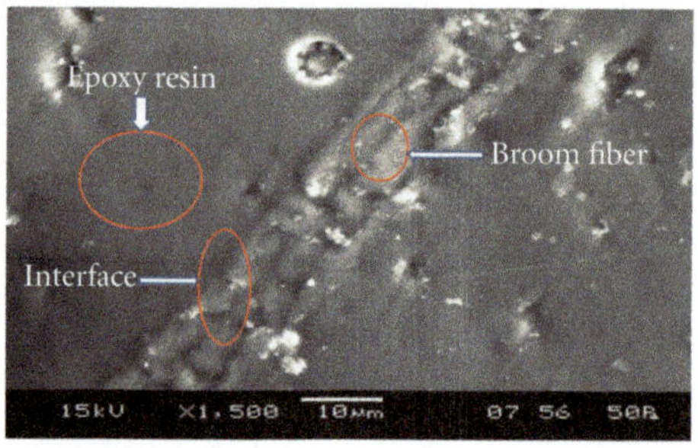

Figure 3: Surface morphology of the composite with 10% of broom fiber reinforcement.

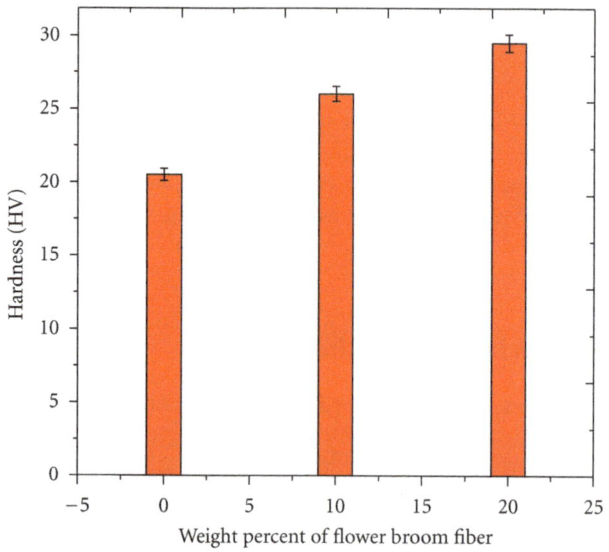

Figure 4: Variation of hardness (HV) of composites with weight percent of flower broom fiber-reinforced epoxy composites.

as fillers in epoxy matrix reduces the density of polymer composite and hence makes it lightweight. This may be due to presence of the high air content.

With addition of flower broom fiber in epoxy resin, the volume fraction of voids is increased as shown in Figure 5. The voids significantly affect some of the mechanical properties and even the performance of the composites in the place of use.

Figure 6 shows the comparison of flexural strengths of the flower broom-reinforced epoxy composites. There is an increase in the flexural strength that is associated with an increase in wt% of short fiber flower broom in composites. This may be due to the presence of stronger interfacial bonds (as shown in SEM figure). Impregnations of natural fibre help in the interface bonding which improve the structural properties of the composite. The micromechanical events that occur for a long fiber-reinforced composite are not the same as those observed for a short fiber reinforced composites. In a short fiber, there are variations in stress distribution along the fiber matrix interface, and end effects can be neglected in the case of long fibers, but they can be very important in the case of short fiber-reinforced composites.

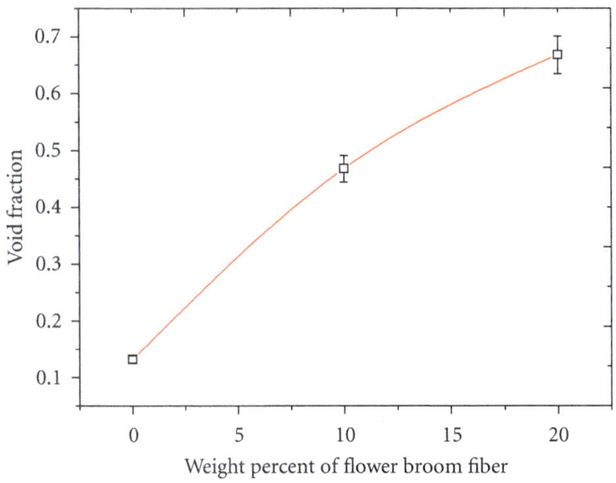

FIGURE 5: Variation of void fraction of composite with weight % of broom fiber-reinforced epoxy composites.

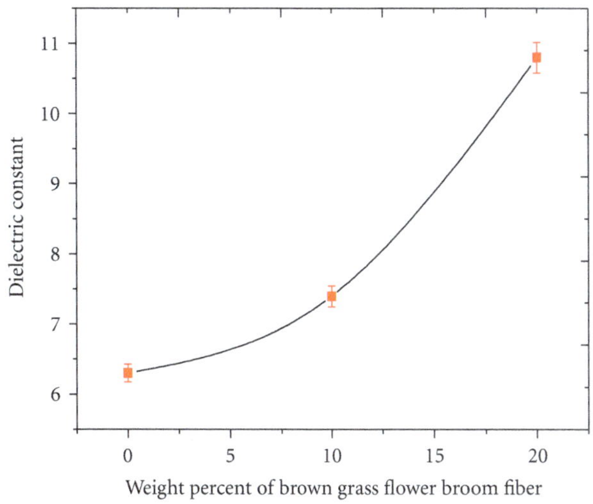

FIGURE 7: Variation of dielectric constant with amount of fibre reinforcement.

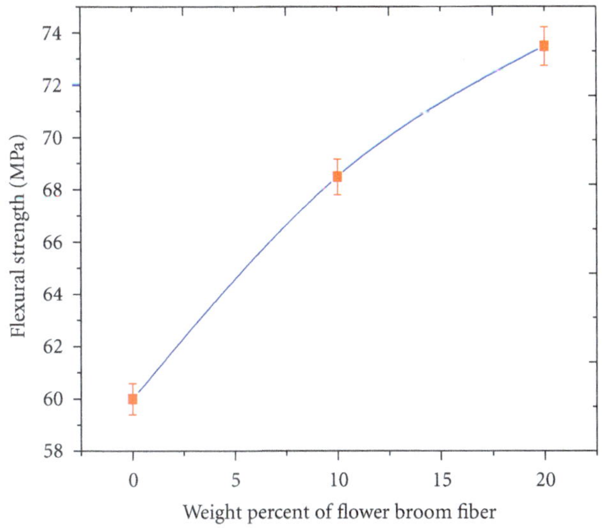

FIGURE 6: Variation of flexural strength of composites with weight percent of flower broom fiber-reinforced epoxy composites.

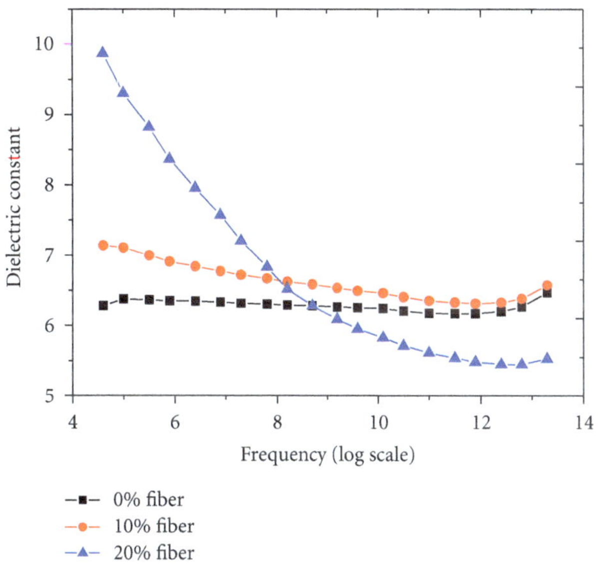

FIGURE 8: Variation of dielectric constant with frequency.

It is observed from Figure 7 that the dielectric constant (at 100 Hz frequency) increases with the volume fraction reinforcement. However, from Figures 8 and 9, it is inferred that the dielectric constant and the dielectric loss initially reduce and attain a steady state with an increase in frequency. This may be due to the fact that (i) dielectric behavior is dependent on porosity, (ii) on material properties, and also on (iii) interface bonding in case of composite materials. So in this study the materials used are having many diverse physico-mechanical properties. However, it is found that making a composite with these wastes, using a polymer binder, is best suited for providing good mechanical strength without sacrificing its dielectric property. Impregnation of natural fibre helps in the interface bonding and distribution of absorbed moisture in the material which may be one of the reasons for change in dielectric properties. With increase in frequency, the dielectric constant of the composites decreases due to dielectric relaxation. From structural point of view, the dielectric relaxation involves oriental polarization which in turn depends on molecular arrangement of the dielectric material. At high frequency, the rotational motion of polar molecule is not sufficiently rapid for attainment of equilibrium with applied field, hence dielectric constant decreases. As the reinforcement content increases, the dielectric constant also increases. Dielectric loss of the composite shows a stabilizing trend with an increase in frequency which appears to be a beneficial from application point of view.

4. Conclusions

Brown grass flower broom fiber-reinforced epoxy composites can be prepared easily with different weight percentage of reinforcement. The polymer composite cost will be low as

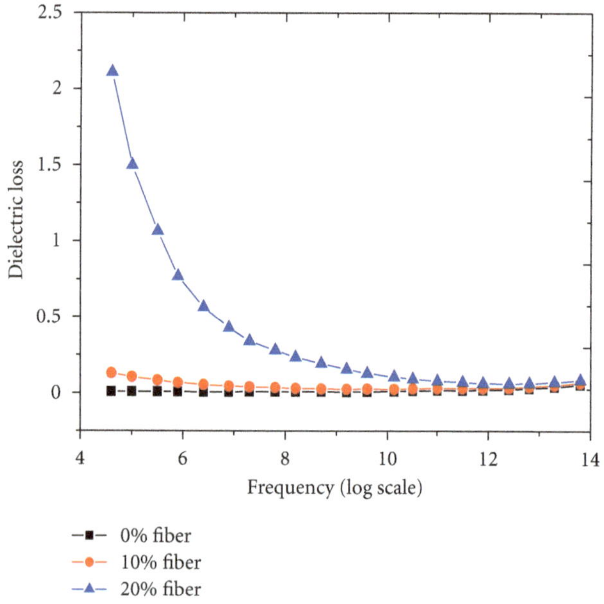

FIGURE 9: Variation of dielectric loss with frequency.

the raw material is plant fibre based, and the processing technique is simple. This composite has the lowest porosity and homogeneous surface structure, and a strong interface bonding of the broom fiber with epoxy resin is present. The void fraction of the composites increases slightly with the increase in the reinforcement content. It might be due to the presence of pores/cavities in the fibre. There is an increase in the flexural strength along with the increase in weight percent of short fiber flower broom in composites. Moreover the material is of light weight and possesses high strength. With the increase in reinforcement content, the density of the composite decreases, and the hardness of the composite increases, and hence make it of light weight and high of strength. Dielectric properties (relative permittivity and loss factor) of pure epoxy resin and composites with different amount of broom fiber-reinforced polymer composite have been studied in the frequency range from 100 Hz to 1 MHz. The experimental results indicate that the dielectric constant and dielectric loss factor decrease with the increasing frequency. It may/can be due to the orientation polarization, which has increased with increasing temperature, due to greater movement of polar molecular dipoles appears to be beneficial in electronic industry and can be processed by village artisans also. As the composite is made using bio-materials from local resources, its cost is less, compared to other polymer composites available today. This can further open up a new frontier for the industrialization at rural sectors.

References

[1] H. Peltola, B. Madsen, R. Joffe, and K. Nättinen, "Experimental study of fiber length and orientation in injection moulded natural fiber/starch acetate composites," *Advances in Materials Science and Engineering*, vol. 2011, Article ID 891940, 7 pages, 2011.

[2] M. H. Ab Ghani and S. Ahmad, "The comparison of water absorption analysis between counterrotating and corotating twin-screw extruders with different antioxidants content in wood plastic composites," *Advances in Materials Science and Engineering*, vol. 2011, Article ID 406284, 4 pages, 2011.

[3] A. Gupta, A. Kumar, A. Patnaik, and S. Biswas, "Effect of different parameters on mechanical and erosion wear behavior of bamboo fiber reinforced epoxy composites," *International Journal of Polymer Science*, vol. 2011, Article ID 592906, 10 pages, 2011.

[4] S. Kalia, L. Avérous, J. Njuguna, A. Dufresne, and B. M. Cherian, "Natural fibers, bio- and nanocomposites," *International Journal of Polymer Science*, vol. 2011, Article ID 735932, 2 pages, 2011.

[5] T. Williams, M. Hosur, M. Theodore, A. Netravali, V. Rangari, and S. Jeelani, "Time effects on morphology and bonding ability in mercerized natural fibers for composite reinforcement," *International Journal of Polymer Science*, vol. 2011, Article ID 192865, 9 pages, 2011.

[6] S. C. Mishra, "Low cost polymer composites with rural resources," *Journal of Reinforced Plastics and Composites*, vol. 28, no. 18, pp. 2183–2188, 2009.

[7] S. K. Acharya and S. C. Mishra, "Weathering behavior of fly-ash jute polymer composite," *Journal of Reinforced Plastics and Composites*, vol. 26, no. 12, pp. 1201–1210, 2007.

[8] J. George, M. S. Sreekala, and S. Thomas, "A review on interface modification and characterization of natural fiber reinforced plastic composites," *Polymer Engineering and Science*, vol. 41, no. 9, pp. 1471–1485, 2001.

[9] S. Varghese, B. Kuriakose, and S. Thomas, "Stress relaxation in short sisal-fiber-reinforced natural rubber composites," *Journal of Applied Polymer Science*, vol. 53, no. 8, pp. 1051–1060, 1994.

[10] Y. Mi, X. Chen, and Q. Guo, "Bamboo fiber-reinforced polypropylene composites: crystallization and interfacial morphology," *Journal of Applied Polymer Science*, vol. 64, no. 7, pp. 1267–1273, 1997.

[11] F. M. B. Coutinho, T. H. S. Costa, and D. L. Carvalho, "Polypropylene-wood fiber composites: effect of treatment and mixing conditions on mechanical properties," *Journal of Applied Polymer Science*, vol. 65, no. 6, pp. 1227–1235, 1997.

[12] H. K. Mishra and S. C. Mishra, "Erosion wear behaviour of coir dust reinforced polymer composite," *Orissa Journal of Physics*, vol. 18, no. 1, pp. 97–108, 2011.

[13] S. Joseph, M. S. Sreekala, Z. Oommen, P. Koshy, and S. Thomas, "A comparison of the mechanical properties of phenol formaldehyde composites reinforced with banana fibres and glass fibres," *Composites Science and Technology*, vol. 62, no. 14, pp. 1857–1868, 2002.

[14] N. Sombatsompop, K. Chaochanchaikul, C. Phromchirasuk, and S. Thongsang, "Effect of wood sawdust content on rheological and structural changes, and thermo-mechanical properties of PVC/sawdust composites," *Polymer International*, vol. 52, no. 12, pp. 1847–1855, 2003.

[15] H. Djidjelli, J. J. Martinez-Vega, J. Farenc, and D. Benachour, "Effect of wood flour content on the thermal, mechanical and dielectric properties of poly(vinyl chloride)," *Macromolecular Material Engineering*, vol. 287, no. 9, pp. 611–618, 2002.

[16] H. Djidjelli, M. Kaci, J. J. Martinez-Vega, and D. Benachour, "Effects of hydrothermal ageing on the thermal behaviour of poly(vinyl chloride) filled with wood flour," *Polymer International*, vol. 53, no. 11, pp. 1760–1765, 2004.

[17] S. C. Mishra, N. B. Nayak, and A. Satapathy, "Investigation on bio-waste reinforced epoxy composites," *Journal of Reinforced Plastics and Composites*, vol. 29, no. 19, pp. 3016–3020, 2010.

[18] S. C. Mishra and H. Aireddy, "Evaluation of dielectric behavior of bio-waste reinforced polymer composite," *Journal of Reinforced Plastics and Composites*, vol. 30, no. 2, pp. 134–141, 2011.

[19] S. C. Mishra and N. B. Nayak, "An investigation of dielectric properties of chicken feather reinforced epoxy matrix composite," *Journal of Reinforced Plastics and Composites*, vol. 29, no. 17, pp. 2691–2697, 2010.

[20] A. Arbelaiz, B. Fernández, G. Cantero, R. Llano-Ponte, A. Valea, and I. Mondragon, "Mechanical properties of flax fibre/polypropylene composites. Influence of fibre/matrix modification and glass fibre hybridization," *Composites Part A*, vol. 36, no. 12, pp. 1637–1644, 2005.

16

On the Prediction of Strength from Hardness for Copper Alloys

S. Chenna Krishna, Narendra Kumar Gangwar, Abhay K. Jha, and Bhanu Pant

Materials Processing and Characterization Group, Vikram Sarabhai Space Centre, Trivandrum 695 022, India

Correspondence should be addressed to S. Chenna Krishna; chenna.sk@gmail.com

Academic Editor: Steve Bull

Hardness and strength values of over 55 copper alloys strengthened by solid solution strengthening, precipitation hardening, cold working, and dispersion strengthening were compiled. The yield strength (YS) and ultimate tensile strength (UTS) values of the copper alloys examined ranged between 50 to 1300 MPa and 200 to 1400 MPa, respectively. The compiled values were classified based on strain-hardening potential an indirect method to understand the effect of strain-hardening characteristics. Least squares regression analysis was employed to establish correlations between strength and Vickers hardness values. Strain-hardening potential showed a significant effect on the correlations. In all the cases, a linear relation was obtained for both YS and UTS with hardness for the entire range of values under analysis. Simple empirical equations were proposed to estimate the strength using bulk hardness. The proposed correlations obtained for the entire range of values were verified with experimental values. A good agreement was observed between experimental and predicted values.

1. Introduction

Ever since indentation hardness testing has come into existence, there were studies to estimate other mechanical properties especially ultimate tensile strength and yield strength from bulk hardness measurement [1–3]. Over the last 70 years, a number of researchers have worked on experimental techniques and theoretical relations to determine the strength from hardness. Such relations can be useful in the design, where direct measurement of tensile properties is not viable. These relations are always attractive as they bring down the number of tests to be conducted to ensure the quality of the materials [4, 5]. As these methods are fast and relatively non-destructive in nature they are effectively used in failure analysis [5, 6].

Tabor [1] has shown that the ratio of ultimate tensile strength to the Vickers hardness is related to strain-hardening coefficient. The relation between these parameters was expressed using

$$\text{UTS} = \left(\frac{H}{C}\right)(1-n)\left(\frac{12.5n}{(1-n)}\right)^{n}, \quad (1)$$

where UTS is the ultimate tensile strength, H is the Vickers hardness, n is the strain-hardening coefficient, and C is a constant which has a value of 2.9 for steels and 3.0 for copper alloys.

This relation was further improved by Cahoon [2] in the form of

$$\text{UTS} = \left(\frac{H}{2.9}\right)\left(\frac{n}{0.217}\right)^{n}, \quad (2)$$

where H is the Vickers hardness and n is the strain-hardening coefficient.

Cahoon et al. [3] proposed relation to estimate the yield strength using simple hardness measurement for copper, aluminum, and steel in the form of

$$\text{YS} = \left(\frac{H}{3}\right)\left(\frac{1}{10}\right)^{n}. \quad (3)$$

All the expressions mentioned previously require the knowledge of strain-hardening coefficient, which can be directly determined from uniaxial tensile or compression test and indirectly through Meyers hardness measurement and empirical relations [4]. Apart from these, few methods were proposed for estimation of strength for metals using an indentation technique in the literature [4–9].

Pavlina and Van Tyne [10] suggested simple linear relationships to estimate the ultimate tensile strength and yield strength using the Vickers hardness number for steels as follows:

$$YS = -90.7 + 2.876\,Hv, \quad (4)$$

$$UTS = -99.8 + 3.734\,Hv, \quad (5)$$

where strength has units of MPa and Hv has units of kg/mm^2. These relationships do not require the knowledge of any other parameter than hardness for estimation of strength.

Even though extensive work was conducted to estimate the ultimate tensile strength and yield strength from hardness measurements for certain metals and alloys, there appears that no attempts were made to obtain such relations for copper alloys. Most of the correlations available were obtained by analysis of data for steels [1–10]. Therefore, an attempt has been made to propose correlations to estimate the strength from hardness for copper alloys.

2. Materials and Testing

In the present investigation, hardness and strength values for copper alloys were collected from the literature. For comparison, few hardness values were converted from HB and HRB to Vickers hardness as per ASTM E140-07 [11]. All the hardness and strength values compiled were divided based on strain-hardening potential as low, medium, and high UTS/YS ratios. Least squares regression analysis was employed to obtain simple expressions, to predict the strength from hardness.

In addition to the literature values, hardness and strength were evaluated for some copper alloys for validation of empirical relations obtained. These measurements were made in different heat treatment, cold-worked, and hot-worked conditions. Samples for hardness were ground and polished with emery paper prior to hardness measurement. Seven readings were taken for each condition and average is reported. The hardness was measured using a Vickers hardness tester (make: FIE FE-20) with a load of 10 kgf. Tensile properties were evaluated with dog-bone shaped flat tensile specimens (25 mm gauge length) using an INSTRON 5500R UTM at a strain rate of $10^{-3}\,s^{-1}$. Three tests were performed for each condition and average is reported.

3. Results

3.1. Entire Data Range. The relationship between compiled ultimate tensile strength and hardness values (120 data points) for different copper alloys is shown in Figure 1. A linear correlation could be observed in most of the hardness ranges of materials tested, except at low hardness. The linear trend line is made to pass through the origin as shown in Figure 1. Regression analysis of the data points yielded a linear relation as

$$UTS = 3.353 * VHN, \quad (6)$$

where ultimate tensile strength is in MPa and VHN is Vickers hardness number. Regression analysis for (6) gives a coefficient of determination (R^2) of 0.933. Even though a linear correlation has been obtained from the entire range of the hardness, a deviation was observed for hardness lower than 110 VHN.

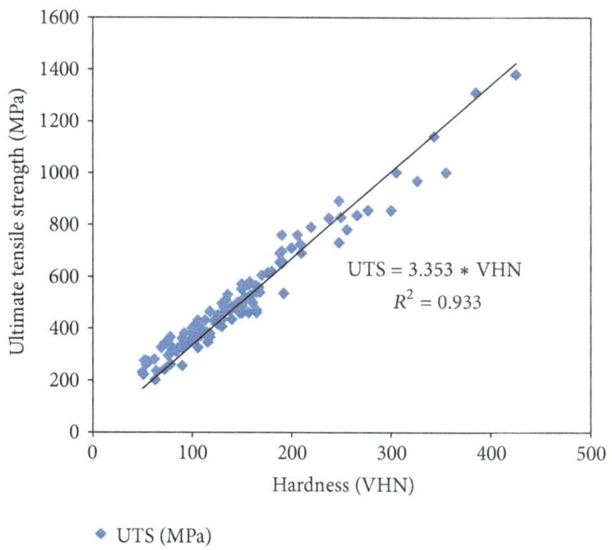

FIGURE 1: Correlation between hardness and ultimate tensile strength for copper alloys [12–28].

Correlation obtained for yield strength and hardness values for copper alloys also depicted a linear relationship as shown in Figure 2. Regression analysis of the data points gave a linear correlation for ultimate tensile strength as

$$YS = 2.874 * VHN, \quad (7)$$

where YS is in MPa and VHN is Vickers hardness number. Coefficient of determination (R^2) for (7) has a value of 0.916. From Figure 2, it can be ascertained that yield strength has a linear correlation with hardness, with correlation tending to nonlinearity at hardness values less than 110 VHN.

3.2. Data Classified by UTS to YS Ratio. Since the compiled data taken from the literature comprises a large number of alloys, strain-hardening coefficient could not be obtained for all the alloys. Therefore, strain-hardening potential determined as the ratio of UTS to YS, which is a measure of maximum strengthening that can occur in a material beyond its yield strength and continues until UTS, was used in the study [10]. The ratio of these two parameters (UTS/YS) was employed to study the effect of strain-hardening on the material indirectly. To determine the effect of strain hardening on the compiled values, the data were classified into three groups: (i) low UTS/YS ratio (UTS/YS < 1.20), (ii) medium UTS/YS ratio (1.20 < UTS/YS < 1.52), and (iii) high UTS/YS ratio (UTS/YS > 1.52). Figures 3(a)–3(c) show the plots of strength and hardness values with the trend line obtained by regression analysis. The values of the regression coefficient and R^2 are given in Table 1. Even though the data were fitted by a linear equation for all three plots in Figure 3, a variation was observed in the values of R^2. It is noteworthy to mention that best fit was observed in the case of low UTS/YS

TABLE 1: Results of regression analysis for hardness-strength data of alloys.

Data type	Property	Regression equation	R^2	Valid range (VHN)	Data points
All	YS	YS = 2.874 * VHN	0.916	50–425	120
	UTS	UTS = 3.353 * VHN	0.933	50–425	120
Medium UTS/YS	YS	YS = 2.594 * VHN	0.887	50–220	30
	UTS	UTS = 3.389 * VHN	0.934	50–220	30
High UTS/YS	YS	YS = 1.97 * VHN	0.845	50–142	26
	UTS	UTS = 3.865 * VHN	0.665	50–142	26
Low UTS/YS	YS	YS = 3.018 * VHN	0.933	78–425	64
	UTS	UTS = 3.301 * VHN	0.917	78–425	64

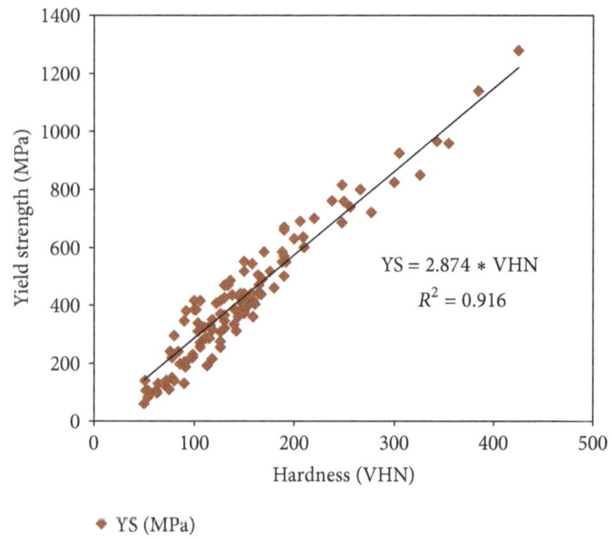

FIGURE 2: Correlation between hardness and yield strength for copper alloys [12–28].

ratio with a coefficient of determination of 0.933 and 0.917 for UTS-hardness and YS-hardness, respectively.

4. Discussion

The results of the regression analysis obtained for different data groups are given in Table 1. It is seen that all the data analyzed could be modelled by linear equations. However, some deviation from the linear trend line was observed in all the cases as shown in Figures 1–3. This can be attributed to the fact that a wide range of copper alloys with different thermomechanical and thermal history were analyzed in the present work. In addition, it may be due to the fact that the hardness measurement is done on samples with varying surface roughness using different indentation load, as both affect the measured hardness. It is noteworthy to mention that the lower range of hardness showed a deviation from linearity for both yield strength and ultimate tensile strength in the data groups analyzed as shown in Figures 1–3. It can be observed that UTS-hardness showed better correlation with R^2 value of 0.9333 with regression constant of 3.05. On the other hand, YS-hardness showed a regression coefficient and coefficient of determination of 2.874 and 0.916, respectively. The regression coefficient obtained for copper alloys in the case of YS-hardness is similar to that reported for steels with a value of 2.876 [10]. Zinkle et al. [23] proposed linear correaltions for AMZIRC and AMAX-MZC between YS and hardness with regression constant of 3.03 and 3.0, respectively.

Empirical relations obtained for hardness strength classified by strain-hardening potential showed a significant effect on the correlations obtained for the data as evident from Table 1. Copper alloys with high strain-hardening potential showed a large deviation from the linear trend line for both YS and UTS as shown in Figure 3(a). This data group to a larger extent consists of alloys in solution treated condition and solid solution strengthened alloys in annealed condition. The hardness values of the alloy analyzed for this group were in the range of 50–140 VHN. On the other hand, data groups with with medium and low UTS/YS showed better correlations for strength hardness with minimal deviations from the linear trend line as shown in Figures 3(b) and 3(c). Among the three data groups analyzed, low UTS/YS group showed the best correlation with regression coefficient of 3.301 and 3.018 for UTS-hardness and YS-hardness, respectively. From the plot, it is also apparent that both trend lines lie very close and have similar behavior. This data group consists alloys that are subjected to cryorolling, cold rolling, and cold rolling + aging. The observed effect of strain hardening on the correlations may be due to wide differences in cold-working that occurs before hardness testing and during the test itself for annealed alloys. An annealed material with high strain-hardening potential will harden much more during the hardness test than a cold-worked metal. Therefore, cold-worked materials yield better correlations for hardness strength.

The predicted values of UTS and YS from hardness using (6) and (7) are compared with the experimental values as shown in Figure 4. The predicted values are plotted as a line and experimental values are indicated by symbols. It can be seen that a reasonably good agreement is observed between experimental and predicted values for UTS whereas, experimental YS showed a reasonable deviation from predicted values at hardness lower than 120 VHN. Nevertheless, the experimental UTS values showed deviation at higher hardness as evident from Figure 4. It was observed that

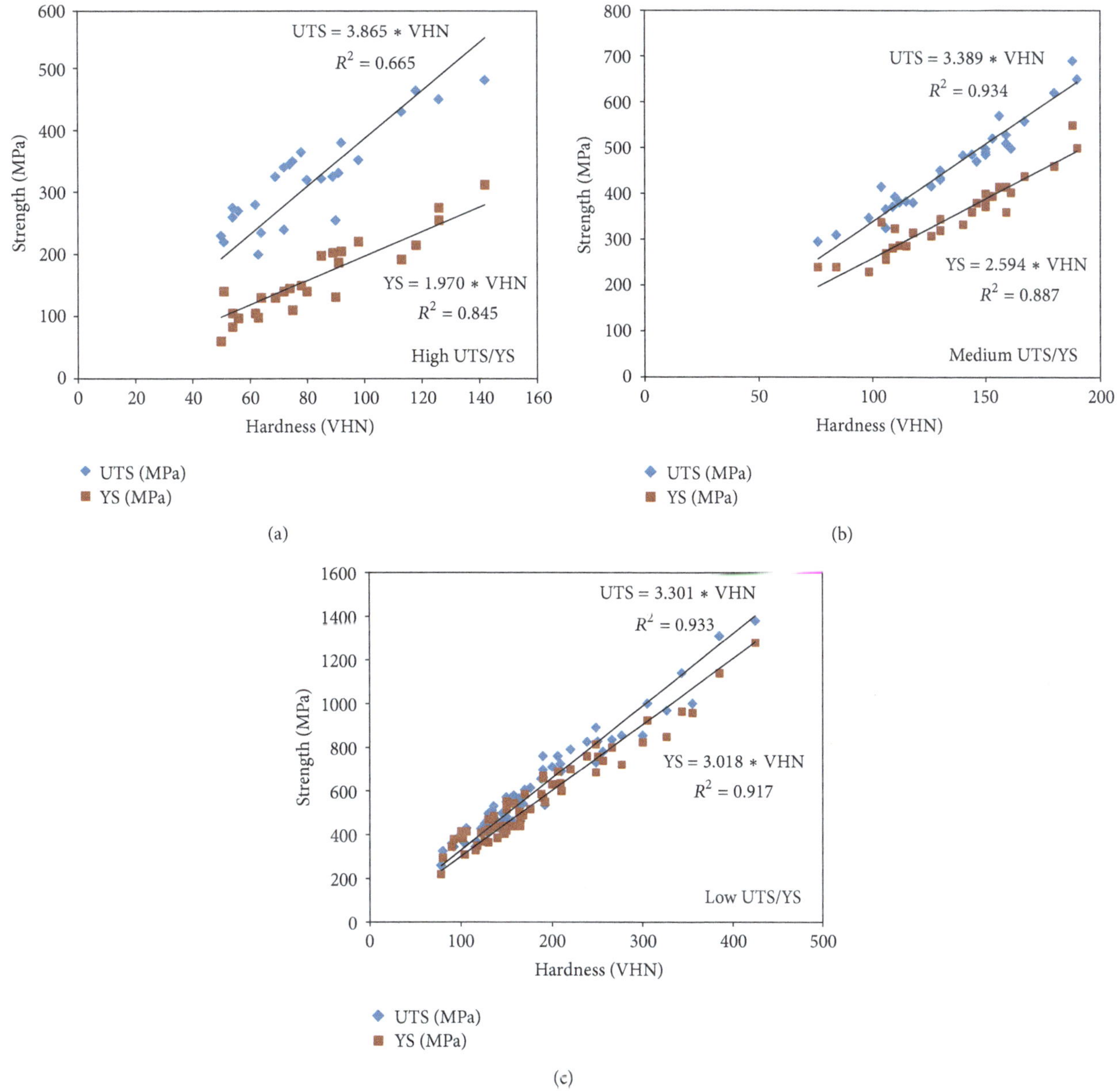

FIGURE 3: Relationship between strength and hardness of various alloys: (a) high UTS/YS ratio, (b) medium UTS/YS ratio, and (c) low UTS/YS ratio [12–28].

(6) estimated the UTS with less than 20% deviation at hardness lower than 150 VHN. Whereas reasonable deviation was observed at higher hardness. On the other hand, YS was estimated within 10%–25% using (7) with large deviation at hardness lower than 140 VHN. These deviations are quite reasonable considering the inherent errors in measuring the hardness and strength [3]. Hence, the correlations can be utilized to estimate the strength of copper alloys from Vickers hardness.

5. Conclusion

The results of the present study indicate that yield strength and ultimate tensile strength of the copper alloys can be determined with fairly good precision from Vickers hardness using simple linear correlations. However, the strength can be estimated with better accuracy using correlations for alloys with low and medium strain-hardening potential, that is, for fully cold-worked, thermomechanically treated, and aged alloys. Alloys with lower hardness and strength have shown deviation from linear correlation. The equations proposed for all the data were validated with the experimental data and a good agreement was noticed for both yield strength and ultimate tensile strength. Therefore, it is concluded that these correlations can be used for predicting the strength of copper alloys covering a wide range (YS-50 to 1200 MPa and UTS-200 to 1400 MPa).

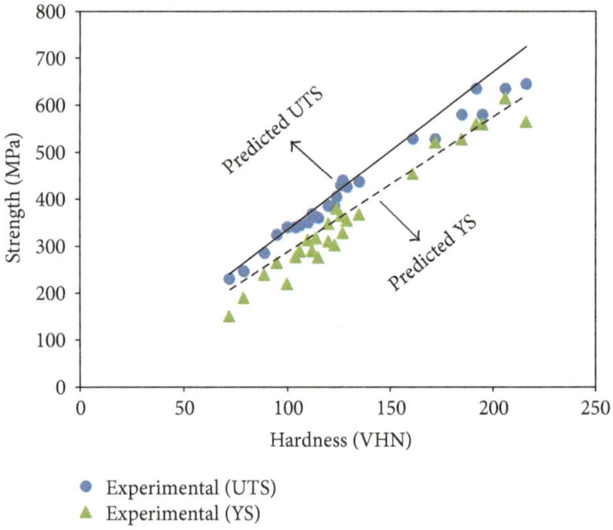

FIGURE 4: Comparison of estimated strength ((6) and (7)) and experimental strength for copper alloys.

Acknowledgments

The authors would like to thank Group Director (MMG) and Deputy Director VSSC (MME) for their support and technical guidance. The authors would like to express sincere gratitude to the Director, VSCC, for his kind permission to publish this work.

References

[1] D. Tabor, "The hardness and strength of metals," *Journal Institute of Metals*, vol. 79, pp. 1–18, 1951.

[2] J. R. Cahoon, "An improved equation relating hardness to ultimate strength," *Metallurgical Transactions*, vol. 3, no. 11, p. 3040, 1972.

[3] J. R. Cahoon, W. H. Broughton, and A. R. Kutzak, "The determination of yield strength from hardness measurements," *Metallurgical Transactions*, vol. 2, no. 7, pp. 1979–1983, 1971.

[4] J. Moteff, R. K. Bhargava, and W. L. McCullough, "Correlation of the hot-hardness with the tensile strength of 304 stainless steel to temperatures of 1200∘C," *Metallurgical Transactions A*, vol. 6, no. 5, pp. 1101–1104, 1975.

[5] M. O. Lai and K. B. Lim, "On the prediction of tensile properties from hardness tests," *Journal of Materials Science*, vol. 26, no. 8, pp. 2031–2036, 1991.

[6] S. C. Chang, M. T. Jahn, C. M. Wan, J. Y. M. Lee, and T. K. Hsu, "The determination of tensile properties from hardness measurements for Al-Zn-Mg alloys," *Journal of Materials Science*, vol. 11, no. 4, pp. 623–630, 1976.

[7] J. Datsko, L. Hartwig, and B. McClory, "On the tensile strength and hardness relation for metals," *Journal of Materials Engineering and Performance*, vol. 10, no. 6, pp. 718–722, 2001.

[8] E. G. Herbert, W. C. Oliver, and G. M. Pharr, "On the measurement of yield strength by spherical indentation," *Philosophical Magazine*, vol. 86, no. 33–35, pp. 5521–5539, 2006.

[9] G. Pintaude, A. Hoechele, and G. Cipriano, "Relation between strain hardening exponent of metals and residual profiles of deep spherical indentation," *Materials Science and Technology*, vol. 28, no. 9-10, pp. 9–10, 2012.

[10] E. J. Pavlina and C. J. Van Tyne, "Correlation of Yield strength and Tensile strength with hardness for steels," *Journal of Materials Engineering and Performance*, vol. 17, no. 6, pp. 888–893, 2008.

[11] Annual Book of ASTM Standards, *Standard Hardness Conversion Tables for Metals Relationship among Brinell Hardness, Vickers Hardness, Rockwell Hardness, Superficial Hardness, Knoop Hardness, and Scleroscope, E140-07*, American Society for Testing and Materials, West Conshohocken, Pa, USA, 2007.

[12] S. C. Krishna, K. Thomas Tharian, B. Pant, and R. S. Kottada, "Age-hardening characteristics of Cu-3Ag-0.5 Zr alloy," *Materials Science Forum*, vol. 710, pp. 563–568, 2012.

[13] I. S. Batra, G. K. Dey, U. D. Kulkarni, and S. Banerjee, "Microstructure and properties of a Cu-Cr-Zr alloy," *Journal of Nuclear Materials*, vol. 299, no. 2, pp. 91–100, 2001.

[14] H. Ehsanian Mofrad, S. Raygan, B. Amin Forghani, K. Hanaei, and F. K. Ahadi, "Effect of cold-working and aging processes on the microstructure, mechanical properties and electrical conductivity of Cu-13. 5%Mn-4%Ni-1. 2%Ti alloy," *Materials and Design*, vol. 41, no. 0, pp. 182–191, 2012.

[15] S. C. Krishna, K. Radhika, K. T. Tharian, G. S. Rao, M. S. Kiranmayee, and B. Pant, "Effect of simulated brazing cycle on the microstructure and mechanical properties of Cu-Cr-Zr-Ti alloy," *Materials Science Forum*, vol. 710, pp. 626–631, 2012.

[16] R. Markandeya, S. Nagarjuna, and D. S. Sarma, "Precipitation hardening of Cu-Ti-Cr alloys," *Materials Science and Engineering A*, vol. 371, no. 1-2, pp. 291–305, 2004.

[17] S. Nagarjuna, U. Chinta Babu, and P. Ghosal, "Effect of cryo-rolling on age hardening of Cu-1.5Ti alloy," *Materials Science and Engineering A*, vol. 491, no. 1-2, pp. 331–337, 2008.

[18] K. Oishi, I. Sasaki, and J. Otani, "Effect of silicon addition on grain refinement of copper alloys," *Materials Letters*, vol. 57, no. 15, pp. 2280–2286, 2003.

[19] B. N. Singh, D. J. Edwards, and P. Toft, "Effects of neutron irradiation on mechanical properties and microstructures of dispersion and precipitation hardened copper alloys," *Journal of Nuclear Materials*, vol. 238, no. 2-3, pp. 244–259, 1996.

[20] A. Vinogradov, T. Ishida, K. Kitagawa, and V. I. Kopylov, "Effect of strain path on structure and mechanical behavior of ultra-fine grain Cu-Cr alloy produced by equal-channel angular pressing," *Acta Materialia*, vol. 53, no. 8, pp. 2181–2192, 2005.

[21] R. Markandeya, S. Nagarjuna, and D. S. Sarma, "Effect of prior cold work on age hardening of Cu-3Ti-1Cr alloy," *Materials Characterization*, vol. 57, no. 4-5, pp. 348–357, 2006.

[22] R. Markandeya, S. Nagarjuna, and D. S. Sarma, "Characterization of prior cold worked and age hardened Cu-3Ti-1Cd alloy," *Materials Characterization*, vol. 54, no. 4-5, pp. 360–369, 2005.

[23] S. J. Zinkle, D. H. Plantz, A. E. Bair, R. A. Dodd, and G. L. Kulcinski, "Correlation of the yield strength and mlcrohardnesss of high-strength, high-conductivity copper alloys," *Journal of Nuclear Materials C*, vol. 133-134, pp. 685–689, 1985.

[24] A. Guha, "Development a high strength high conductivity Cu-Ni-Be alloy," in *High Conductivity Copper and Aluminum Alloys*, E. Ling and W. Pierre, Eds., pp. 133–145, The Metallurgical Society of AIME, California, Calif, USA, 1984.

[25] S. C. Krishna, J. Srinath, A. Jha, B. Pant, S. C. Sharma, and K. George, "Microstructure and properties of a high-strength Cu-Ni-Si-Co-Zr alloy," *Journal of Materials Engineering and Performance*, 2013.

[26] F. R. Mollar, K. G. Wikle, and A. R. Chaudhry, "Copper-Beryllium for elevated temperature electronic applications," in *High Conductivity Copper and Aluminum Alloys*, E. Ling and W. Pierre, Eds., pp. 147–167, The Metallurgical Society of AIME, California, Calif, USA, 1984.

[27] ASM International Handbook Committee, *ASM Handbook, Volume 02-Properties and Selection: Nonferrous Alloys and Special-Purpose Materials*, ASM International, California, Calif, USA, 1990.

[28] E. G. West, *Copper and Its Alloys*, Ellis Haorwood Limited, Chichester, UK, 1983.

17

Synthesis, Characterization, and Thermal Decomposition of Pure and Dysprosium Doped Yttrium Phosphate System

K. K. Bamzai, Nidhi Kachroo, Vishal Singh, and Seema Verma

Crystal Growth and Material Research Laboratory, Department of Physics & Electronics, University of Jammu, Jammu 180006, India

Correspondence should be addressed to K. K. Bamzai; kkbamz@yahoo.com

Academic Editor: Rodrigo Martins

Yttrium phosphate and dysprosium doped yttrium phosphate were synthesized from aqueous solutions using rare earth chloride, phosphoric acid, and traces of ammonium hydroxide. The synthesized material was then characterized for their structural investigations using powder X-ray diffraction (XRD) analysis and scanning electron microscopy (SEM) supplemented with energy dispersive X-ray analysis (EDAX). The spectroscopic investigations were carried out using Fourier transform infrared (FTIR) spectroscopy. The thermal stability was studied using differential thermogravimetric analysis (DTA), thermogravimetric analysis (TGA), and differential scanning calorimetry (DSC) techniques. X-ray diffraction analysis reveals that both yttrium phosphate and dysprosium doped yttrium phosphate belong to tetragonal system with lattice parameter $a = b = 6.8832$ Å, $c = 6.0208$ Å and $a - b = 6.9987$ Å, $c = 6.0142$ Å, respectively. The stoichiometry of the grown composition was established by energy dispersive X-ray analysis. The EDAX analysis suggests the presence of water molecules. The presence of water molecules along with orthophosphate group and metallic ion group was confirmed by FTIR analysis. Thermogravimetric analysis suggests that decomposition in case of yttrium phosphate takes place in three different stages and the final product stabilizes after 706°C, whereas in case of dysprosium doped yttrium phosphate the decomposition occurs in two different stages, and the final product stabilizes after 519°C.

1. Introduction

Rare earth phosphates belong to the family of rare earth zircons. The crystal of rare earth phosphate crystallizes into tetragonal zircon structure. The general formula for rare earth zircon is RXO_4, where "R" = rare earth and "X" = P, V, or As. Here each "X" atom is surrounded by an oxygen tetrahedron, and each rare earth atom surrounded by eight nearest neighbor oxygen atoms [1]. The rare earth ions (R^{3+}) and phosphorous (P^{5+}) ions occupy sites of tetragonal symmetry. There are four molecules per unit cell. Rare earth phosphates find their place in wide variety of applications such as optical materials including lasers [2], phosphors [3], and more recently as anti-UV materials [4]. Rare earth orthophosphates exhibit certain properties that make them of interest as scintillators for gamma-ray detection [5–7], as thermophosphors for remote measurement of temperature on moving components [8], and as rare earth analytical standards [9]. Rare earth codoping in inorganic materials has a long held transition of facilitating highly desirable optoelectronic properties for their potential applications to the laser industry. Rare earth compounds were extensively applied in luminescent and display, such as lighting, field emission display (FED), cathode ray tubes (CRT), and plasma display panel (PDP) [10–12].

Amorphous rare earth phosphates were synthesized by precipitation method [13]. Guo et al. [14] had reported the sol-gel synthesis of $RePO_4$ (Re = La, Ce, Nd, Eu, Y). Synthesis of rare earth phosphates by wet precipitation method has been reported by different research groups [15–17]. Nedelec et al. [18] reported the dependence of optical properties on the synthesis of YPO_4, whereas the thermal decomposition of $RePO_4 \cdot nH_2O$ (Re = La, Ce, Y) was reported by Lucas et al. [19]. Several other authors reported the synthesis of rare earth phosphate compounds by different methods such as high temperature solid state reaction technique, wet chemical precipitation technique, and sol-gel and hydrothermal synthesis [20–24]. The physicochemical properties of the material, which depends on the synthesis route, that is, chemical composition, grain size, morphology, and the crystalline structure, influence the thermal behavior the end product, and therefore their final physicochemical properties. So these factors are of prime importance in the manufacturing processes. There is no study concerning pure and doped yttrium phosphate through the influence synthesis parameters on the characteristics of the resulting material. This paper, therefore, deals with study of yttrium phosphate and dysprosium doped yttrium phosphate grown by an aqueous solution method using constituent material like yttrium chloride, dysprosium chloride, phosphoric acid, and traces of ammonium hydroxide. To the best of author's knowledge no such detailed work on the synthesis by this method, characterization, and thermal behavior was reported.

2. Materials and Methods

2.1. Materials Preparation. Yttrium phosphate (YPO_4) hereafter abbreviated as (YP) was grown by reacting phosphoric acid (H_3PO_4) with rare earth chlorides (RCl_3, R = Y) using ammonia solution (NH_4OH) to adjust the pH. The chemical used in the present study are: yttrium chloride (YCl_3) and dysprosium chloride ($DyCl_3$) (Indian Rare Earth Ltd. 99%); phosphoric acid (H_3PO_4) and ammonia solution (NH_4OH) (AR grade from S.D.Fine Chemicals). Many researchers have given this technique the name of precipitation method [25] or sol-gel synthesis [14]. However, it is worth noticeable that in all the reported works the synthesized material was later sintered at high temperatures to obtain the crystalline form, whereas in the present work no such sintering or heating of the material was done. In case of YP the material was obtained in the form of crystals at the room temperature, whereas in case of dysprosium doped yttrium phosphate instead of crystals floral shaped growth it was obtained over the condensed gel.

An aqueous solution of 0.5 M of yttrium chloride was mixed with an aqueous solution of 0.5 M of phosphoric acid in the ratio of 1 : 5. Then ammonia solution was added to the mixture drop by drop until a pH of 3 was obtained amid continuous stirring. The solution was then put into the crystallizing dish, then kept undisturbed in the constant temperature bath at ambient temperature (35–40°C). After two weeks, the solvent was evaporated from the crystallizing dish, condensed gel was obtained and cracks were seen on the surface of condensed gel after a few days, and then small size crystals appeared. After the period of 30 days, crystals were taken out and washed under running water. The chemical reaction that took place in the previous process is as follows:

$$YCl_3 + H_3PO_4 + NH_4OH \longrightarrow YPO_4 + 3HCl + NH_3 + H_2O. \quad (1)$$

The preparation of 2% dysprosium doped yttrium phosphate ($Dy_{0.02}Y_{0.98}PO_4$) here after abbreviated as (DyYP) was also based on the same procedure. The solution of yttrium and dysprosium was made in such a way that the only traces of dysprosium nearly 2% appear in the material. An aqueous solution of 0.5 M of yttrium chloride and dysprosium chloride was mixed with an aqueous solution of 0.5 M of phosphoric acid in the ratio of 1 : 5. Then ammonia solution was added to the mixture drop by drop until a pH of 3 was obtained amid continuous stirring. The solution was then kept undisturbed in the constant temperature bath at ambient temperature (35°–40°C). During the period of 40 days, the material in the crystallizer condensed into a gel-like substance and became hard, and after few days, the gel broke into pieces and small flower-like growth appeared over each piece of the condensed gel. The chemical reaction that took place in the previous process is as follows:

$$Dy_{0.02}Y_{0.98}Cl_3 + H_3PO_4 + NH_4OH \longrightarrow Dy_{0.02}Y_{0.98}PO_4 + 3HCl + NH_3 + H_2O. \quad (2)$$

2.2. Characterization. The characterization techniques consisted of X-ray diffraction (XRD), scanning electron microscopy (SEM) supplemented with energy dispersive X-ray analysis (EDAX), Fourier transform infrared spectroscopy (FTIR), thermogravimetry (TGA), differential thermogravimetry (DTA), and differential scanning calorimetry (DSC). Powder X-ray diffraction was performed using Rich Seifert powder X-ray diffractometer (model ISO Debye Flux 2002). Scanning electron microscope model number-JSM6100 supplemented with energy dispersive X-ray analysis was used to study morphology and elemental composition of the grown crystals. To study the presence of phosphate and other groups in the crystals Fourier transform infrared (FTIR) spectrum was obtained on Perkin-Elmer 781 spectrophotometer in the regions from 400 to 4000 cm^{-1} using KBr pellet. TGA and DTA curves were recorded simultaneously on the thermal analyzer (Shimadzu make DTG-60) over the temperature range from 25 to 1000°C at the heating rate of 10°C/min in the N_2 atmosphere at a flow rate of 30 mL/min.

3. Results and Discussion

3.1. Optical Microscopy. Optical microscopy was involved for rapid scanning of the grown crystals. Photomicrograph of YP and DyYP is shown in Figures 1(a) and 1(b), respectively. Yttrium phosphate appears transparent as well as platelet in nature, whereas DyYP shows clear view of the flower-like arrangement.

3.2. Scanning Electron Microscopy (SEM). Electron microscopy is a powerful tool to investigate the microstructure of single crystal. Figure 2 shows the SEM image of YP crystal. From the image, it is clear that it is a platelet-like crystal with clear and smooth surfaces. However, DyYP grew in the form of floral growth over the condensed gel. These floral growths were then scanned to have a closer look on the morphology of the material. Figure 3 gives a clear picture of the SEM image of DyYP. From the image, it is seen that the material has grown in such a way that there are striations all over the grown area. Striations are caused by a crystal alternating between crystal faces, as it grows. Striations are generally classified as positive and negative types. Striations, parallel to the crystal

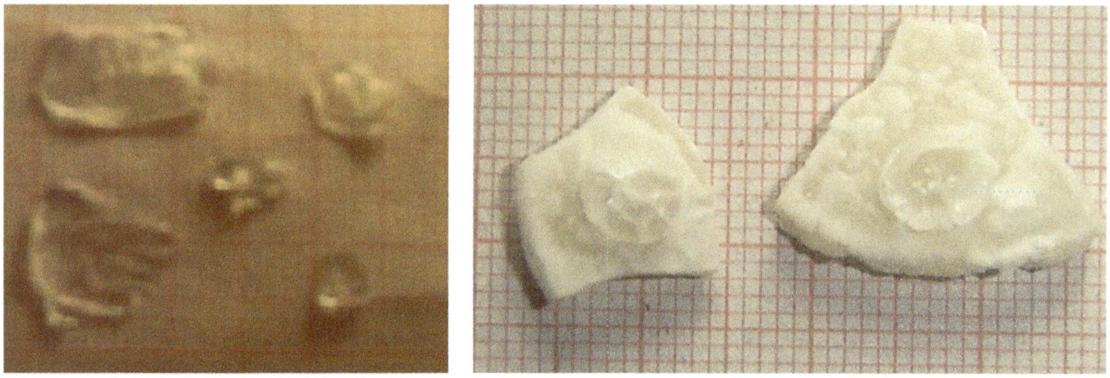

(a) (b)

Figure 1: Photomicrograph as seen under optical microscope for (a) yttrium phosphate crystals; (b) dysprosium doped yttrium phosphate.

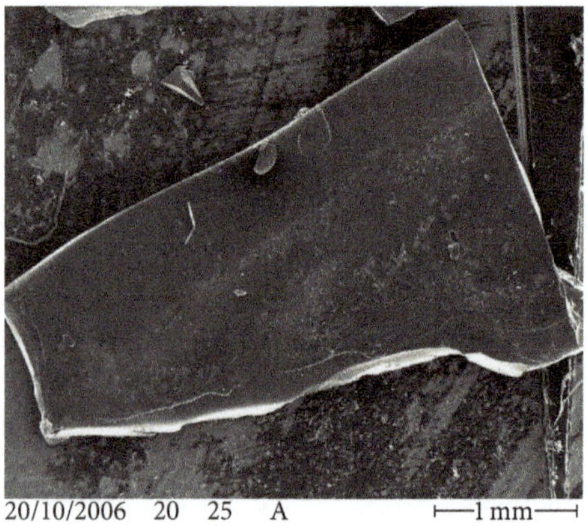

Figure 2: Scanning electron micrograph of pure YP crystal which clearly shows the platelet morphology of the crystal.

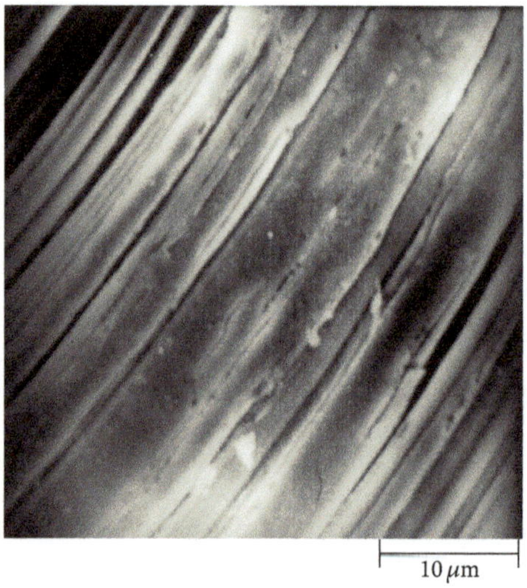

Figure 3: Scanning electron micrograph of floral part of DyYP revealed certain kind of striations at the surface of the grown material.

faces are called positive striations and those perpendiculars to the growth faces are called negative striations [26]. In the present case, since the striations were parallel to the growth surface, they are suggested to be positive striations. Temperature fluctuations are often regarded as the root cause of all types of striations [27].

3.3. X-Ray Diffraction Analysis (XRD). Figure 4(a) shows the XRD graph of YP crystal. The graph consists of high resolved peaks at some specific 2θ Bragg's angles depicting the crystalline nature of the material. The data was compared with JCPDS data number 84–0335 which suggest that crystal belongs to tetragonal system. The cell parameters were found to be $a = b = 6.8832$ Å, $c = 6.0208$ Å. The unit cell volume as calculated by WINPLOTR software comes out to be 285.3 cubic Å. Figure 4(b) shows the diffraction pattern of the DyYP crystal. From the graph, it is clear that like YP, DyYP also has a well-versed crystallinity. The lattice parameters for DyYP comes out to be $a = b = 6.9987$ Å, $c = 6.0142$ Å. The unit cell volume came out to be 294.6 cubic Å.

3.4. Energy Dispersive X-Ray Analysis (EDAX). To study the elemental composition of YP and DyYP, qualitative and quantitative analysis were performed by energy dispersive X-ray analysis. The spectrum obtained from EDAX analyses is shown in Figures 5(a) and 5(b). EDAX pattern shows peaks corresponding to all the major elements present in the grown crystals as should be expected from YP system. The spectra corresponding to doped DyYP shows peaks corresponding to all the major elements, that is, yttrium, phosphorous, and oxygen along with dysprosiumby suggesting that Dy has entered into the lattice of YP system. However, along with these elements some trace impurities in the form of chlorine and nitrogen were observed. The experimental and theoretical calculated atomic and weight percentages of elements in YP and DyYP is given in Tables 1(a) and 1(b), respectively. For YP crystals the theoretical values were calculated as per the formula $YPO_4{\cdot}H_2O$, whereas for DyYP system the values were calculated as per the formula $Dy_{0.02}Y_{0.98}PO_4{\cdot}2H_2O$. However, it is important to mention here that EDAX analysis does not give experimental values of atomic and weight percentage for lighter elements like hydrogen. Therefore, in

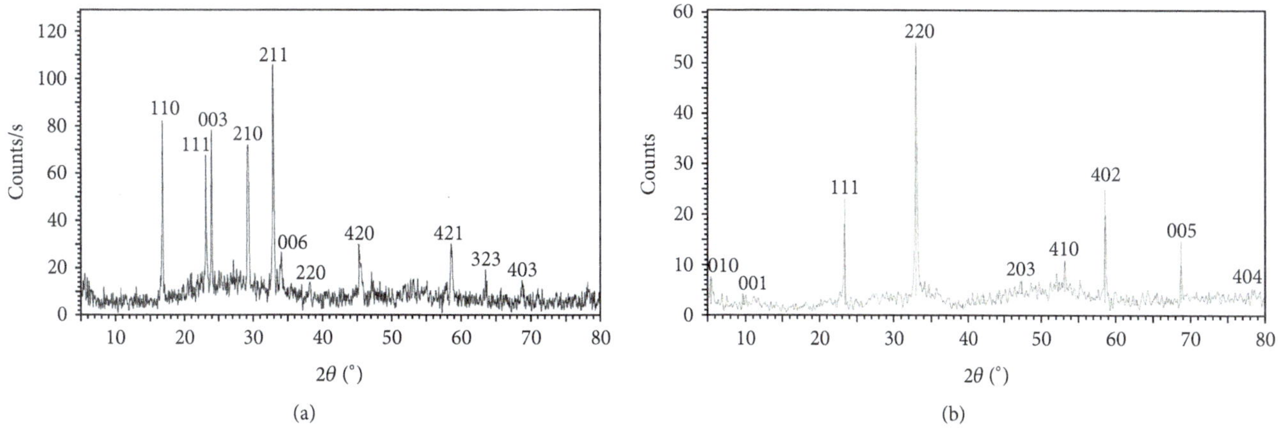

FIGURE 4: XRD diffractograms which clearly depict the presence of crystallinity in the material, along with indexing of all the prominent peaks for (a) yttrium phosphate (YP); (b) dysprosium doped yttrium phosphate (DyYP).

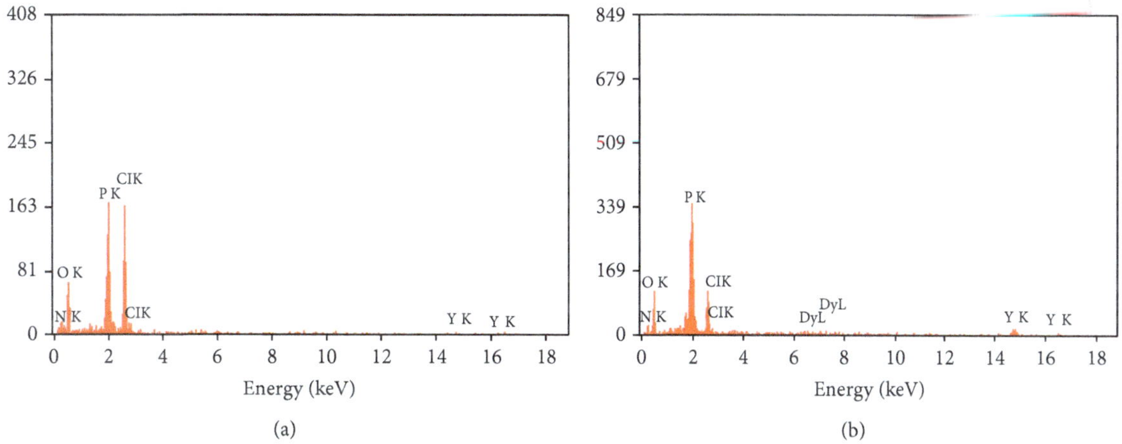

FIGURE 5: EDAX spectra showing the presence of suggested elements along with some impurity elements for (a) Yttrium phosphate (YP); (b) Dysprosium doped yttrium phosphate (DyYP).

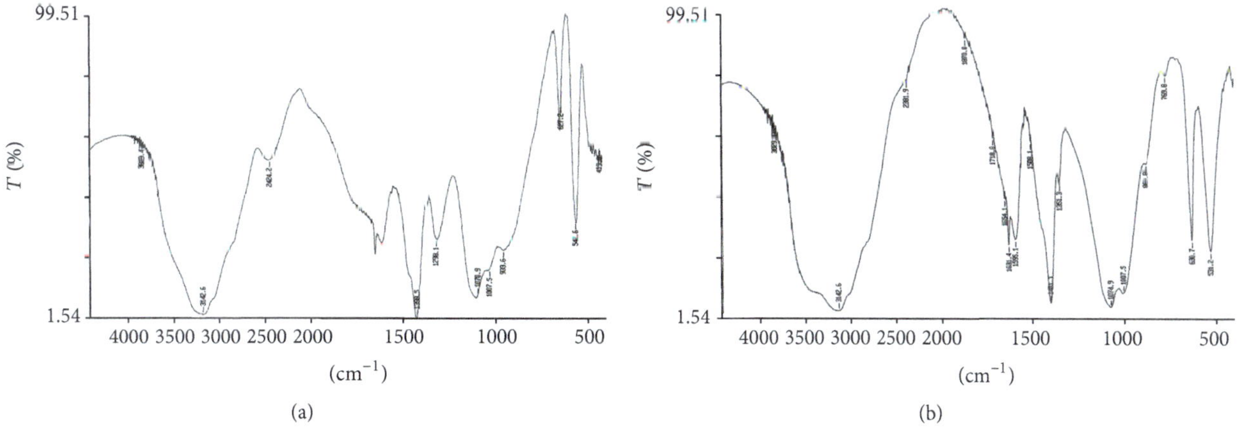

FIGURE 6: FT-IR spectra depicting the various functional groups present in (a) yttrium phosphate (YPO_4); (b) dysprosium doped yttrium phosphate ($Dy:YPO_4$).

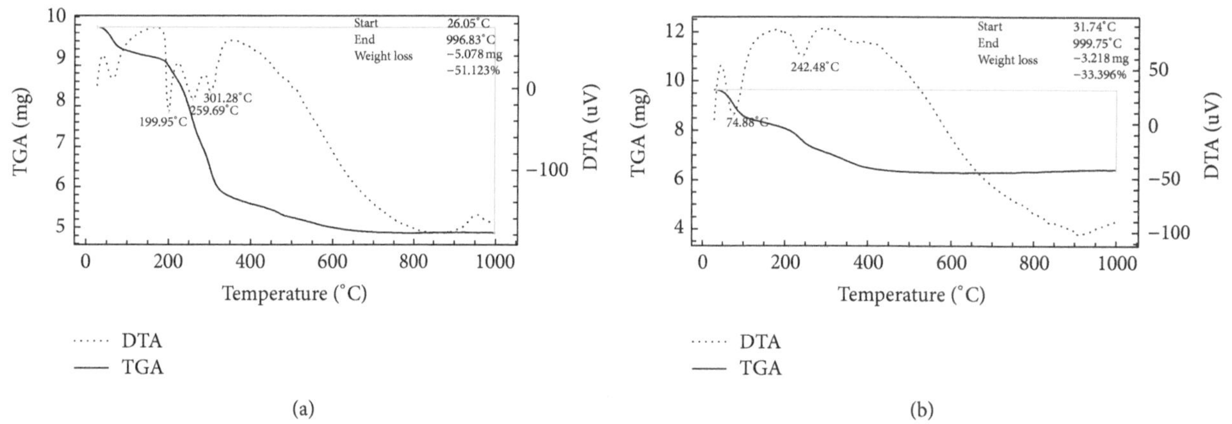

FIGURE 7: (a) Thermograph depicting simultaneous recording of TGA and DTA curves which clearly shows that pure YPO_4 crystal is thermally unstable as the temperature increases from room temperature to higher values. (b) Thermograph of Dy:YPO_4 depicting the TGA and DTA curves shows that it is thermally more stable as compared to the pure crystals of YPO_4.

the table theoretical values of the lighter element (H) based on the formula have been given, and it is found that the experimental and theoretical values are in close agreement with each other within the experimental error. The presence of H_2O molecules in YP and DyYP was further confirmed by FTIR and thermoanalytical analysis.

3.5. Fourier Transform Infrared Spectroscopy (FT-IR). The infrared spectrum is formed as a consequence of the absorption of electromagnetic radiation at frequencies that correlate with the vibration of specific sets of chemical bonds from a molecule. Thus, the vibrational spectrum of a molecule is considered to be a unique physical property and is characteristic of the molecule. Figures 6(a) and 6(b) shows the FTIR spectrum for YP and DyYP, respectively. On analyzing the spectrum, it was observed that grown crystals show the presence of water molecules, orthophosphate group and metallic ions group. For the water vibration in addition to the wide bands associated with different types of OH groups extending from 3829.4 to 2424.2 in case of YP and from 3829.4 to 2381.9 in case of DyYP the presence of two bands at 1631.6, 1596 in YP and 1631.4, 1595.1 in DyYP is indicative of the characteristics of coordinated water molecule [19], that is, the hydrated water molecule in the as-synthesized sample are chemically bonded to the rare earth ions [28, 29]. The bands from orthophosphate functional group were observed at 1074.9, 1070.9, 1007.5, and 933.6 cm^{-1} in both the cases with the difference that for doped yttrium phosphate the values of transmittance decreases which can be attributed to the doping effect [30, 31]. The band around this wave number was attributed to the symmetric stretching mode (ν_4) and asymmetric stretching mode (ν_3) of PO_4 tetrahedron [32, 33]. The bands at 630.7, 627.2, 542.6, and 531.2 cm^{-1} show the presence of metallic ions. Tables 2(a) and 2(b) give complete information about the type of functional groups present in YP and DyYP along with their frequency bands and transmittance percentage.

3.6. Thermal Decomposition. Thermogravimetric analysis is a technique to assess the stability of various substances.

TABLE 1: Experimental and theoretical calculated composition obtained from energy dispersive X-ray analysis (EDAX) of various constituent elements present in the following cases.

(a) Yttrium phosphate (YPO_4)

Element	Experimental values		Theoretical values	
	Wt%	At%	Wt%	At%
Oxygen	32.92	56.34	30.8	55.42
Phosphorous	11.08	12.03	11.95	11.08
Yttrium	33.15	09.23	34.31	11.08
Hydrogen	—	—	22.94	22.42

(b) Dysprosium doped yttrium phosphate (Dy:YPO_4)

Element	Experimental values		Theoretical values	
	Wt%	At%	Wt%	At%
Oxygen	40.12	68.07	42.15	70.89
Phosphorous	11.29	12.19	13.60	11.82
Dysprosium	01.13	00.68	01.43	00.24
Yttrium	38.40	10.10	38.26	11.58
Hydrogen	—	—	4.56	5.47

Figures 7(a) and 7(b) show the simultaneously recorded thermogravimetric analysis (TGA) and differential thermogravimetric analysis (DTA) curve for YP and DyYP crystal. Thermograms were first analyzed to obtain information about the percentage mass loss at different temperatures and hence about the thermal stability and kinetics of dissociation of crystals. YP and DyYP when heated at a uniform rate of 10°C/min, mass was found to lose continuously, as a function of temperature applied. For YP starts to decompose at 39°C, and the decomposition went up to 706°C in three different stages. However, in case of DyYP the decomposition starts at 43°C, and the thermal stability was acquired by the material at 519°C in two different stages. During these temperature ranges, the mass loss was 50.9% and 32.6% for YP and DyYP respectively. Therefore, the mechanism involved in the decomposition of the constituent material is different for pure

TABLE 2: Presence of various functional groups along with their frequency bands and transmittance percentage for the following.

(a) Yttrium phosphate (YPO_4)

Assignments of bands/peaks	IR bands (cm^{-1})	Transmittance percentage (%)
PO_4	1070.9	10%
	1007.5	18%
	933.6	20%
OH	3829.4	60%
	3142.6	2%
Metallic ion	627.2	30%
	542.6	60%

(b) Dysprosium doped yttrium phosphate (Dy:YPO_4)

Assignments of bands/peaks	IR bands (cm^{-1})	Transmittance percentage (%)
PO_4	1074.9	5%
	1007.5	10%
OH	3829.4	60%
	3142.6	5%
Metallic ion	630.7	25%
	531.2	20%

and doped one. Our group has earlier reported [34, 35] the growth and thermal kinetics of pure and cadmium-doped barium/calcium phosphate single crystal.

From Figure 7(a), YP system is thermally stable up to a temperature of 39°C and thereafter starts decomposing. The whole process of decomposition completes in three steps. The first stage of decomposition begins from 39°C and continues up to a temperature of 176°C resulting in a weight loss of 7.9% of the total weight. First step of decomposition involves conversion of hydrated YP crystal into anhydrous in nature. The second stage of decomposition starts from 176°C and ends at a temperature of 437°C leading to weight loss of 37%. This weight loss in the second stage of decomposition corresponds to the conversion of anhydrous YPO_4 to Y_2O_3. The third stage of decomposition starts from 437°C and goes up to 706°C resulting in weight loss of 6% during which the intermediate product Y_2O_3 decomposes to form YO with the release of half a molecule of oxygen as the final product. This type of fractional release of oxygen molecule has also been reported earlier by Brown [36] during the thermal decomposition of inorganic solids. Table 3(a) gives the compiled summary of the decomposition process of $YPO_4{\cdot}H_2O$. It can be seen that the calculated weight loss is in close proximity with the observed values. Based on these thermal analyses, we confirm that the grown crystal is having a composition of $YPO_4{\cdot}H_2O$.

From the thermogram of DyYP (Figure 7(b)), it is clear that the doped crystal is thermally stable up to temperature of 43°C, which means that doped crystal is more stable than pure one. In case of DyYP, the decomposition takes place in two steps. In the first step from 43 to 176°C, two water molecules attached to the doped system get decomposed. In the second stage of the decomposition, doped orthophosphate reduces to pyrophosphate with the release of phosphorous oxide and oxygen. Table 3(b) gives detailed summary of the decomposition of $Dy_{0.02}Y_{0.98}PO_4{\cdot}2H_2O$ along with observed as well as calculated weight losses. In this case, the calculated and observed weight losses are in close agreement with each other. It is worth mentioning here that the temperature for the formation of stable product after decomposition in case of pure one is 706°C, whereas in case of doped one the stable product is formed at a temperature of 519°C. This means that the temperature for the formation of end product decreases with dysprosium substitution.

The thermal decomposition of YPO_4 into yttrium oxide through different stages has been accomplished with the release of H_2O, P_2O_5 and O_2, whereas for doped Dy:YPO_4 the decomposition was accompanied by the release of H_2O, PO_3, and O_2. Corresponding to each stage of decomposition, there are endothermic and exothermic peaks in the DTA curve. As seen from DTA curve in case of YP (Figure 7(a)) and DyYP (Figure 7(b)) there is well-marked endothermic and exothermic peak corresponding to each stage of decomposition. Since peaks in DTA curve correspond to weight loss in TGA curve, thereby suggesting that some changes takes place in the material because of the weight loss in the material.

From the thermal analysis of the system, we can therefore confirm that product formed in the pure form, that is, yttrium phosphate is associated with one water molecules having composition $YPO_4{\cdot}H_2O$, whereas the doped one that is, dysprosium doped yttrium phosphate, is associated with two water molecules having composition $Dy_{0.02}Y_{0.98}PO_4{\cdot}2H_2O$. These compositions were further supported by other analyses like energy dispersive X-ray analysis (EDAX) and Fourier transform infrared (FTIR) spectroscopy.

4. Conclusions

From the research work carried out over the synthesis and characterization of the pure and dysprosium doped yttrium phosphate, the following conclusions can be drawn.

(1) Pure yttrium phosphate with composition $YPO_4{\cdot}H_2O$ is obtained in the form of platelet like crystals, whereas the dysprosium doped yttrium phosphate having composition $Dy_{0.02}Y_{0.90}PO_4{\cdot}2H_2O$ is obtained in the form of floral growth on the condensed gel.

(2) Yttrium phosphate and dysprosium doped yttrium phosphate belong to tetragonal system. The lattice parameters obtained in case of YP are $a = b = 6.8832$ Å, $c = 6.0208$ Å. Similarly the lattice parameters in case of doped one, that is, DyYP, come out to be $a = b = 6.9987$ Å, $c = 6.0142$ Å. Thus, the crystal structure of yttrium phosphate remains unaffected by modification of its composition by dysprosium.

(3) Scanning electron microscopy (SEM) studies give a clear picture about the morphology of the grown crystals. The qualitative and quantitative elemental analyses employing EDAX technique confirm the

TABLE 3: Results of thermal decomposition for different temperature ranges with observed and calculated weight loss in the following cases.

(a) Yttrium phosphate (YP)

Stage	Temperature (°C)	Decomposition steps	Weight loss (%)	
			Observed	Calculated
First	39–176	$YPO_4 \cdot H_2O \rightarrow YPO_4 + H_2O$	7.9	8.9
Second	176–437	$2[YPO_4] \rightarrow Y_2O_3 + P_2O_5(g)$	37	38.7
Third	437–706	$Y_2O_3 \rightarrow YO + 0.5O_2(g)$	6	7

(b) Dysprosium doped yttrium phosphate (DyYP)

Stage	Temperature (°C)	Decomposition steps	Weight loss (%)	
			Observed	Calculated
First	43–176	$Dy_{0.02}Y_{0.98}PO_4 \cdot 2H_2O \rightarrow Dy_{0.02}Y_{0.98}PO_4 + 2H_2O$	13.65	16.28
Second	176–519	$3[Dy_{0.02}Y_{0.98}PO_4] \rightarrow (Dy_{0.02}Y_{0.98})_2P_2O_7 + PO_3(g) + O_2$	19.04	19.95

presence of major elements in the grown material. The stoichiometric composition of the grown system as established by EDAX technique is $YPO_4 \cdot H_2O$ and $Dy_{0.02}Y_{0.98}PO_4 \cdot 2H_2O$.

(4) Fourier transformed infrared studies on YP and DyYP confirm the presence of water molecules, orthophosphate functional group and metallic ion group. The frequency bands within the range of 3829.4 to 2424.2 cm^{-1} are attributed to the presence of water molecules. The effect of doping can be seen clearly in FTIR spectrum where the transmittance percentage decreases in case of DyYP.

(5) The thermal studies carried out on YP and DyYP reveal that pure yttrium phosphate is less stable as compared to doped yttrium phosphate. YPO_4 starts to decompose early and becomes stable at a later stage in comparison to its doped counterpart.

(6) The thermal decomposition of YP into the final stable product of yttrium monoxide underwent three different stages with the release of H_2O, P_2O_5, and O_2 whereas for DyYP the decomposition was accompanied by the release of H_2O, PO_3, and O_2, and a final product of doped yttrium pyrophosphate was obtained.

References

[1] Z. A. Kazei, N. P. Kolmakova, and O. A. Shishkina, "Magnetoelastic contribution to thermal expansion of rare-earth zircons," *Physics B*, vol. 245, no. 2, pp. 164–172, 1998.

[2] V. Mehta, G. Aka, A. L. Dawarb, and A. Mansingh, "Optical properties and spectroscopic parameters of Nd^{3+}-doped phosphate and borate glasses," *Optical Materials*, vol. 12, pp. 53–63, 1999.

[3] K. Riwotzki, H. Meyssamy, A. Kornowski, and M. Haase, "Liquid-phase synthesis of doped nanoparticles: colloids of luminescing $LaPO_4$:Eu and $CePO_4$:Tb particles with a narrow particle size distribution," *Journal of Physical Chemistry B*, vol. 104, no. 13, pp. 2824–2828, 2000.

[4] N. Imanaka, T. Masui, H. Hirai, and G. Y. Adachi, "Amorphous cerium-titanium solid solution phosphate as a novel family of band gap tunable sunscreen materials," *Chemistry of Materials*, vol. 15, no. 12, pp. 2289–2291, 2003.

[5] W. W. Moses, M. J. Weber, S. E. Derenzno, D. Perry, P. Berdahl, and L. A. Boatnor, "Prospects for dense, infrared emitting scintillators," *IEEE Transactions on Nuclear Science*, vol. 45, pp. 462–466, 1998.

[6] A. J. Wojtowicz, D. Wisniewski, A. Lempicki, and L. A. Boatner, "Scintillation mechanisms in rare earth orthophosphates," *Radiation Effects and Defects in Solids*, vol. 135, no. 1, pp. 305–310, 1995.

[7] A. Lempicki, E. Berman, A. J. Wojtowicz, M. Balcerzyk, and L. A. Boatner, "Cerium-doped orthophosphates: new promising scintillators," *IEEE Transactions on Nuclear Science*, vol. 40, no. 4, pp. 384–387, 1993.

[8] S. W. Allison, L. A. Boatner, and G. T. Gillies, "Characterization of high-temperature thermographic phosphors: spectral properties of $LuPO_4$:Dy(1%),Eu(2%)," *Applied Optics*, vol. 25, pp. 5624–5627, 1995.

[9] E. Jarosewich and L. A. Boatner, "Rare-earth element reference samples for electron microprobe analysis," *Geostandards Newsletter*, vol. 15, pp. 397–399, 1991.

[10] J. Dhanaraj, R. Jagannathan, T. R. N. Kutty, and C. H. Lu, "Photoluminescence characteristics of Y_2O_3:Eu^{3+} nanophosphors prepared using sol-gel thermolysis," *Journal of Physical Chemistry B*, vol. 105, no. 45, pp. 11098–11105, 2001.

[11] Z. Wei, L. Sun, C. Liao, C. Yan, and S. Huang, "Fluorescence intensity and color purity improvement in nanosized YBO_3:Eu," *Applied Physics Letters*, vol. 80, no. 8, pp. 1447–1449, 2002.

[12] R. S. Meltzer, S. P. Feofilov, B. Tissue, and H. B. Yuan, "Dependence of fluorescence lifetimes of Y_2O_3:Eu^{3+} nanoparticles on the surrounding medium," *Physical Review B*, vol. 60, no. 20, pp. R14012–R14015, 1999.

[13] H. Hirai, T. Masui, N. Imanaka, and G. Y. Adachi, "Characterization and thermal behavior of amorphous rare earth phosphates," *Journal of Alloys and Compounds*, vol. 374, no. 1-2, pp. 84–88, 2004.

[14] Y. Guo, P. Woznicki, A. Barkatt, E. E. Saad, and I. G. Talmy, "Sol-gel synthesis of microcrystalline rare earth orthophosphates," *Journal of Materials Research*, vol. 11, no. 3, pp. 639–649, 1996.

[15] D. Bregiroux, S. Lucas, E. Champion, F. Audubert, and D. Bernache-Assollant, "Sintering and microstructure of rare earth phosphate ceramics $REPO_4$ with RE = La, Ce or Y," *Journal of the European Ceramic Society*, vol. 26, no. 3, pp. 279–287, 2006.

[16] J. Zhu, W. D. Cheng, D. S. Wu et al., "Crystal and band structures, and optical characterizations of sodium rare earth phosphates $NaLnP_2 2O_7$ and $NaLn(PO_3)_4$ (Ln = Ce, Eu)," *Journal of Alloys and Compounds*, vol. 454, no. 1-2, pp. 419–426, 2008.

[17] H. Lai, A. Bao, Y. Yang et al., "UV luminescence property of YPO_4:RE (RE = Ce^{3+}, Tb^{3+})," *Journal of Physical Chemistry C*, vol. 112, no. 1, pp. 282–286, 2008.

[18] J. M. Nedelec, D. Avignant, and R. Mahiou, "Soft chemistry routes to YPO_4-based phosphors: dependence of textural and optical properties on synthesis pathways," *Chemistry of Materials*, vol. 14, no. 2, pp. 651–655, 2002.

[19] S. Lucas, E. Champion, D. Bernache-Assollant, and G. Leroy, "Rare earth phosphate powders $RePO_4 \cdot nH_2O$ (*Re*=La, Ce or Y) II. Thermal behavior," *Journal of Solid State Chemistry*, vol. 177, no. 4-5, pp. 1312–1320, 2004.

[20] R. S. Fugelson, "Synthesis and single-crystal growth of rare-earth orthophosphates," *Journal of the American Ceramic Society*, vol. 47, pp. 257–258, 1964.

[21] N. N. Chudinova, L. P. Shklover, and G. M. Balagina, "Reactions of lanthanum oxide with phosphoric acids at 100–500°C," *Inorganic Materials*, vol. 11, pp. 590–593, 1975.

[22] N. Arul-Dhas and K. C. Patel, "Synthesis of $AlPO_4$, $LaPO_4$ and $KTiOPO_4$ by flash combustion," *Journal of Alloys and Compounds*, vol. 202, pp. 137–141, 1993.

[23] H. Onoda, H. Nariani, H. Maki, and I. Motooka, "Mechanochemical effects on synthesis of Rhabdophane-type neodymium and cerium phosphates," *Materials Chemistry and Physics*, vol. 78, pp. 400–404, 2002.

[24] O. Terra, N. Dacheux, R. Podar, and N. Clavier, "Preparation and characterization of lanthanum-gadolinium monazites as ceramics for radioactive waste storage," *New Journal of Chemistry*, vol. 27, pp. 957–967, 2003.

[25] S. V. Ushakov, K. B. Helean, A. Navrotsky, and L. A. Boatner, "Thermochemistry of rare-earth orthophosphates," *Journal of Materials Research*, vol. 16, no. 9, pp. 2623–2633, 2001.

[26] Y. Endo and I. Sunagawa, "Positive and negative striations in pyrite," *American Mineralogist*, vol. 58, pp. 930–935, 1973.

[27] B. Cockayne and M. P. Gates, "Growth striations in vertically pulled oxide and fluoride single crystals," *Journal of Materials Science*, vol. 2, pp. 118–123, 1967.

[28] W. Di, X. X. Zhao, S. Lu, X. Wang, and H. Zhao, "Thermal and photoluminescence properties of hydrated $YPO_4:Eu^{3+}$ nanowires," *Journal of Solid State Chemistry*, vol. 180, pp. 2478–2484, 2007.

[29] L. Qiong, S. Yiguo, Y. H. Sheng, and H. Wei, "YPO_4 nanocrystals: preparation and size-induced lattice symmetry enhancement," *Journal of Rare Earths*, vol. 26, no. 4, pp. 495–500, 2008.

[30] S. Lucas, E. Champion, C. Penot, G. Leroy, and D. Bernache-Assollant, "Synthesis and characterization of rare earth phosphate powders," *Key Engineering Materials*, vol. 206–213, pp. 47–50, 2001.

[31] A. Hezel and S. D. Ross, "Forbidden transitions in the infra-red spectra of tetrahedral anions-III. Spectra-structure correlations in perchlorates, sulphates and phosphates of the formula MXO_4," *Spectrochimica Acta*, vol. 22, pp. 1949–1961, 1966.

[32] D. K. Breitinger, G. Brehm, J. Mohr et al., "Vibrational spectra of synthetic crandallite-type minerals—optical and inelastic neutron scattering spectra," *Journal of Raman Spectroscopy*, vol. 37, no. 1-3, pp. 208–216, 2006.

[33] M. J. Bushiri, R. S. Jayasree, M. Fakhfakh, and V. U. Nayar, "Raman and infrared spectral analysis of thallium niobyl phosphates: $Tl_2NbO_2PO_4$, $Tl_3NaNb_4O_9(PO_4)_2$ and $TlNbOP_2O_7$," *Materials Chemistry and Physics*, vol. 73, pp. 179–185, 2002.

[34] S. Suri, K. K. Bamzai, and V. Singh, "Growth and thermal kinetics of pure and cadmium doped barium phosphate single crystal," *Journal of Thermal Analysis and Calorimetry*, vol. 105, no. 1, pp. 229–238, 2011.

[35] K. K. Bamzai, S. Suri, and V. Singh, "Synthesis, characterization, thermal and dielectric properties of pure and cadmium doped calcium hydrogen phosphate," *Materials Chemistry and Physics*, vol. 135, pp. 158–167, 2012.

[36] M. E. Brown, "Quantitative thermoanalytical studies of the kinetics and mechanisms of the thermal decomposition of inorganic solids," *Thermochimica Acta*, vol. 110, pp. 153–158, 1987.

Sago Starch-Mixed Low-Density Polyethylene Biodegradable Polymer: Synthesis and Characterization

Md Enamul Hoque,[1] Tan Jie Ye,[1] Leng Chuan Yong,[1] and KhairulZaman Mohd Dahlan[2]

[1] *University of Nottingham Malaysia Campus, 43500 Semenyih, Selangor, Malaysia*
[2] *Malaysian Nuclear Agency, 43000 Kajang, Selangor, Malaysia*

Correspondence should be addressed to Md Enamul Hoque; enamul.hoque@nottingham.edu.my

Academic Editor: Guoping Chen

This research focuses on synthesis and characterization of sago starch-mixed LDPE biodegradable polymer. Firstly, the effect of variation of starch content on mechanical property (elongation at break and Young's modulus) and biodegradability of the polymer was studied. The LDPE was combined with 10%, 30%, 50%, and 70% of sago for this study. Then how the cross-linking with trimethylolpropane triacrylate (TMPTA) and electron beam (EB) irradiation influence the mechanical and thermal properties of the polymer was investigated. In the 2nd study, to avoid overwhelming of data LDPE polymer was incorporated with only 50% of starch. The starch content had direct influence on mechanical property and biodegradability of the polymer. The elongation at break decreased with increase of starch content, while Young's modulus and mass loss (i.e., degradation) were found to increase with increase of starch content. Increase of cross-linker (TMPTA) and EB doses also resulted in increased Young's modulus of the polymer. However, both cross-linking and EB irradiation processes rendered lowering of polymer's melting temperature. In conclusion, starch content and modification processes play significant roles in controlling mechanical, thermal, and degradation properties of the starch-mixed LDPE synthetic polymer, thus providing the opportunity to modulate the polymer properties for tailored applications.

1. Introduction

The development of innovative biopolymers has been underway for a number of years and continues to be an area of interest for the modern scientists. In 1996, shipments from the Canadian Plastic Industry increased by 10.6% from 1995 levels to $9.1 billion [1, 2]. As reported by Fomin [3], worldwide production of synthetic polymers reached 130 million tons per year at the end of the 20th century, while the demand for biodegradable polymer is reported to be growing by 30% each year [4, 5]. Meanwhile, Europe has an estimated average usage of 100 kg of polymer per person each year [6]. Polyethylene (PE) is widely used in medical devices and pharmaceutical packaging [7]. Different types of polymers such as high-density PE (HDPE), low-density PE (LDPE), or linear low-density PE (LLDPE) offer different benefits and functionalities. HDPE provides stiffness, chemical resistance, and barrier properties, while LDPE offers resistance to stress cracking and excellent impact properties [8]. In addition, its capability of being sterilized is a favourable and an important aspect. Cross-linking is a broadly used method for modifying the polymer properties. This process involves the formation of tridimensional structures that cause substantial changes [9]. The mechanical property of biomaterial is equally important as biocompatibility since an inadequate performance or premature failure of the implants might impose serious health issues to the patients [10–15]. Besides, most of the body implants undergo dynamic loading that might affect the mechanical behaviour (e.g., creep or relaxation) of the implants. The implants should maintain their stability in the body for a prescribed time as required [16]. Therefore, the durability or long-term mechanical behaviour of new biomaterials has become a prime concern for their adoption to medical devices.

Starch is a hydrocolloid biopolymer that can be found in a variety of agricultural feedstocks such as wheat, corn, rice, beans, and potatoes [17, 18]. Starch consists of one branched and linear polymer [19] and is usually found in the

TABLE 1: Relative percentage of LDPE and starch of polymer samples.

Sample ID	% LDPE	% Starch
LDPE00-Starch0	100	0
LDPE90-Starch10	90	10
LDPE70-Starch30	70	30
LDPE50-Starch50	50	50
LDPE30-Starch70	30	70

form of granules. Different starches and percentages could be employed to modulate polymer properties. A number of studies have reported the development of biodegradable polymer by incorporating starch into LDPE and the characterization of these synthesized polymers [20–27]. These studies mostly focused on investigating the influences of starch content and so forth on the mechanical and degradation properties of the polymer. However, too little attention had been paid so far to evaluate the effects of modification processes (e.g., addition of cross-linker and electron beam irradiation) on the mechanical and thermal properties of the polymer. In this study, the effects of cross-linking with trimethylolpropane triacrylate (TMPTA) and electron beam (EB) irradiation on mechanical and thermal properties of the polymer were investigated besides the effect of starch content on the mechanical and degradation characteristics of the sago starch-incorporated LDPE polymer.

TABLE 2: Polymer composition with TMPTA cross-linker and EB irradiation doses.

Sample ID	% LDPE	% Starch	% TMPTA	EB irradiation (kGy)
LDPE50-STARCH50	50	50	0	0
TMPTA1-DOSE0	50	50	1	0
TMPTA3-DOSE0	50	50	3	0
TMPTA5-DOSE0	50	50	5	0
TMPTA1-DOSE10	50	50	1	10
TMPTA3-DOSE10	50	50	3	10
TMPTA5-DOSE10	50	50	5	10
TMPTA1-DOSE30	50	50	1	30
TMPTA3-DOSE30	50	50	3	30
TMPTA5-DOSE30	50	50	5	30
TMPTA1-DOSE50	50	50	1	50
TMPTA3-DOSE50	50	50	3	50
TMPTA5-DOSE50	50	50	5	50
TMPTA1-DOSE70	50	50	1	70
TMPTA3-DOSE70	50	50	3	70
TMPTA5-DOSE70	50	50	5	70
TMPTA1-DOSE100	50	50	1	100
TMPTA3-DOSE100	50	50	3	100
TMPTA5-DOSE100	50	50	5	100

2. Materials and Experimental Methods

2.1. Synthesis of Polymer. In the 1st phase of the experiment, five types of LDPE polymer samples were produced in combination with various percentages of starch. Low-density polyethylene (LDPE) and sago starch (Igan Sago Industries Sdn. Bhd.) were weighed by a toploader balance (AND, model GF-3000) with various percentages of LDPE and starch as reported in Table 1. The weighed LDPE and starch were premixed and blended in a brabender mixer for six minutes at 135°C. The blends produced from the brabender were then transferred to a hydraulic machine (Lab Tech) for compression. The compression process involved four steps: preheating, venting, pressing, and cooling. First three steps involved hot pressing, while the last step was cold pressing. The samples were placed in the middle of a square steel frame which was then clipped between two other steel plates. There were two thinner and smoother steel plates between these three layers of thicker plates. Polyethylene terephthalate (PET) sheets were used in between polymer surface and steel plate to prevent sticking the polymer samples to the steel plate. The molding temperature and pressure were 150°C and 10 MPa, respectively. Hot press and cold press processes were conducted using the same machine but different moulds. Upon hot pressing that was conducted at the upper part of the machine, samples were transferred to the bottom part for cold pressing. Mechanical and biodegradation tests were performed over these 1st phase samples.

In the 2nd phase, the effects of crosslinking with trimethylolpropane triacrylate (TMPTA) and electron beam (EB) irradiation on the mechanical and thermal properties of the polymer samples were investigated. To avoid overwhelming of data, the polymer samples used in this second phase experiments consisted of 50% LDPE and 50% starch. Various percentages of TMPTA were added to the LDPE-starch blends during the blending process and then irradiated accordingly with various doses of EB as presented in Table 2. Upon crosslinking, the polymer samples were pressed under the hydraulic machine and irradiated with EB at the doses of 10 kGy, 30 kGy, 50 kGy, 70 kGy, and 100 kGy. The energy and current were set at 1.5 MeV and 30 mA, respectively.

2.2. Sample Preparation: Dumb-Bell-Shaped Cutting. Upon completion of synthesis process, the polymer samples were cut into dumb-bell-shaped by a dumb-bell cutter (model SDL-100; Dumb-Bell Co., Ltd.) according to ASTMD 1882L standard. The dumb-bell-shaped samples had smooth surface especially, in the neck section that avoided stress concentration during mechanical test.

2.3. Mechanical Characterization: Tensile Test. Tensile test was carried out to investigate the influence of starch content and further the effect of modification processes, namely, crosslinking with TMPTA and EB irradiation on the polymer's mechanical property. The dumb-bell-shaped sample thicknesses were measured by a thickness gauge (Mitutoyo, model EMD-57B-11M) and were keyed in the test system. Tensile test was performed by using a uniaxial testing system (Instron 3365) and a 5 kN load cell (Canton, MA, USA) according to ASTMD1882L standard. The cross-head speed was set at 10 mm/min. For each type of polymer, five samples were tested. Young's modulus was calculated as stress divided by strain to evaluate the mechanical property of the polymer samples. A Student's t-test was performed to compare mean values from all independent sample groups using a Minitab statistics software version 12.2 (Minitab Inc., State College, PA, USA) at a significance level of 0.05.

2.4. Degradation Study. The in vitro degradation study was performed on the dumb-bell-shaped LDPE polymer samples containing various percentages of starch (10%, 30%, 50%, and 70%) by burying the polymer samples at the exterior under the soil at a depth of 2 feet for a period of one month. Each type of polymer had five samples. This study was to primarily investigate the variation of degradation kinetics due to the variation of starch content. The degradation phenomenon was evaluated through mass loss and change in surface morphology. Prior to burial, the polymer samples were characterized by weighing and recording their initial mass using an electronic balance with a resolution of 0.1 mg. After one month of burial, the samples were dug out and cleaned to ensure complete removal of soil/mud. Samples were then placed in an area with sufficient ventilation for natural drying. The dried degraded samples were weighed using the same electronic balance as carried out before starting degradation. Subsequently, the percentage of mass loss of respective sample was measured as follows:

$$\% \text{ Mass Loss} = \frac{M_i - M_f}{M_i} \times 100, \quad (1)$$

where, M_i is the initial mass (i.e., mass before degradation) and M_f is the final mass (i.e., mass after degradation).

2.5. Thermal Analysis: Differential Scanning Calorimetry (DSC). The thermal characteristic of cross-linked and EB irradiated polymer sample was determined by differential scanning calorimeter (TA Instruments DSC 2910, New Castle, DE, USA). An indium standard was used to calibrate the instrument. The sample weight of 5 mg was taken for scanning. All samples were placed in aluminum pans and scanned from −70°C to 200°C at a rate of 10°C/min, using argon as purge gas. Five samples for each type of polymer were analyzed. The DSC analysis provided the melting point of the polymer.

3. Results and Discussion

3.1. Mechanical Characteristics. The mechanical experiment (tensile test) initially investigated the influence of starch content and further the effect of modification processes, namely, cross-linking with TMPTA and EB irradiation on the polymer's mechanical property. The starch content and modification process (crosslinking and EB irradiation) had direct influences on the mechanical characteristics of the starch-incorporated LPDE polymer. The elongation at break of the polymer samples decreased with the increase of starch content, while Young's modulus increased with the increase of starch content as demonstrated in Figures 1 and 2. The results are in agreement with the findings of some other

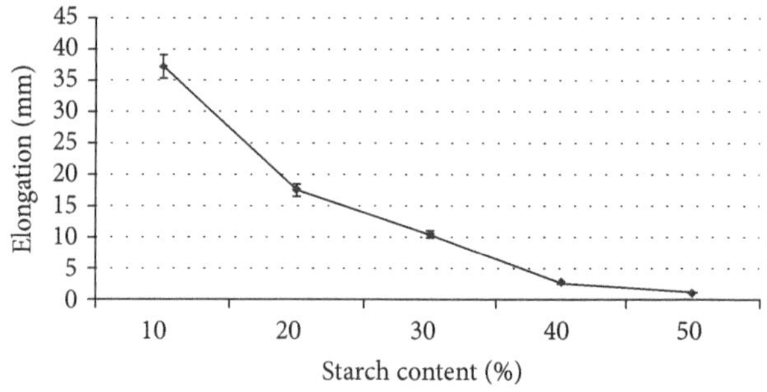

FIGURE 1: Variation of elongation with starch content of LDPE/starch blends without modification.

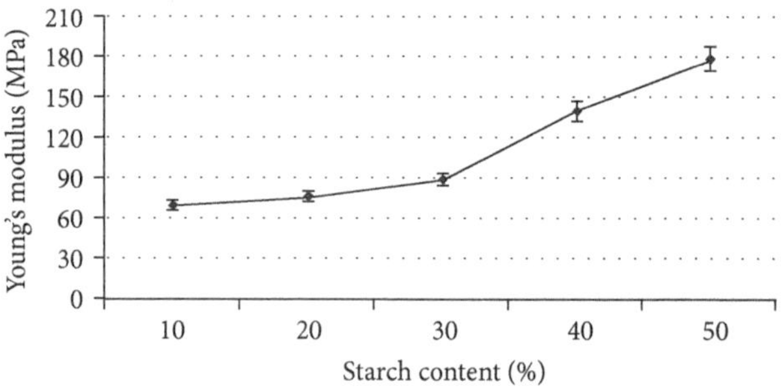

FIGURE 2: Variation of Young's modulus with starch content of LDPE/starch blends without modification.

studies [18–23]. The elongation at break which decreased with the starch content could be attributed to the phenomenon that the starch granule containing hydroxyl groups on its surface is highly hydrophilic, whereas LDPE is nonpolar. In such system, strong interfacial bond (e.g., hydrogen bond) does not form between LDPE and starch. Furthermore, the presence/absorption of moisture by the starch at the LDPE-starch interface weakens the weak interfacial adhesion. Therefore, the starch granules do not elongate along with the LDPE and thus give rise to easy crack generation and/or propagation eventually resulting in fracture of the polymer sample at lower values of elongation. Figure 1 shows that the drop in elongation at break of the polymer became more prominent as the starch content increased to higher percentage. This could be due to the reason that, at higher starch contents, filler-filler interaction becomes more pronounced than filler-matrix interaction which reduces the effective cross-sectional area of the polymer sample caused by the presence of starch particles. The applied stress is not transferred accordingly from the polymer matrix to the rigid starch particles, and hence the effective stress experienced by the matrix is essentially higher [19].

Unlike the elongation, Young's modulus of the synthesized polymer increased with the increase of starch content (Figure 2). As mentioned, with the increase of starch content the filler-filler interaction becomes more pronounced than filler-matrix interaction, which leads to the agglomeration of starch granules that are inherently more rigid or stiffer than the LDPE matrix [28]. Rigid filler is generally known to increase the modulus of a blend even in the case where true

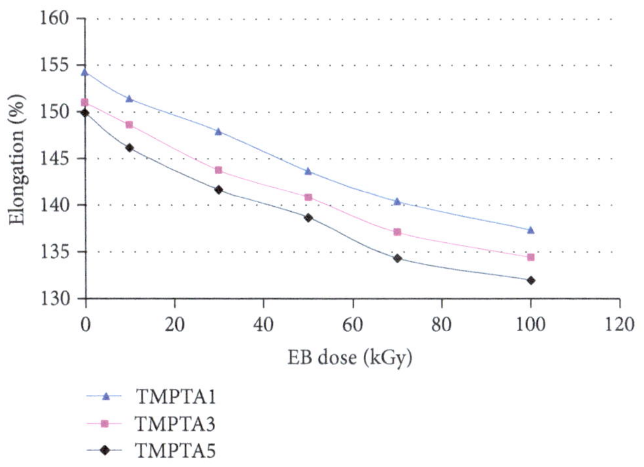

Figure 3: Variation of elongation with TMPTA cross-linker and electron beam irradiation.

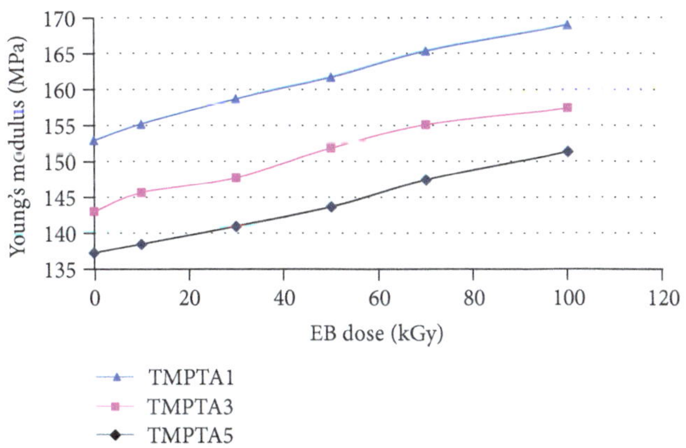

Figure 4: Variation of Young's modulus with TMPTA cross-linker and electron beam irradiation.

reinforcement does not occur [24]. Therefore, the increase in modulus of the polymer blend with the increase of starch content is considered to be due to the higher stiffness of the starch granules.

It was observed that the increase of TMPTA cross-linker resulted in decrease of elongation and consequently, increased Young's modulus as shown in Figures 3 and 4. Likewise, the increase of EB irradiation dose also decreased the elongation and thus increased Young's modulus. It is believed that the variation of starch content, cross-linker, and EB irradiation influenced the bond strength of the polymer. As a result, the polymer's mechanical properties were accordingly affected. For example, the elongation became tougher, thus caused higher Young's modulus. However, the detail explanations of these phenomena could be confirmed through intensive bond strength analysis that is out of scope in this preliminary study.

3.2. Thermal Property. The effects of TMPTA cross-linker and EB irradiation on the thermal property (melting temperature) of the polymer were determined via DSC analysis.

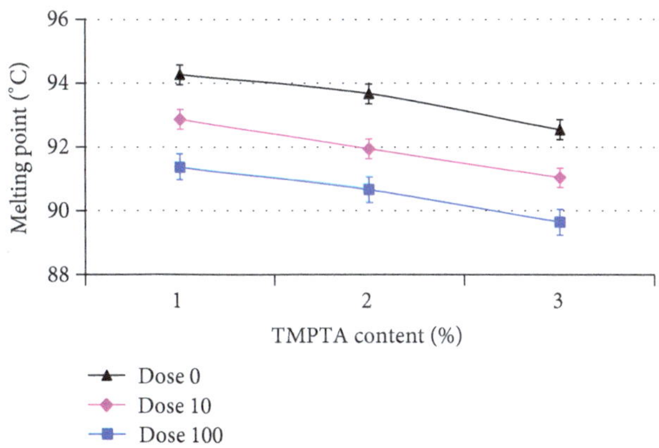

Figure 5: Influence of TMPTA cross-linker and electron beam irradiation on the polymer's melting temperature.

Table 3: The mass of LPDE polymer samples (before and after degradation) containing various percentages of starch.

Starch Content (%)	Mass (before degradation) (g)	Mass (after degradation) (g)	Mass loss (%)
0	0.464	0.464	0.00
10	0.452	0.452	0.05
30	0.492	0.496	3.81
50	0.53	0.528	5.37
70	0.632	0.562	11.07

The overall influence of TMPTA cross-linker and EB irradiation on the polymer's melting temperature is presented in Figure 5. The DSC result shows that the increase of EB dosage lowered the polymer's melting points. Similarly, the increase of TMPTA cross-linker also lowered the melting points. The graph illustrates that all the polymer samples modified with TMTA crosslinker and EB irradiation had lower melting points compared to the samples without any modification. As discussed in the previous section, the cross-linker and EB irradiation could have influenced the bond strength of the polymer which, consequently, affected the polymer's melting temperature.

3.2. Thermal Property. Technically, all polymers regardless of their chemical structure or origin degrade under appropriate conditions. However, the term "nondegradable polymer" is meant to indicate the polymer that does not degrade during use or even after very long-term use (e.g., decades to centuries) rather than not to degrade at all [29–31]. The primary rationale of this in vitro degradation study was to understand the degradation behavior and kinetics of the synthesized LPDE polymer and investigate the variation of degradation kinetics due to the variation of starch content in the polymer. The measured masses before and after degradation of the polymer samples are presented in Table 3 from which ultimately the mass loss (i.e., degradation) is calculated.

Pure LDPE and low starch containing (e.g. 10%) LDPE showed no apparent mass loss. This result is in agreement

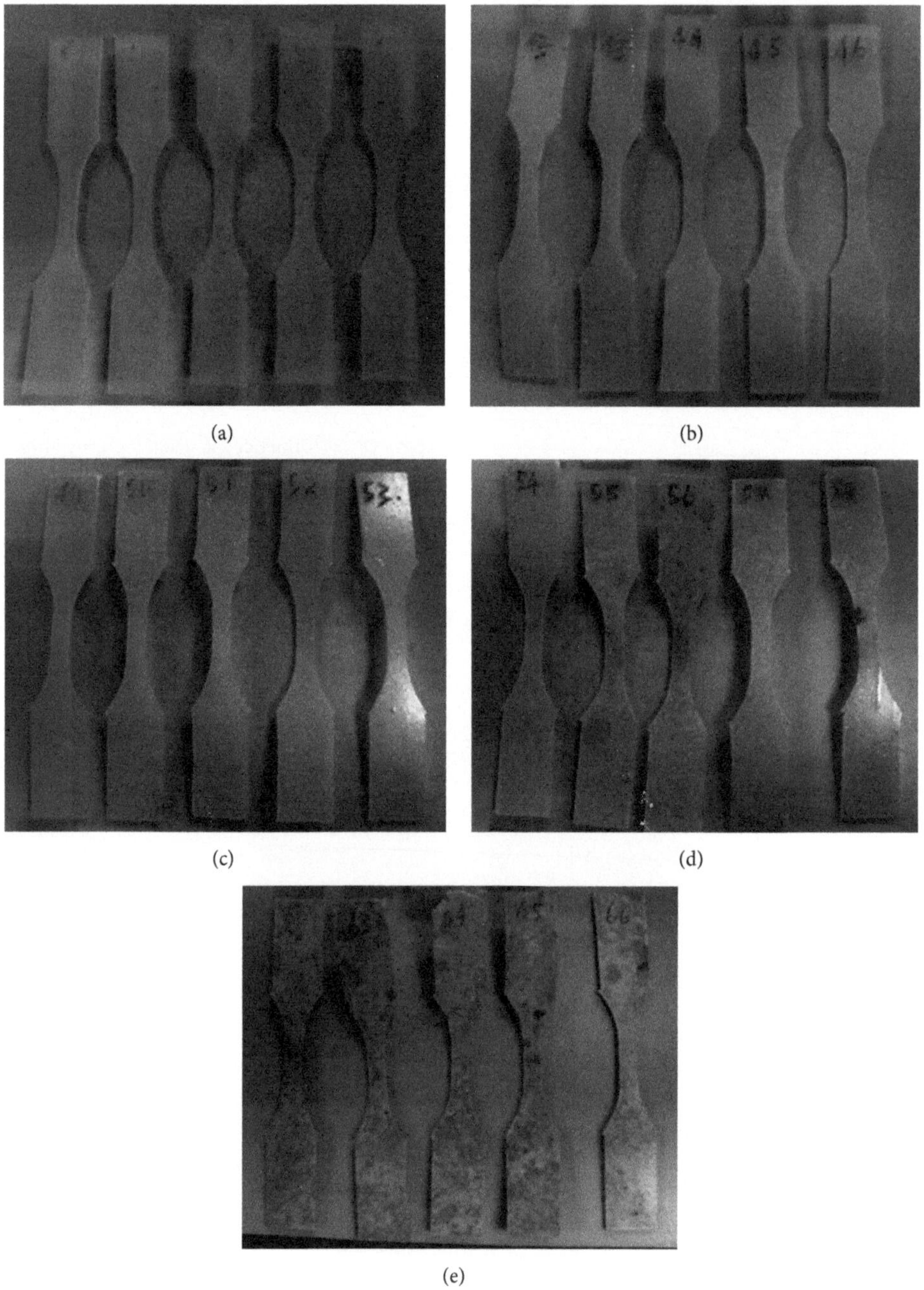

FIGURE 6: Physical appearances of LDPE polymer samples containing various percentages of starch after a one-month long burial that demonstrate different levels of degradation: (a) Pure LDPE, (b) LDPE 90 and Starch 10, (c) LDPE 70 & Starch 30, (d) LDPE 50 & Starch 50, and (e) LDPE 30 & Starch 70.

with the fact that the pure LDPE is considered to be technically non-degradable, and there should not be any mass loss. Virtually the LDPE polymer sample containing 10% starch should demonstrate some mass loss (i.e., degradation). However, due to time constraints (e.g., too short burial duration) the degradation might not be observed appreciably. Besides, the presence of starch (only 10%) might be too low to render the polymer degrade. The LDPE polymer samples containing 30%, 50%, and 70% starch demonstrated mass loss of 3.81%, 5.37%, and 11.07%, respectively. These results reveal that higher starch content enhances the degradation kinetics and thus increases mass loss. This could be due to the hydrophilic nature of starch that has been argued in another study [6]. The starch being hydrophilic in nature retains moisture that contributes to the degradation of the polymer. The higher the starch content in the polymer, the higher the moisture content that renders faster degradation. The polymer sample's gross morphology was observed to be changed physically; for example, the surface roughened over the degradation period as seen in Figure 6. The increased starch content resulted in higher surface roughness of the samples due to increased degradation. The weight loss and change in physical appearance (i.e., surface roughness) of the sample in the soil could be considered as an evidence of biodegradation of this polymer in the landfills or natural environment. The results indicate that the incorporation of hydrophilic starch into hydrophobic LDPE enhances the hydrophilicity and degradability of the overall polymer. Similar degradation phenomena of the LDPE incorporated with starch were observed in some other studies [20, 25, 27]. Therefore, the degradation characteristic of the starch-mixed LDPE polymer could be modulated by manipulating the starch content in the polymer. Indeed, the polymer should be developed with essentially a controlled degradation characteristic while maintaining the required strength of the polymeric object during its designed life time for a particular application.

4. Conclusions

The LDPE polymer was successfully synthesized incorporating sago starch and utilizing some modification processes like, TMPTA cross-linker and EB irradiation. The overall process provides the freedom to develop a large variety of polymers with various mechanical, thermal, and degradation properties. The characterization results indicate that the variations of starch content, percentage of TMPTA crosslinker, and EB irradiation dose are very effective means to modulate the polymer's mechanical, thermal, and degradation properties. The polymer's strength (i.e., Young's modulus) increased with the increase of starch content, percentage of TMPTA, and EB irradiation dose, while the ductility of the polymer decreased with the increase of the said parameters. The melting temperature decreased with the increase of TMPTA percentage and EB irradiation dose. The degradation of the polymer was enhanced with the increase of starch content. In conclusion, the starch-incorporated LDPE polymer properties could be modulated by manipulating the starch content and modification processes (e.g., TMPTA crosslinker and EB irradiation dose) for tailored applications.

Conflict of Interests

The authors would like to clarify that there is no direct financial relation with the software and corporation mentioned in this paper that might lead to a conflict of interests for any of the authors.

References

[1] G. Swift, "Requirements for biodegradable water-soluble polymers," *Polymer Degradation and Stability*, vol. 59, no. 1–3, pp. 19–24, 1998.

[2] P. M. Subramanian, "Plastics recycling and waste management in the US," *Resources, Conservation and Recycling*, vol. 28, no. 3-4, pp. 253–263, 2000.

[3] V. A. Fomin, "Biodegradable polymers, their present state and future prospects," *Progress in Rubber and Plastics Technology*, vol. 17, no. 3, pp. 186–204, 2001.

[4] R. Leaversuch, "Biodegradable polyesters: packaging goes green," *Plastics Technology*, vol. 48, no. 9, pp. 66–78, 2002.

[5] P. Mukhopadhyay, "Emerging trends in plastic technology," *Plastics Engineering*, vol. 58, pp. 28–35, 2002.

[6] K. F. Mulder, "Sustainable consumption and production of plastics?" *Technological Forecasting and Social Change*, vol. 58, no. 1-2, pp. 105–124, 1998.

[7] J. Chen, Y. Maekawa, M. Asano, and M. Yoshida, "Double crosslinked polyetheretherketone-based polymer electrolyte membranes prepared by radiation and thermal crosslinking techniques," *Polymer*, vol. 48, no. 20, pp. 6002–6009, 2007.

[8] B. Ratner, A. Hoffman, F. Schoen, and J. Lemons, *Biomaterials Science: An Introduction to Materials in Medicine*, Academic Press, San Diego, Calif, USA, 1996.

[9] M. Cerrada, R. Benavente, M. Fernández-García, E. Pérez, J. Campos, and M. Ribeiro, "Crosslinking in metallocene ethylene-*co*-5,7-dimethylocta-1,6-diene copolymers initiated by electron-beam irradiation," *Polymer*, vol. 50, pp. 1095–1102, 2009.

[10] C. Götz, U. A. Handge, M. Piatek, M. El Fray, and V. Altstädt, "Influence of e-beam irradiation on the dynamic creep and fatigue properties of poly(aliphatic/aromatic-ester) copolymers for biomedical applications," *Polymer*, vol. 50, no. 23, pp. 5499–5507, 2009.

[11] A. Magoń and M. Pyda, "Study of crystalline and amorphous phases of biodegradable poly(lactic acid) by advanced thermal analysis," *Polymer*, vol. 50, no. 16, pp. 3967–3973, 2009.

[12] S. Chang, H. Ismail, and Q. Ashan, "Effect of maleic anhydride on kenaf dust filled polycaprolactone/thermoplastic sago starch composites," *Bioresources*, vol. 7, pp. 1594–1616, 2012.

[13] M. Wahab, H. Ismail, and N. Othman, "Compatibilization effects of PE-g-MA on mechanical, thermal and swelling properties of high density polyethylene/natural rubber/thermoplastic tapioca starch blends," *Polymer-Plastics Technology and Engineering*, vol. 51, pp. 298–303, 2012.

[14] A. Zuraida, Y. Yusliza, H. Anuar, and R. Mohd Khairul Muhaimin, "The effect of water and citric acid on sago starch bio-plastics," *International Food Research Journal*, vol. 19, no. 2, pp. 715–719, 2012.

[15] A. Zuraida, Y. Yusliza, O. Nurizan, H. Anuar, H. Zahurin, and S. Noorasikin, "Properties of montmorillonite-reinforced thermoplastic sago starch composites," *Advanced Materials Research*, vol. 445, pp. 469–474, 2012.

[16] D. Wise, *Biomaterial and Bioengineering Handbook*, Marcel Dekker, New York, NY, USA, 2000.

[17] E. M. Salmoral, M. E. Gonzalez, and M. P. Mariscal, "Biodegradable plastic made from bean products," *Industrial Crops and Products*, vol. 11, no. 2-3, pp. 217–225, 2000.

[18] O. Martin, E. Schwach, L. Avérous, and Y. Couturier, "Properties of biodegradable multilayer films based on plasticized wheat," *Starch*, vol. 53, pp. 372–380, 2001.

[19] R. Chandra and R. Rustgi, "Biodegradable polymers," *Progress in Polymer Science*, vol. 23, no. 7, pp. 1273–1335, 1998.

[20] M. Borghei, A. Karbassi, S. Khoramnejadian, A. Oromiehie, and A. H. Javid, "Microbial biodegradable potato starch based low density polyethylene," *African Journal of Biotechnology*, vol. 9, no. 26, pp. 4075–4080, 2010.

[21] S. M. Goheen and R. P. Wool, "Degradation of polyethylene-starch blends in soil," *Journal of Applied Polymer Science*, vol. 42, no. 10, pp. 2691–2701, 1991.

[22] J. L. Willett, "Mechanical properties of LDPE/granular starch composites," *Journal of Applied Polymer Science*, vol. 54, no. 11, pp. 1685–1695, 1994.

[23] B. G. Kang, S. H. Yoon, S. H. Lee, J. E. Yie, B. S. Yoon, and M. H. O. Suh, "Studies on the physical properties of modified starch-filled HDPE film," *Journal of Applied Polymer Science*, vol. 60, no. 11, pp. 1977–1984, 1996.

[24] I. Danjaji, R. Nawang, U. Ishiaku, H. Ismail, and Z. Ishak, "Sago starch-filled linear low-density polyethylene (LLDPE) films: their mechanical properties and water absorption," *Journal of Applied Polymer Science*, vol. 79, pp. 29–37, 2001.

[25] I. D. Danjaji, R. Nawang, U. S. Ishiaku, H. Ismail, and Z. Ishak, "Degradation studies and moisture uptake of sago-starch-filled linear low-density polyethylene composites," *Polymer Testing*, vol. 21, no. 1, pp. 75–81, 2002.

[26] N. Sharmaa, H. Ismaila, and Z. Ishaka, "Mechanical, water absorption and enzymatic degradation properties of sago starch filled linear low density polyethylene," *International Journal of Polymeric Materials*, vol. 47, pp. 351–366, 2000.

[27] W. A. W. Abdul Rahman, R. R. Ali, and N. Zakaria, "tudies on biodegradability, morphology and mechanical properties of low density polyehtylene/sago based blends," in *Proceedings of the 1st International Conference on Natural Resources Engineering & Technology*, Putrajaya, Malaysia, July 2006.

[28] R. Nawang, I. D. Danjaji, U. S. Ishiaku, H. Ismail, and Z. A. Mohd Ishak, "Mechanical properties of sago starch-filled linear low density polyethylene (LLDPE) composites," *Polymer Testing*, vol. 20, no. 2, pp. 167–172, 2001.

[29] A. Gopferich, "Mechanisms of polymer degradation and erosion," *Biomaterials*, vol. 17, pp. 103–114, 1996.

[30] J. Pan, X. Han, W. Niu, and R. E. Cameron, "A model for biodegradation of composite materials made of polyesters and tricalcium phosphates," *Biomaterials*, vol. 32, no. 9, pp. 2248–2255, 2011.

[31] R. E. Sallach, W. Cui, F. Balderrama et al., "Long-term biostability of self-assembling protein polymers in the absence of covalent crosslinking," *Biomaterials*, vol. 31, no. 4, pp. 779–791, 2010.

Parameter Optimization in the Synthesis of BZT Ceramics to Achieve Good Dielectric Properties

A. Frattini,[1,2] A. Di Loreto,[1,2] and O. De Sanctis[2]

[1] *Área Física, Departamento de Química-Física, FCByF, UNR, Suipacha 531, 2000 Rosario, Argentina*
[2] *Laboratorio de Materiales Cerámicos, FCEIyA, IFIR, UNR, Pellegrini 250, 2000 Rosario, Argentina*

Correspondence should be addressed to A. Di Loreto; diloreto@fceia.unr.edu.ar

Academic Editor: Antoni Morawski

The powder synthesis of barium zirconate titanate (BZT) ($BaZr_xTi_{(1-x)}O_3$) from the mechanochemical activation of $BaCO_3$, ZrO_2, and TiO_2 was studied. The grinding effect, by using a planetary ball milling, on the crystallization temperature of BZT powders ($x = 0.05$) was analyzed. X-ray diffractometry, differential thermal analysis, thermogravimetric analysis, and scanning electronic microscopy (SEM) were used as characterization methods. The crystallization behavior of powders activated by high-energy grinding and the effect of grinding time on the BZT crystallization were analyzed. After grinding by 4 h, the $BaZr_{(0.05)}Ti_{(0.95)}O_3$ sample was almost fully crystallized at 800°C. The results of dielectric and ferroelectric properties show that high-energy ball milling is a practical and promising way to prepare BZT ceramics.

1. Introduction

Ferroelectric materials based in Pb have been studied for many years for its important applications in piezoelectric, pyroelectric, and optical devices [1–5]. Lead-free compositions can be of great interest for environmentally friendly applications such as actuators and dielectrics for capacitors. $BaTiO_3$ is the most widely used ferroelectric material in ceramic capacitors and is known for its large electrochemical coupling factor [2, 3]. Within $BaTiO_3$ family, barium zirconate titanate ($Ba(Zr_xTi_{(1-x)})O_3$, BZT) has become one of the most attractive materials because it has been reported that the zirconium substitutions into the titanium lattices enhance the dielectric and piezoelectric properties [6, 7]. Furthermore, Curie temperature can be shifted below room temperature by doping with Zr [8, 9]. However, under common preparation conditions, BZT ceramics have very high sintering temperature that does not fit the industrial requirements. In order to decrease the sintering temperature, it is necessary to produce powders with fine particle sizes and homogeneous distribution.

Mechanochemical activation by mixing of high energy (high-energy ball milling) has become important to produce solid-state reactions at lower temperatures than those used in conventional methods of powder preparation [10–14]. This mechanical technique is unique because of its advantages: (i) low cost of starting materials, (ii) low operating temperature, and (iii) higher sinterability of milled powders.

The purpose of this work is to reach low energy cost in the synthesis of BZT ceramics without expense of their electrical properties. The effects of increasing the milling time of the powders and, on the other hand, of lowering the powder calcination temperature and/or sintering temperature of the pellets are analyzed.

2. Experimental

BZT ceramics were prepared via conventionally mixed-oxide method by high-energy ball milling. Commercial powders of barium carbonate ($BaCO_3$) (99.98% Sigma-Aldrich), titanium dioxide (TiO_2) (99.98% Sigma-Aldrich), and zirconium oxide (ZrO_2) (99.9% Sigma-Aldrich), were used as starting materials with the nominal composition of $BaZr_{0.05}Ti_{0.95}O_3$. The precursors were ground in a planetary ball milling (Torrey Hills Technologies ND 0.4L) during 1, 2, and 4 h, into a zirconia jars. Simultaneous differential thermal analysis (DTA) and thermogravimetric analysis (TGA) measurements were carried out on the as-obtained powder from 20°C to 1200°C at 10°C/min heating rate, with 50 mL/min of O_2 flux into alumina crucibles, using a Shimadzu DTG-60 analyzer. The mixed powders were calcined between 600°C and 1200°C for 2 h in the air. Afterwards, powders were analyzed by X-ray diffraction (XRD-Philips PW1700 diffractometer), scanning electron microscopy (SEM Philips 505), and high-resolution SEM (FEI QUANTA 200).

The powders with two weight percent PVB binder were uniaxially pressed (200 MPa) into green pellets with 10 mm of diameter and 1 mm of thickness. The sintering behavior of these pellets was monitored at constant heating rate (5°C/min) in a Theta Dilatronic dilatometer. Structural analyses of sintered samples at 1350°C and 1400°C were performed by XRD and SEM.

To measure the electric properties, the surfaces were polished, and silver electrodes were painted on both sides of the pellets. The capacitance and dielectric loss of the samples were measured with a HP 4192A impedance analyzer, and the hysteresis loops were examined by using a conventional Sawyer-Tower tester at 50 Hz.

3. Results and Discussion

Figures 1 and 2 show the DTA-TGA curves of the oxides mixture milled for 1 and 4 h, respectively, when they were heated from room temperature to 1200°C. Three endothermic peaks (labeled as (1), (2), and (3)) are observed in both DTA curves. The first two peaks match with weight losses, whereas no weight loss is seen for the last one. It is well known [15, 16] that the first two peaks correspond to reactions of $BaCO_3$ decomposition and synthesis of $BaTiO_3$, including the formation of a transient Ba_2TiO_4 phase. These reactions are as follows:

$$BaCO_3 + TiO_2 \longrightarrow BaTiO_3 + CO_2 \quad (1)$$

where the CO_2 emission causes the first weight loss realized in the TGA graph (around 600°C),

$$BaCO_3 + BaTiO_3 \longrightarrow Ba_2TiO_4 + CO_2 \quad (2)$$

where again there is weight loss by the CO_2 emission, and

$$Ba_2TiO_4 + TiO_2 \longrightarrow 2BaTiO_3 \quad (3)$$

which has no weight loss associated.

By comparing both figures, it is noted that the peak (1) is broader and deeper in Figure 2 which means further development of reaction (1) in the sample milled by 4 h and matches with the greater mass loss (about double) observed in the TGA curve. This behavior also agrees with the smaller depth of peak (2) in Figure 2 since more $BaCO_3$ has decomposed and also with the 800°C XRD patterns in Figures 3(a) and 3(b), where it is quite clear that there is fewer remaining $BaCO_3$ in the sample milled for 4 h.

It is noteworthy that the formation of Ba_2TiO_4 phase advances at the temperature range between 600°C and 900°C due to the diffusive nature of this reaction [15]. The peak (2) in Figure 1 is quite sharp since superposes the $\gamma \rightarrow \beta$ allotropic transition of $BaCO_3$ (at 827°C) [17] with reaction (2). As it is observed, this peak in Figure 2 is not so deep because that transition is hindered by the lattice deformation resulting of the greater milling time.

On the other hand, Figure 2 shows an increase in the mass loss rate (see the derivative of the TGA curve from the inset) because of the higher reactivity of the powders and the faster kinetic of the reaction. Furthermore, the mass loss finishes at 885°C in the BZT-4 h sample and at 920°C for the BZT-1 h one. By revising DTA graphs, only reaction (3) could be present from these temperatures until the maximum of hump (4). This temperature defines the synthesis of $BaTiO_3$ after accomplishment of all the reactions above mentioned.

Berbenni et al. [18] have obtained a final plateau in its TG curve at 730°C by applying high-energy milling for 159 h on BZT powders. By analyzing the temperatures where the mass loss is over, Berbenni's results and ours indicate that crystallization starts at lower temperatures if more energy

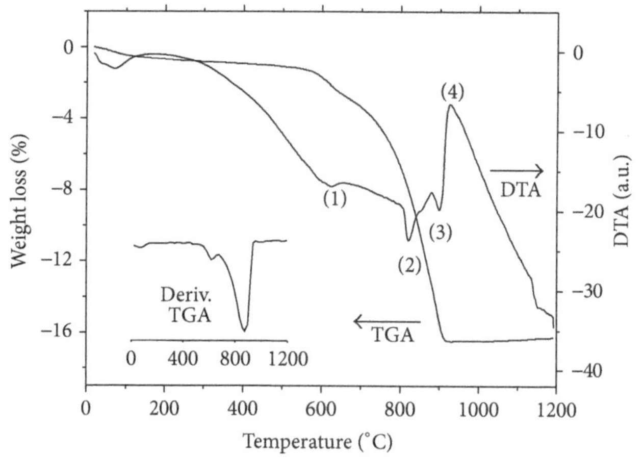

FIGURE 1: DTA-TG of BZT powder milled for 1 h (10°C/min heating rate, sample weight = 30.158 mg).

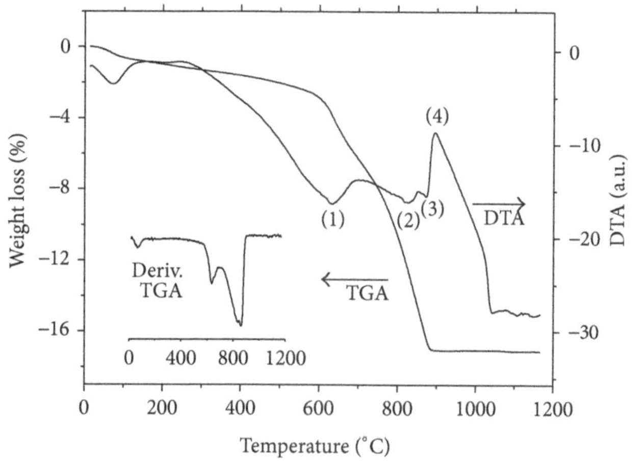

FIGURE 2: DTA-TG of BZT powder milled for 4 h (10°C/min heating rate, sample weight = 28.135 mg).

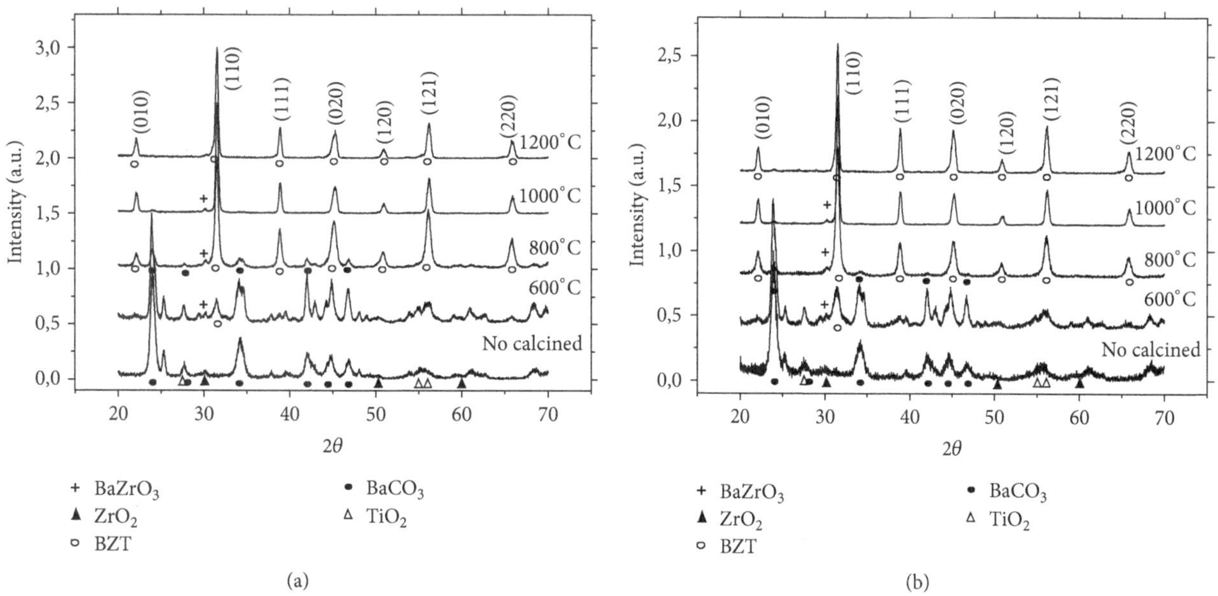

FIGURE 3: (a) XRD of 1 h milled powders as calcination temperature function. (b) XRD of 4 h milled powders as calcination temperature function.

from milling is delivered. That is to say, by increasing the milling time the powders will have more reactivity because of their smaller particle size and hence higher surface area. By comparing the end of hump (4) in both DTA curves, it is clear that the sample ground for 4 h has completely crystallized at 1040°C. The sample milled for 1 h seems to be completely crystallized at 1200°C. Finally, from Figures 1 and 2, it can be expected that Ba_2TiO_4 and BZT coexist only in samples calcined in temperatures ranging between 700°C and 900°C.

XRD patterns of both samples (Figures 3(a) and 3(b)) are on good agreement with the DT and TG analysis. The diffractogram of 600°C for the 4 h milled sample shows the main peak of BZT sharper and higher than the one corresponding to the 1 h milled sample. The 800°C pattern for the 1 h ground sample (Figure 3(a)) shows clear presence of $BaCO_3$ peaks yet (23.89°, 27.71°, 34.08°, 34.58°, 41.98°, 42.95°, and 46.77°) while they have disappeared in the 4 h ground sample (Figure 3(b)).

Furthermore, in the 800°C-4 h ground XRD pattern (Figure 3(b)), the BZT seems to be practically crystallized; just some peaks of $BaCO_3$ and the strongest of cubic $BaZrO_3$ ($2\theta = 30.14°$) are weakly seen. At 1000°C, there only remains the main peak of $BaZrO_3$, whereas, at 1200°C, the sample is fully crystallized. Even though the DTA and TGA graphics have suggested possible presence of Ba_2TiO_4 phase between 800°C and 900°C, this phase has not been detected in the XRD patterns. It is worth noting that XRD measurements were carried out on samples previously calcined for 2 h, while thermal measurements have a quite different thermal evolution (10°C/min). The calcination for 2 h could have increased the Ba_2TiO_4 diffusion, and there it will be transformed completely into $BaTiO_3$ [15].

Figure 4 shows the milling time effect on powders calcined at 1200°C for 2 h. The peaks are broad, showing poorly crystallized phases with amorphous phases for unmilled

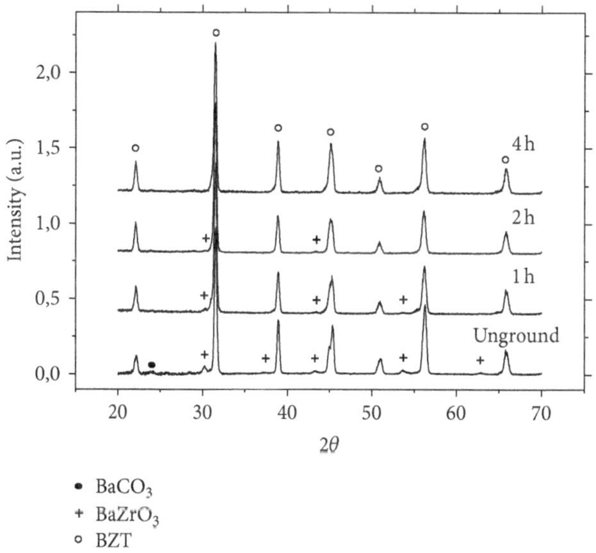

FIGURE 4: XRD of BZT powders milled during different times and calcined at 1200°C for 2 h.

powders. They become stronger and sharper when the samples are milled. It is observed that the unground sample presents peaks corresponding to $BaCO_3$ ($2\theta = 24.09°$) and $BaZrO_3$ ($2\theta = 30.14°$, 37.14°, 43.14°, 53.54°, and 62.67°). After grinding by 1 hour, the samples did not exhibit $BaCO_3$, but, however, the $BaZrO_3$ content decreased gradually as the milling time increased. Finally, a BZT structure free of spurious phases was obtained after 4 h of grinding and 2 h of calcination at 1200°C. The crystallite sizes of powders calcined at different times were calculated from the X-ray broadening of (110) peaks, using the Scherrer equation [19]. Crystallite sizes for powders treated at 1 h, 2 h, and 4 h are 32.9, 33.8, and

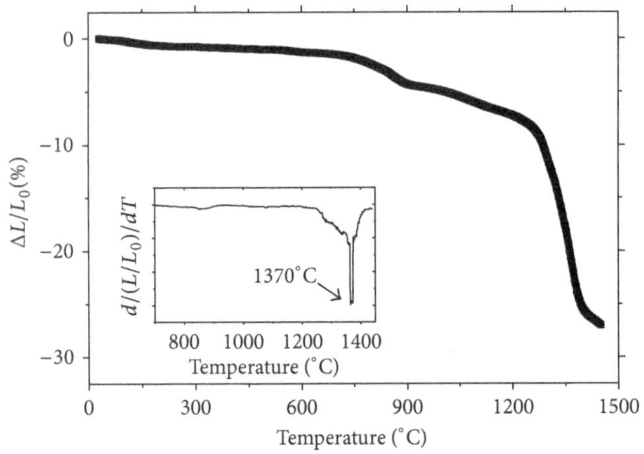

Figure 5: Linear shrinkage and linear shrinkage rate (inset) as a temperature function. The pellets were fabricated from powder calcined at 800°C.

35.7 nm, respectively. This indicates that BZT crystallizes and grows as the milling time increases.

The curves of linear shrinkage and its derivative with respect to temperature ($d(\Delta L/L_0)/dT$), both as functions of temperature, are shown in Figure 5. The first contraction between 750°C and 900°C indicates powder agglomeration during the sintering due to processes inside the agglomerates and between them [20]. Densification then increases sharply (~20%) in the range from 1250°C up to 1400°C. From these data, the 1350°C and 1400°C values were set as suitable temperatures for the sintering of ceramic pellets.

The SEM from Figure 6(a) indicates that the average grain size of ceramic is 250 nm, which means that grains have not grown in the sintering process with respect to the initial size of the crystallites in the powders (Figure 6(b)). However, certain porosity can be seen in the micrograph presumably induced by a binder evaporation and/or by a residual inhomogeneity in the shape of the calcined crystallites. Indeed, if the calcined crystallites are irregularly shaped prior to pressing, greater porosity will result in the sintered ceramic.

Figure 7 shows the XRD patterns of BZT pellets sintered at 1350°C and 1400°C. BZT pellets sintered at 1400°C display full crystallization in a perovskite phase, while small peaks of secondary phase are noted in the sample sintered at 1350°C. Moreover, larger grain growth is detected in the 1400°C sample. Precisely for both ceramics, crystallite sizes of 56.3 nm and 58.9 nm were calculated from half-width of (011) peaks.

The sintering temperature has significant influence on the electrical properties as displayed in Figure 8. The behavior of the dielectric constant (ε_r) with frequency [1 kHz–1 MHz] is similar in both samples (1350°C and 1400°C), but their values are higher, in all this range, for the sample sintered at 1400°C. The comparison between both samples in the range 1 kHz–10 kHz is reported at the inset. Furthermore, the dielectric loss (D) in both samples decreases with the increasing frequency in this range, this value being lower for the sample sintered at 1400°C. These two behaviors come from a further densification at higher temperature.

Figure 6: (a) SEM of the ceramic sintered at 1350°C. (b) HRSEM of ground powders by 4 h.

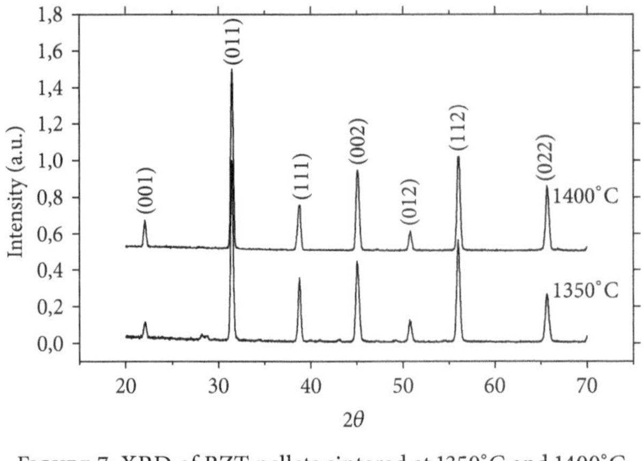

Figure 7: XRD of BZT pellets sintered at 1350°C and 1400°C.

Figure 9(a) shows the dielectric constant evolution of the BZT sample (sintered at 1400°C) with the temperature, where the three curves correspond to measurements at 10, 50, and 100 kHz. The ceramic presents two abrupt changes in dielectric constant corresponding to the orthorhombic-tetragonal (~60°C) and tetragonal-cubic (~125°C) phase transitions, displaying a typical ferroelectric behavior [21–23].

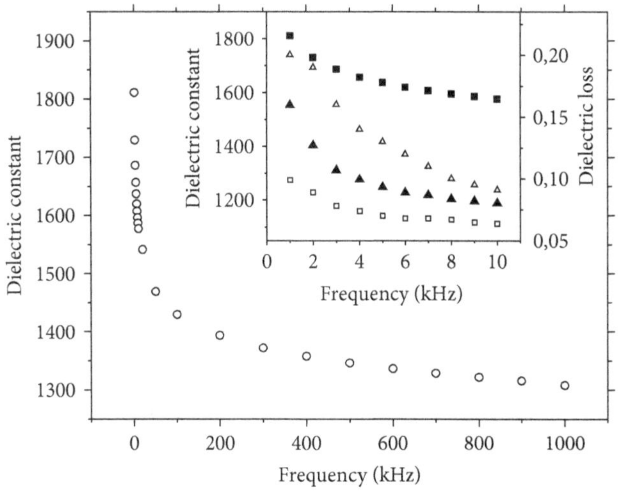

FIGURE 8: Dielectric constant (ε_r) of BZT sintered at 1400°C. Inset: zoom for {1–10 kHz} range, Δ: ε_r for sample sintered at 1400°C, ▲: ε_r for 1350°C sintered BZT, □: dielectric loss for 1350°C sintered BZT, ■: dielectric loss for 1400°C sintered BZT.

Besides, it is not observed a T_C shift (Curie temperature) with frequency, according to a nonrelaxor ferroelectric.

The Curie temperature increases slightly as does the sintering temperature; it can be seen in Figure 9(b). Moreover, the orthorhombic to tetragonal transition phase around 60°C is more evident in the sample sintered at 1400°C.

The hysteresis cycle measured at 50 Hz and room temperature, with the sample inside of silicon oil, is shown in Figure 10. Typical performances of normal ferroelectrics, according to the designed ceramic composition, were observed. The sample sintered at 1400°C shows a remnant polarization of 35 $\mu C/cm^2$ and a coercive electric field of about 2.5 kV/cm, while the 1350°C sintered ceramic shows lower values in its ferroelectric properties.

These results can be compared to the dielectric constant and the ferroelectric P-E hysteresis loops previously reported. Nanakorn et al. [23] have obtained 25 $\mu C/cm^2$ for remnant polarization and about 3000 as dielectric constant (ε_r) in ceramics of BZT ($x = 0.05$). Their procedure involved ball milling in acetone during 24 h, calcination at 1200°C for 2 h, milling for 24 h in ethanol, and sintering at 1300°C–1450°C for 2 h. Similar method has been used by Yu et al. [24] (except for sintering at 1500°C–1560°C from 2 to 15 h) and they have found values of 13.3 $\mu C/cm^2$ and $\varepsilon_r = 1600$ for $BaZr_{(0.05)}Ti_{(0.95)}O_3$ ceramics.

4. Conclusions

The purpose of reaching low energy cost in the synthesis of BZT ceramics was achieved. By high-energy milling, suitable crystallization of $BaZr_{0.05}Ti_{0.95}O_3$ powders was achieved at 800°C due to the increasing in their reactivity. From these powders, BZT ceramics were sintered for 2 h at 1350°C and 1400°C. BZT ceramics with high values of remnant polarization (35 $\mu C/cm^2$) and good values of dielectric constant (1700

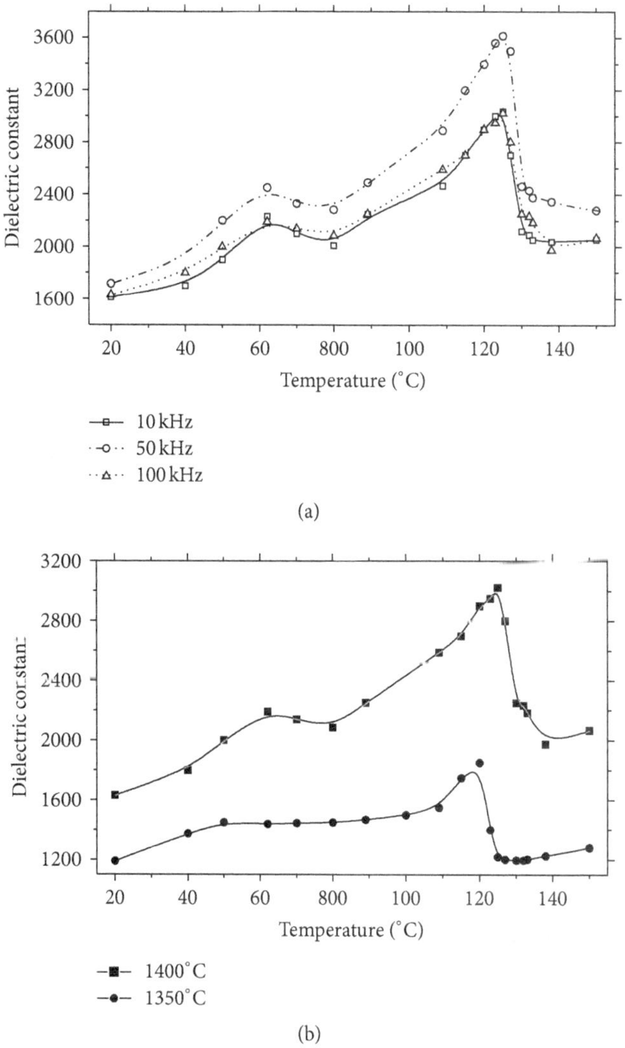

FIGURE 9: (a) Dielectric constant versus temperature and frequency for the sample sintered at 1400°C, measured at room temperature. (b) Dielectric constant versus temperature for ceramics sintered at 1350°C and 1400°C, measured at 100 kHz and room temperature.

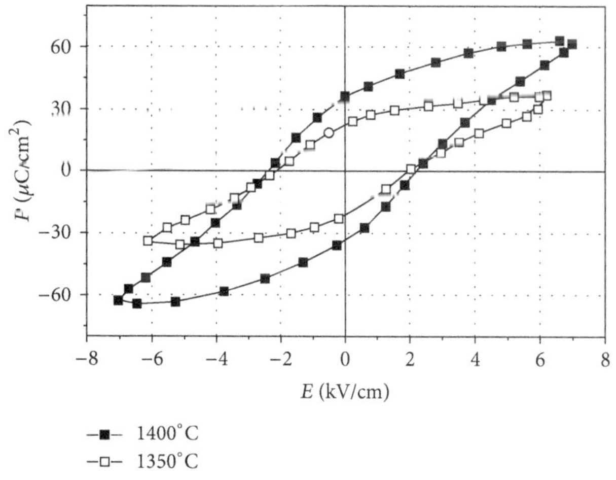

FIGURE 10: Polarization versus electric field for BZT sintered at 1400°C, measured at 50 Hz and room temperature.

at room temperature) were obtained. These results show an efficient way in the synthesis of BZT ceramics as compared to other proceedings reported in the literature.

Moreover, several common features of these ceramics were found in our samples, summarized as follows. The dielectric constant drops when the frequency rises at the range from 1 kHz to 1 MHz. The Curie temperature does not change with frequency but grows slightly with sintering temperature increasing. Furthermore, sintering at 1400°C improves the dielectric properties (rise of ε and decreasing of D) with regard to sintering at 1350°C.

References

[1] Y. Xu, *Ferroelectric Materials and Their Applications*, North Holland, 1991.

[2] M. T. Sebastian, *Dielectric Materials for Wireless Comunications*, Elsevier, 2008.

[3] R. Frank, *Understanding Smart Sensors*, Artech House, 2000.

[4] B. Jaffe, W. R. Cook, and H. Jaffe, *Piezoelectric Ceramics*, Academic Press, 1971.

[5] A. Moulson and J. Herbert, *Electroceramics: Materials-Properties-Applications*, John Wiley & Sons, 2003.

[6] U. Weber, G. Greuel, U. Boettger, S. Weber, D. Hennings, and R. Waser, "Dielectric properties of $Ba(Zr,Ti)O_3$-based ferroelectrics for capacitor applications," *Journal of the American Ceramic Society*, vol. 84, pp. 759–766, 2001.

[7] K.-i. Mimura, T. Naka, T. Shimura, W. Sakamoto, and T. Yogo, "Synthesis and dielectric properties of $(Ba,Ca)(Zr,Ti)O_3$ thin films using metal-organic precursor solutions," *Thin Solid Films*, vol. 516, no. 23, pp. 8408–8413, 2008.

[8] A. Simon, J. Ravez, and M. Maglione, "Relaxor properties of $Ba_{0.9}Bi_{0.067}(Ti_{1-x}Zr_x)O_3$ ceramics," *Solid State Sciences*, vol. 7, no. 8, pp. 925–930, 2005.

[9] S. Lee, M. Kwak, S. Moon, H. Ryu, Y. Kim, and K. Kang, *Integrated Ferroelectrics*, vol. 77, pp. 93–99, 2005.

[10] C. Suryanarayana, *Progress in Materials Science*, vol. 46, pp. 1–184, 2001.

[11] J. Lee, M. Choi, N. Hung et al., *Ceramics International*, vol. 33, pp. 1283–1286, 2007.

[12] R. Rajkhowa, L. Wang, and X. Wang, "Ultra-fine silk powder preparation through rotary and ball milling," *Powder Technology*, vol. 185, no. 1, pp. 87–95, 2008.

[13] C. B. Reid, J. S. Forrester, H. J. Goodshaw, E. H. Kisi, and G. J. Suaning, "A study in the mechanical milling of alumina powder," *Ceramics International*, vol. 34, no. 6, pp. 1551–1556, 2008.

[14] C. Gomez-Yañez, C. Benitez, and H. Balmori-Ramirez, "Mechanical activation of the synthesis reaction of $BaTiO_3$ from a mixture of $BaCO_3$ and TiO_2 powders," *Ceramics International*, vol. 26, no. 3, pp. 271–277, 2000.

[15] A. Beauger, J. C. Mutin, and J. C. Niepce, "Synthesis reaction of metatitanate $BaTiO_3$—part 1 Effect of the gaseous atmosphere upon the thermal evolution of the system $BaCO_3$-TiO_2," *Journal of Materials Science*, vol. 18, no. 10, pp. 3041–3046, 1983.

[16] L. B. Kong, J. Ma, H. Huang, R. F. Zhang, and W. X. Que, "Barium titanate derived from mechanochemically activated powders," *Journal of Alloys and Compounds*, vol. 337, no. 1-2, pp. 226–230, 2002.

[17] L. Templeton and J. Pask, "Formation of $BaTiO_3$ from $BaCO_3$ and TiO_2 in air and in CO_2," *Journal of the American Ceramic Society*, vol. 42, pp. 212–216, 1959.

[18] V. Berbenni, A. Marini, and G. Bruni, "Effect of mechanical milling on solid state formation of $BaTiO_3$ from $BaCO_3$-TiO_2 (rutile) mixtures," *Thermochimica Acta*, vol. 374, no. 2, pp. 151–158, 2001.

[19] A. West, *Solid State Chemistry and Its Applications*, John Wiley & Sons, 1992.

[20] F. Moura, A. Z. Simões, B. D. Stojanovic, M. A. Zaghete, E. Longo, and J. A. Varela, "Dielectric and ferroelectric characteristics of barium zirconate titanate ceramics prepared from mixed oxide method," *Journal of Alloys and Compounds*, vol. 462, no. 1-2, pp. 129–134, 2008.

[21] A. Ioachim, M. I. Toacsan, M. G. Banciu et al., "Synthesis and properties of $Ba(Zn_{1/3}Ta_{2/3})O_3$ for microwave and millimeter wave applications," *Thin Solid Films*, vol. 516, no. 7, pp. 1558–1562, 2008.

[22] T. Maiti, R. Guo, and A. S. Bhalla, "Structure-property phase diagram of $BaZr_xTi_{1-x}O_3$ system," *Journal of the American Ceramic Society*, vol. 91, no. 6, pp. 1769–1780, 2008.

[23] N. Nanakorn, P. Jalupoom, N. Vaneesorn, and A. Thanaboonsombut, "Dielectric and ferroelectric properties of $Ba(Zr_xTi_{1-x})O_3$ ceramics," *Ceramics International*, vol. 34, no. 4, pp. 779–782, 2008.

[24] Z. Yu, C. Ang, R. Guo, and A. S. Bhalla, "Piezoelectric and strain properties of $Ba(Ti_{1-x}Zr_x)O_3$ ceramics," *Journal of Applied Physics*, vol. 92, no. 3, 1489.

PVA/Polysaccharides Blended Films: Mechanical Properties

Fábio E. F. Silva, Maria Carolina B. Di-Medeiros, Karla A. Batista, and Kátia F. Fernandes

Laboratório de Química de Proteínas, Departamento de Bioquímica e Biologia Molecular, ICB 2, Universidade Federal de Goiás, Caxia Postal 131, 74001-970 Goiânia, GO, Brazil

Correspondence should be addressed to Kátia F. Fernandes; katia@icb.ufg.br

Academic Editor: Peter Chang

Blends of polyvinyl alcohol (PVA) and angico gum (AG) and/or cashew gum (CG) were used to produce films by casting method. Morphological and mechanical properties of these films were studied and compared to the properties of a commercial collagen membrane of bovine origin (MBO). The films presented thickness varying from 70 to 140 μm (PVA/AG) and 140 to 200 μm (PVA/CG). Macroscopic analysis showed that a PVA/CG film was very similar to MBO regarding the color and transparency. The higher values of tensile strength (TS) and elastic modulus (EM) were observed in the PVA/AG_{HCl} film. On the other hand, PVA/CG and PVA/CG-AG presented the highest value of percentage of elongation (E%). Pearson's Correlation Analysis revealed a positive correlation ($r = 0.83$) between TS and EM and a negative correlation ($r = -0.79$) between E% and EM. The PVA/CG film presented mechanical properties very similar to MBO, with the advantage of a higher E% (11.96) than MBO (2.94). The properties of the PVA blended films depended on the polysaccharide added in the blend, as well as the acid used as a catalyst. However, all produced films presented interesting mechanical characteristics which enables several biotechnological applications.

1. Introduction

There has been an expanding search for new materials with high performance at affordable costs in recent years [1]. Particularly important are the efforts to develop new materials with film-forming capacity for biomedical applications. Materials for biomedical applications or biomaterials have been defined as those intended to interface with biological systems to evaluate, treat, augment, or replace any tissue, organ, or function of the body [2]. Currently, bioresorbable and biodegradable membranes have been highlighted in the medical market, owing to their inherent biocompatibility [2, 3].

Polyvinyl alcohol (PVA) has excellent film-forming properties and has been frequently blended with synthetic and natural polymers due to its water-soluble, biodegradable, innocuous, noncarcinogenic, and biocompatible character [4, 5]. Blended materials assemble the properties of each material on its own. In general, blended materials are made of a material that acts as a matrix, while the other is imbedded in the matrix material improving its mechanical and physical properties. In this sense, the final properties of the blended material depend on the properties of the imbedded material [6].

In this scenario, blends of proteins or polysaccharides with polyvinyl alcohol (PVA) are promising biodegradable materials since the combination of these polymers may have beneficial effects on the characteristics of the blended material. However, the use of animal proteins in the formulation of biological membranes requires a highly purified material to avoid allergic reactions and increased inflammatory response. The laborious and time-consuming techniques of protein purification result in materials with a high cost.

On the other hand, polysaccharides have been explored as biomaterials, particularly when used as scaffolds for tissue regeneration and controlled drug releasing. Several plants produce gummous exudates mainly composed of high-molecular polysaccharides. Studies of several different gummous materials showed that they consist of complex branched polysaccharides with highly varied chemical and structural composition. Red angico (*Anadenanthera macrocarpa*) produces gummous exudates composed of arabinose (67%), galactose (24%), rhamnose (2%), and glucuronic acid (7%) [7, 8]. Cashew (*Anacardium occidentale*) also produces gummous exudates, constituted of galactose (72–61%), glucose (14–8%), arabinose (4.6–14%), rhamnose (3.2–7%), and glucuronic acid (4.7–5%) [9–11]. These gummous exudates are inexpensive, nontoxic, biocompatible, and biodegradable polymers with good rheological properties. These factors enable their use for biodegradable membrane fabrication [7, 9].

In this research, biodegradable films were manufactured by blending cashew gum and/or angico gum with PVA. The blended films were characterized according to their morphology and mechanical properties. Additionally, in order to evaluate the potential of these blended films, their properties were compared with those presented by a commercially available biomembrane containing collagen (bovine origin).

2. Material and Methods

2.1. Materials. Samples of the cashew exudates (nodules) (*Anacardium ocidentale*) were collected at CIALNE (Companhia de Alimentos do Nordeste) farm, Pacajus, CE, Brazil. Samples of the angico exudates (nodules) (*Anadenanthera macrocarpa*) were collected at Anápolis-GO, Brazil. The commercial membrane of bovine origin (MBO) used in this work was purchased from GenDerm (Baumer, Brazil).

2.2. Origin and Purification of Polysaccharides. The gum from cashew exudates was extracted according to the methodology described by Silva et al. [9]. Briefly, the nodules were milled, immersed in distilled water in a proportion of 20% (w/v), and kept at room temperature (25°C), for 24 hours. The solution was filtered to remove fragments of bark and then precipitated with absolute ethanol, in the proportion of 1 : 3 (v/v), for 24 hours. The precipitated cashew gum (CG) was separated by filtration, washed with ethanol, dried at room temperature (25°C), and stored at room temperature in hermetically closed vessels.

The gums from angico exudates were milled, immersed in distilled water in a proportion of 20% (w/v), and kept at room temperature (25°C), for 24 hours. The solution was filtered to remove fragments of bark and then precipitated with absolute ethanol, in the proportion of 1 : 3 (v/v), for 24 hours. Alternatively, after filtration, the angico gum solution was acidified to pH 4.0 with acetic acid followed by precipitation with cold ethanol. The precipitated angico gum (AG or AG_{ac}) was separated by filtration, washed with ethanol, dried at room temperature (25°C), and stored at room temperature in hermetically closed vessels.

2.3. The Preparation of the Films. In order to obtain a film with mechanical properties similar to the commercial MBO, several formulations were tested using cashew gum (CG), angico gum (AG), or a mixture of cashew and angico gum.

The PVA/CG film was prepared by casting on the glass molds a solution containing a mixture of 7.5 mL of 3% (w/v) acidic PVA solution (pH 2.0), 7.5 mL of 3% (w/v) cashew gum aqueous solutions, 0.75 mL of 0.1 mol L^{-1} sodium metaperiodate as a cross-linker, and 0.75 mL of 0.1 mol L^{-1} sulfuric acid as catalyst. The solvent evaporation was left to occur at room temperature (25°C) for 24 hours. The dry films were stored in plastic bags before all subsequent characterization procedures.

The PVA/AG films were prepared by casting on the glass molds a solution containing a mixture of 7.5 mL of 3% (w/v) acidic PVA solution (pH 2.0), 7.5 mL of 3% (w/v) angico gum aqueous solutions, 0.75 mL of 0.1 mol L^{-1} sodium metaperiodate as a cross-linker, and 0.75 mL of 0.1 mol L^{-1} phosphoric acid (PVA/AG_{PO_4}) or 0.1 mol L^{-1} hydrochloric acid (PVA/AG_{HCl}) as a catalyst. The PVA/AG_{ac} film was prepared by casting on the glass molds a solution containing a mixture of 7.5 mL of 3% (w/v) acidic PVA solution (pH 2.0), 7.5 mL of 3% (w/v) AG_{ac} aqueous solution, 0.75 mL of 0.1 mol L^{-1} sodium metaperiodate as a cross-linker, and 0.75 mL of 0.1 mol L^{-1} sulfuric acid as a catalyst. The solvent evaporation was left to occur at room temperature (25°C) for 24 hours. The dry films were stored in plastic bags before all subsequent characterization procedures.

The PVA/CG-AG film was prepared by casting on the glass molds a solution containing a mixture of 7.5 mL of 3% (w/v) acidic PVA solution (pH 2.0), 3.25 mL of 3% (w/v) cashew gum aqueous solutions, 3.25 mL of 3% (w/v) angico gum aqueous solutions, 0.75 mL of 0.1 mol L^{-1} sodium metaperiodate as a cross-linker, and 0.75 mL of 0.1 mol L^{-1} sulfuric acid as catalyst. The solvent evaporation was left to occur at room temperature (25°C) for 24 hours. The dry films were stored in plastic bags before all subsequent characterization procedures.

2.4. Film Characterization

2.4.1. Films Thickness and Morphology. The thickness of the blended films was determined using a manual micrometer (Mitutoyo, São Paulo, Brazil). Final thickness was determined by means of ten random determinations in all film area. The films were analyzed according to morphological overall characteristics such as color, transparency, and homogeneity.

2.4.2. Mechanical Properties. Mechanical properties were determined in an Instron M3345 (Instron Co., Norwood, MA, USA), with a 50 N load cell equipped with tensile grips. Samples of the film were cut into strips of 10 mm wide and 30 mm long, according to the ASTM D-638M-93 standard [12]. The grip separation was set at 10 mm, with a crosshead speed of 5 mm min^{-1}. Each sample used was previously inspected and those containing any defect such as air bubbles, holes, and tears or showing average thickness variation superior to 5% were rejected. All samples were tested with curing time greater than 30 days, in dehydrated state.

Tensile strength (TS), percentage of elongation ($E\%$) at break, and elastic modulus (EM) were evaluated using the Software Blue Hill 2 (Instron Co.). The percentage of elongation ($E\%$) at break was calculated as follows:

$$\text{Elongation } (E\%) = \frac{(L_f - L_i) \times 100}{L_i}, \quad (1)$$

where L_i is the initial length and L_f is the final length.

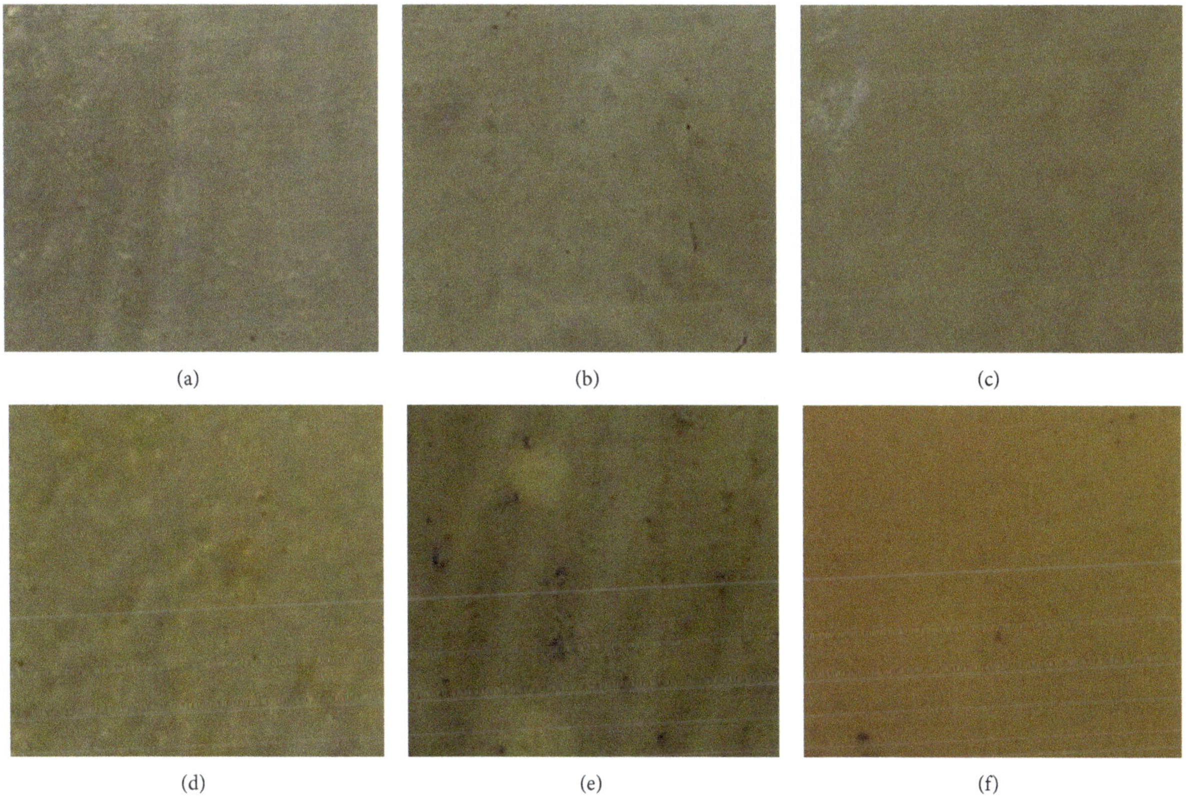

FIGURE 1: Macroscopic morphology of the films: (a) MBO; (b) PVA/CG film; (c) PVA/CG-AG film; (d) PVA/AG_{PO_4} film; (e) PVA/AG_{HCl} film; and (f) PVA/AG_{ac} film.

The modulus of elasticity (Young's modulus) was determined according to the following equation:

$$\text{Modulus of elasticity } (Y) = \frac{FL_i}{A_0 \Delta L}, \quad (2)$$

where F is the force exerted on an object under tension; A_0 is the original cross-sectional area through which the force is applied; ΔL is the amount by which the length of the object changes; and L_i is the original length of the object.

2.5. Statistical Analysis. All experiments were expressed as average ± standard deviation ($\bar{X} \pm$ SD). The variance analysis and Tukey's test were used to define differences in mean values of the data from 3 to 6 replicates. The variance analyses were carried out using the software Statistica 6.0 (StatSoft, Inc., Tulsa, OK, USA, 1997). The means of the treatment with the same letters are not significantly different ($P \leq 0.05$).

3. Results and Discussion

3.1. Appearance of the Films. The PVA/CG films presented thickness values between 140 to 200 μm, while PVA/AG films presented thickness around 70 and 140 μm. Figure 1 shows samples of formulated films and the membrane of bovine origin (MBO). As can be observed, the produced films presented a high diversity of color and transparency. Among them, the PVA/CG films presented the higher transparency and slightly white color, very similar to MBO.

TABLE 1: Mechanical properties of blended films.

	Tensile strength (kgf)	Elongation (%)	Elastic modulus (kgf)
MBO	193.61^b	2.94^b	6568.14^b
PVA/CG	151.84^b	11.96^a	1305.25^c
PVA/CG-AG	218.17^b	13.24^a	1800.85^c
PVA/AG_{PO_4}	227.97^b	4.87^b	4773.82^b
PVA/AG_{HCl}	426.46^a	3.80^b	11590.02^a
PVA/AG_{ac}	221.30^b	3.18^b	6972.95^b

^{a}, ^{b}, and ^{c}: The results are the means of three determinations. Within columns, means with the same superscript letter are not significantly different ($P > 0.05$).

3.2. Mechanical Properties. Artificial membranes have focus of special attention owing to their potential application in the medical fields, such as drug delivery systems, artificial organs, and tissue engineering [13]. To every application, specific mechanical features are required. In general, the mechanical properties of the films produced by blending PVA with CG were similar to those of the MBO (Table 1), whereas films with AG or AG_{ac} had mechanical properties slightly different.

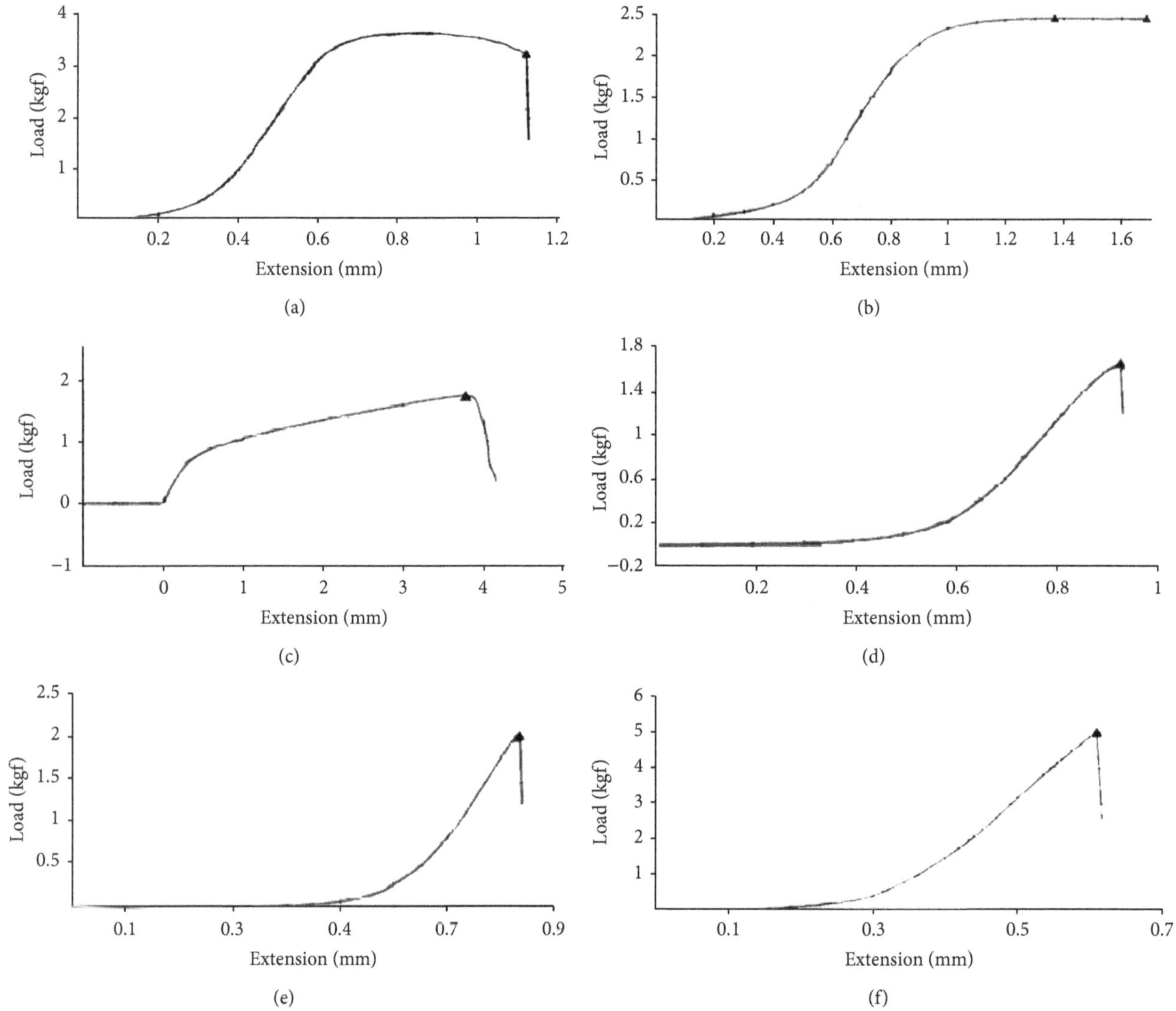

Figure 2: Typical behavior of films under tension tests: (a) MBO; (b) PVA/CG film; (c) PVA/CG-AG film; (d) PVA/AG_{PO_4} film; (e) PVA/AG_{HCl} film; and (f) PVA/AG_{ac} film.

The behavior of the films under tension test is depicted in Figure 2. Analyzing the behavior of the MBO under tension test (Figure 2(a)), it is possible to observe three zones—the first characterized by an increase in the load accompanied by extension of the film, followed by a zone in which the extension occurs without increases in the load and the point of rupture of the film. As can be seen, the extension of MBO film occurred at a maximum load of 3.65 kgf. Similar behavior was observed for PVA/CG film (Figure 2(b)) although the maximum load to extend the film has been 2.49 kgf.

Regarding PVA/AG (Figures 2(d) and 2(e)) or PVA/AG_{ac} films (Figure 2(f)), the tension test showed an absence of an elastic zone. It seems that the acid used as a catalyst may influence the load necessary to extend the film. The PVA/AG_{ac} film required the highest load to rupture (5.0 kgf).

The films produced by blending PVA/CG-AG presented the best set of features regarding the properties of strength and flexibility (Figure 2(c)).

The different behavior observed for the blended films may be ascribed to the chemical structure of the polysaccharides. The property of starches to produce film is related to the relative amount of amylose and amylopectin present in the starch granule. The branching degree and the relative amount of amylopectin present in the starch granule will determine the elastic property of the produced film [14]. In this sense, the different behavior of the blended films may be ascribed to different branching degree or the higher molar mass of AG (3.7×10^6 g mol^{-1}) compared to CG (6.41×10^4 g mol^{-1}) [7, 15].

The mechanical characteristics of PVA blended films are shown in Table 1. The tensile strength (TS) is defined as the capacity of resistance to rupture presented by the material when submitted to pressure force [16]. On the other hand, the percentage of elongation (*E*%) is related to the elasticity of a material since it is measured through the extension under traction. The blended films showed TS and *E*% quite diverse.

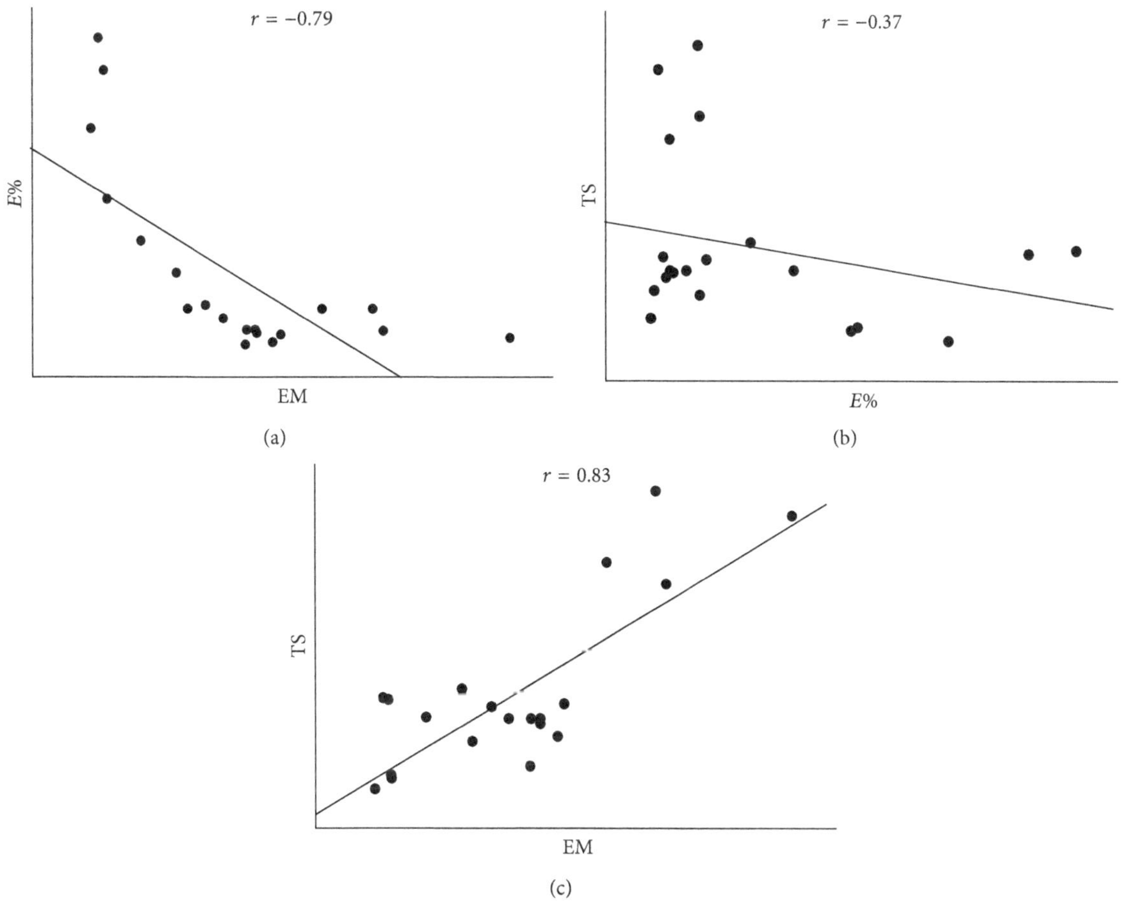

Figure 3: Graphic representation of Pearson's correlation between (a) percentage of elongation (*E*%) and elastic modulus (EM), (b) tensile strength (TS) and percentage of elongation (*E*%), and (c) tensile strength (TS) and elastic modulus (EM).

The film that presented the higher TS was PVA/AG_{HCl} with a value twofold higher than MBO. However, the value of *E*% of PVA/AG_{HCl} did not differ from that of MBO. From these results, it is possible to infer that the polymers chains in the blended film are tightly bounded, with a very low degree of flexibility.

On the other hand, the PVA/CG film showed a remarkable percentage of elongation (*E*%), fourfold higher than MBO. It seems that the polymers chains of the blended film, although covalently bounded, preserved a certain degree of chain flexibility.

Regarding the elastic modulus (EM), three groups with distinct properties were observed: (i) PVA/AG_{HCl} with the highest value of EM; (ii) MBO, PVA/AG_{PO_4}, and PVA/AG_{ac} with EM around 6900–4700 kgf; and (iii) PVA/CG and PVA/CG-AG, with values of EM around 1300–1800 kgf (Table 1).

The results of the correlation analysis evidenced a positive relationship between TS and EM ($r = 0.84$), whereas the *E*% had a negative correlation with EM ($r = -0.79$). This observation is evidenced in Figure 3.

Considering the macroscopic features and the mechanical properties of the blended films, the PVA/CG showed characteristics that enable this material to substitute the MBO with the advantage of higher *E*%. Furthermore, the composition of PVA/CG is advantageous regarding the safety of the material used, since it uses a plant polysaccharide instead of an animal protein.

Although the films produced with angico gum presented inferior macroscopic characteristics, the PVA/AG_{HCl} film showed remarkable mechanical properties such as the high values of TS and EM. In this sense, this material is also promising for biotechnological applications.

4. Conclusion

The films produced by blending PVA with AG and/or CG presented quite different macroscopic and mechanical properties. Among them, the PVA/CG film presented the macroscopic and mechanical properties similar to MBO film, with the advantage of a higher *E*% (11.96) than MBO (2.94). On the other hand, the PVA/AG_{HCl} presented higher values of TS and EM. The inclusion of CG in the blend PVA/AG-CG improved the *E*% of this film, compared to the other PVA/AG films. Finally, all produced films presented interesting mechanical characteristics, which enables several biotechnological applications.

Conflict of Interests

The authors do not have any conflict of interests with the content of the paper.

Acknowledgment

F. E. F. Silva, M. C. B. Di-Medeiros, and K. A. Batista thank CAPES for the fellowship support.

References

[1] K. G. Satyanarayana, G. G. C. Arizaga, and F. Wypych, "Biodegradable composites based on lignocellulosic fibers—an overview," *Progress in Polymer Science*, vol. 34, no. 9, pp. 982–1021, 2009.

[2] K. Y. Lee and D. J. Mooney, "Alginate: properties abd biomedical applications," *Progress in Polymer Science*, vol. 37, pp. 106–126, 2012.

[3] E. Khor, H. Wu, L. Y. Lim, and C. M. Guo, "Chitin-methacrylate: preparation, characterization and hydrogel formation," *Materials*, vol. 4, pp. 1728–1746, 2011.

[4] S. Tripathi, G. K. Mehrotra, and P. K. Dutta, "Physicochemical and bioactivity of cross-linked chitosan-PVA film for food packaging applications," *International Journal of Biological Macromolecules*, vol. 45, no. 4, pp. 372–376, 2009.

[5] E. S. Costa-Júnior, E. F. Barbosa-Stancioli, A. A. P. Mansur, W. L. Vasconcelos, and H. S. Mansur, "Preparation and characterization of chitosan/poly(vinyl alcohol) chemically crosslinked blends for biomedical applications," *Carbohydrate Polymers*, vol. 76, no. 3, pp. 472–481, 2009.

[6] I. S. Fahim, S. M. Elhaggar, and H. Elayat, "Experimental investigation of natural fiber reinforced polymers," *Materials Sciences and Applications*, vol. 3, pp. 59–66, 2012.

[7] M. A. Oliveira, D. A. Silva, D. E. A. Uchoa et al., "Synthesis and characterization of carboxymethylated red angico (*Anadenanthera macrocarpa*) exudate polysaccharide," *Journal of Applied Polymer Science*, vol. 103, no. 5, pp. 2985–2991, 2007.

[8] R. C. M. De Paula, P. M. Budd, and J. F. Rodrigues, "Characterization of *Anadenanthera macrocarpa* exudate polysaccharide," *Polymer International*, vol. 44, no. 1, pp. 55–60, 1997.

[9] T. M. Silva, P. O. Santiago, L. L. A. Purcena, and K. F. Fernandes, "Study of the cashew gum polysaccharide for the horseradish peroxidase immobilization—structural characteristics, stability and recovery," *Materials Science and Engineering C*, vol. 30, no. 4, pp. 526–530, 2010.

[10] J. S. Maciel, D. A. Silva, H. C. B. Paula, and R. C. M. De Paula, "Chitosan/carboxymethyl cashew gum polyelectrolyte complex: synthesis and thermal stability," *European Polymer Journal*, vol. 41, no. 11, pp. 2726–2733, 2005.

[11] D. M. W. Anderson and P. C. Bell, "Structural analysis of the gum polysaccharide from anacardium occidentale," *Analytica Chimica Acta*, vol. 79, no. C, pp. 185–197, 1975.

[12] ASTM, "Designation: D638M-93, standard test method for tensile properties of thin plastic sheeting," in *Annual Book of ASTM Standards*, ASTM, Philadelphia, Pa, USA, 1995.

[13] D. F. Stamatialis, B. J. Papenburg, M. Gironés et al., "Medical applications of membranes: drug delivery, artificial organs and tissue engineering," *Journal of Membrane Science*, vol. 308, no. 1-2, pp. 1–34, 2008.

[14] Y. U. Nabar, D. Draybuck, and R. Narayan, "Physicomechanical and hydrophobic properties of starch foams extruded with different biodegradable polymers," *Journal of Applied Polymer Science*, vol. 102, no. 1, pp. 58–68, 2006.

[15] S. Bose and M. Biswas, "The structure of the gum of *Anacardium occidentale*," *Acta Horticulturae*, vol. 108, pp. 207–217, 1985.

[16] F. Yamashita, A. Nakagawa, G. F. Veiga, S. Mali, and M. V. E. Grossmann, "Biodegradable active packaging for minimally processed fruits and vegetables," *Brazilian Journal of Food Technology*, vol. 8, pp. 335–343, 2005.

21

Study on the Optoelectronic Properties of UV Luminescent Polymer: ZnO Nanoparticles Dispersed PANI

J. Kayathri, N. RaniMeiyammai, S. Rani, K. P. Bhuvana, K. Palanivelu, and S. K. Nayak

Central Institute of Plastics Engineering & Technology, Chennai 600 032, India

Correspondence should be addressed to K. P. Bhuvana; kpusha27@gmail.com

Academic Editor: Amit Bandyopadhyay

The present work is focused on the synthesis and characterization of a new polymer thin film material suitable for emissive layer in polymer light-emitting diodes. Thin films of polyaniline and ZnO nanoparticles dispersed polyaniline have been prepared and investigated for the suitability of emissive layer in polymer light emitting diodes. ZnO nanoparticles have been synthesized using wet chemical method and characterized using powder X-ray diffraction. Diffraction pattern reveals that the ZnO nanoparticles are grown in hexagonal wurtzite structure with the preferential orientation of (001) plane. Polyaniline was synthesized using potassium persulphate (initiator) at room temperature. ZnO nanoparticles are then dispersed in different doping level (5%, 10%, 15% 20% and 50%) with polyaniline. Thin films of ZnO dispersed polyaniline (PANI/ZnO) were prepared and characterized using scanning electron microscopy in order to investigate the surface morphology of the film. The optical characteristics and the suitability of film as emissive layer were analyzed using UV-visible absorption and fluorescence spectroscopy.

1. Introduction

After the discovery of conductivity in polymers and in certain organic molecules, remarkable progress has been made in synthesizing organic materials and in developing them for use in electronic and optical devices [1, 2]. Currently, polymer-based organic light-emitting diodes (LEDs) [3], photovoltaic cells [4], and field effect transistors [5] are being pushed towards commercialization. So far, a wide range of conjugated polymer systems has been developed, such as poly(1,4-phenylenevinylene) (PPV) [6, 7], poly(p-phenylene) (PPP) [8], polyfluorene (PF) [9], and polythiophene (PT) [10] and their derivatives. Polymer-based optoelectronic devices have gained importance due to the cost-effective nature of polymers. Polymer light-emitting diodes (PLEDs) were first demonstrated by Richard Friend and coworkers on PPV that emitted 3 primary colors. Among the conducting polymers, PANI has been of particular interest because of its cheap monomer, simple synthesis technology, unique electrochemical properties, high conductivity, and environmental stability. PANI differs from most other conducting polymers such as PPV and PT because it possesses readily accessible oxidation states. Zinc oxide has received much attention in recent years due to its diverse properties. It is a direct wide gap semiconductor, and it is a prominent inorganic material for optoelectronic devices. The interest of the researchers has increased rapidly during the last decade in the application of inorganic-organic hybrid polymers for electronic and optoelectronic devices. A new finding of a nanocrystal ZnO retarded gelation of low-density polyethylene (LDPE) has also been reported [11]. It has been reported that nanocrystals of inorganic materials exhibit many properties which are unexpected from the conventional point of view. PLEDs by depositing a p-type polymer, poly(3,4-ethylenedioxythiophene): poly(styrene sulfonate)(PEDOT:PSS) on n-type ZnO nanowires grown by electrodeposition on transparent conducting glass (ITO) substrates have also been fabricated and reported [12]. Incorporation of few wt% of ZnO nanoparticles in polymer matrix has enhanced its emissivity. The present investigation is focused on the hybridization of PANI with ZnO and analyzing its optical properties as an emissive layer of LED.

2. Experimental

Thin films of PANI and PANI/ZnO have been synthesized by solution polymerization technique with different compositions of ZnO (5, 10, 15, 20 and 50 wt %). 0.1 M of zinc acetate was added in starch solution. Starch solution is prepared by adding 2.5 g of starch in 250 mL of distilled water. After complete dissolution of zinc acetate under constant stirring, ammonia (0.2 M) is added dropwise to the aforemention solution. The obtained solution is then filtered and dried in oven at 80°C to get ZnO nanoparticles. PANI was synthesized by solution polymerization technique. Prescribed amount of aniline (5.58 g) is dissolved in 1 N hydrochloric acid under constant stirring for 4 hours. Calculated amount of potassium persulphate, $K_2S_2O_8$ (1.62 g), in 10 mL of distilled water is added to aniline in HCl solution and constantly stirred for 24 hours [13]. The polymerization solution results in the formation of dark green precipitate. The precipitate is then filtered and dried in oven at 110°C to get PANI powder. ZnO nanoparticles at different composition were dispersed into PANI by adding ZnO nanoparticles to the aqueous solution of aniline in HCl after adding the initiator while the solution is in constant stirring. PANI and PANI/ZnO composites are then powdered in mortar and preserved in a desiccator.

Thin films of PANI and PANI/ZnO were prepared by solution method. The PANI/ZnO powdered composite was mixed with polyvinyl alcohol (PVA) solution (1 gm of PVA in 20 mL of distilled water). The PVA solution was used as a binding agent. The obtained solution is coated on a transparent glass substrate by spinning method on glass substrate. The solution was allowed to fall dropwise, and the substrate was placed in a rotating disc of the abrasion tester. The disc was set to rotate with the speed of 100 rpm for 2 hrs. The solution was deposited on the glass substrate. The film was kept in the oven for 2 hrs by maintaining the temperature of 150°C. The thickness of the films was measured using digital foil thickness gauge (Make: Precision, Model: 500-FD, Accuracy: 1.0 μm) and found to be around 310–350 μm.

3. Results and Discussion

3.1. Density. The density of the nanocomposites was determined using the formula $D_C = vD_{ZnO} + (1 - v)D_{PANI}$. The density of ZnO nanoparticles and PANI was determined as 5.59 g/cc. and 1.32 g/cc., respectively. The density of PANI/ZnO nanocomposites with different composition of ZnO was determined using the previously formula, and the values are tabulated in Table 1.

Table 1: Density of PANI/ZnO nanocomposites.

Concentration of ZnO in PANI/ZnO nanocomposites	Density (g./cc.)
0%	1.320
5%	1.534
10%	1.747
15%	1.961
20%	2.174
50%	3.455

3.2. Structural Characteristics

3.2.1. X-Ray Diffraction Studies. The structural properties of as-synthesized ZnO nanoparticles and PANI were investigated using X-ray diffraction (XRD). The XRD pattern was recorded in Shimadzu diffractometer, using the Cu K_α radiation, $\lambda = 1.5418$ Å with 40 kV and 20 mA, at 0.04° scan rate. The measurements were made at room temperature with the diffraction angle (2θ) ranging from 10 to 70 degrees. XRD pattern of ZnO nanoparticles and PANI is depicted in Figure 1. The XRD pattern of PANI shows the amorphous nature of PANI; however the XRD pattern of ZnO nanoparticles reveals that they are grown in hexagonal wurtzite structure with the preferential orientation of (100) plane. The diffraction peaks do not imitate the high crystallinity; hence the size of the particles must be smaller. Inset of Figure 1 shows the Williamson-Hall plot which was to calculate the particle size and the strain induced due to the reduction of size of the particle.

From Williamson-Hall plot it is observed that the particle size (D) and the strain (ε) were calculated as 4.496 nm and −0.0685, respectively. The negative sign represents that the strain induced is compressive. Moreover, the lattice constants "a" and "c" are determined using the following formulae:

$$a = \frac{\lambda}{\sqrt{3}\sin\theta}, \qquad c = \frac{\lambda}{\sin\theta}. \tag{1}$$

The "a" axis lattice constant and "c" axis lattice constant are determined to be $a = 3.1606$ Å and $c = 5.4743$ Å. The bulk ZnO has the lattice constant of $a = 3.25$ Å and $c = 5.206$ Å. Along "a" axis the particle undergoes the compressive strain, and hence the lattice constant is reduced from that of the bulk. However, the c axis lattice constant is increased. The reason is that, when there is a compressive strain on one direction, there will be a tensile strain on the opposite direction. Thus the lattice constant is increased along c axis. The average particle size using Debye-Scherer formula is around 3.93 nm.

3.2.2. SEM Analysis. The distribution of ZnO particles was examined using scanning electron microscopy. The SEM images of PANI/ZnO films coated on glass substrate were depicted in Figure 2. The SEM images have been recorded using Carl Zeiss scanning electron microscope at the magnification of 8.00 Kx. The SEM images clearly reveal the uniform dispersion of ZnO nanoparticles in PANI matrix. Further it is observed from SEM image that the ZnO particles exist in the form of hexagonal rod. Moreover, the distribution increases with the increase of ZnO concentration into the PANI matrix. PANI with 20% of ZnO particles shows the higher degree of uniform distribution. However, at higher concentrations of ZnO (50%), the particles are clustered together due to the stronger intermolecular forces between them.

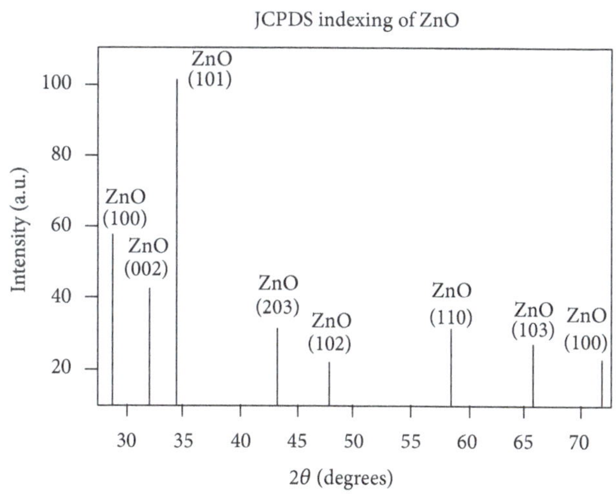

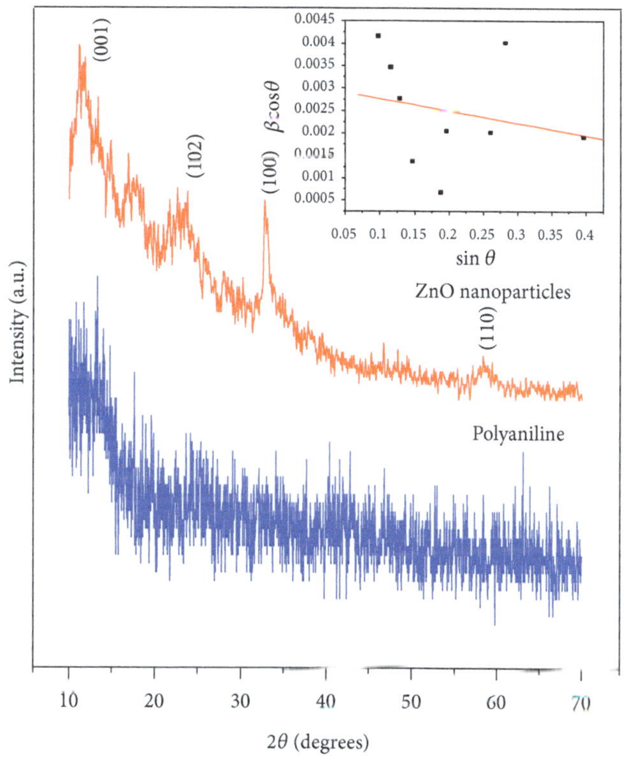

FIGURE 1: XRD pattern of PANI and ZnO nanoparticles. Inset: W-H plot for ZnO nanoparticles.

3.3. Optical Characteristics

3.3.1. Optical Transmittance. Optical transmittance data was observed using Clarity meter (Make: Gardner Laboratory, Inc., USA) which emits the light radiation of wavelength 860 nm. Figure 3 shows the variation of optical transmittance as a function of concentration of ZnO nanoparticles. It is observed that the incorporation of ZnO nanoparticles into the PANI matrix increases the transparency of the PANI film.

3.3.2. Determination of Optical Band Gap. UV-visible absorption spectra were observed using CARY 5E UV-VIS-NIR spectrophotometer at room temperature within the wavelength ranges from 300 nm to 1100 nm in steps of 1 nm. The optical band gap as a function of ZnO nanoparticles concentration has been extracted from UV-Visible spectrum. The transmittance data from the UV spectrum have been used for the further determination of optical parameters like band gap and refractive index.

To determine the optical band gap of the composites, Tauc's plot for pure PANI and PANI/ZnO composites was plotted and depicted in Figure 4. From the graph, it is observed that the optical band gap decreases with respect to the concentration of ZnO. Since ZnO is a direct band gap semiconductor, Tauc's plot is plotted between $(\alpha h\nu)$ and $h\nu$. The band gap is determined by extrapolating the linear region

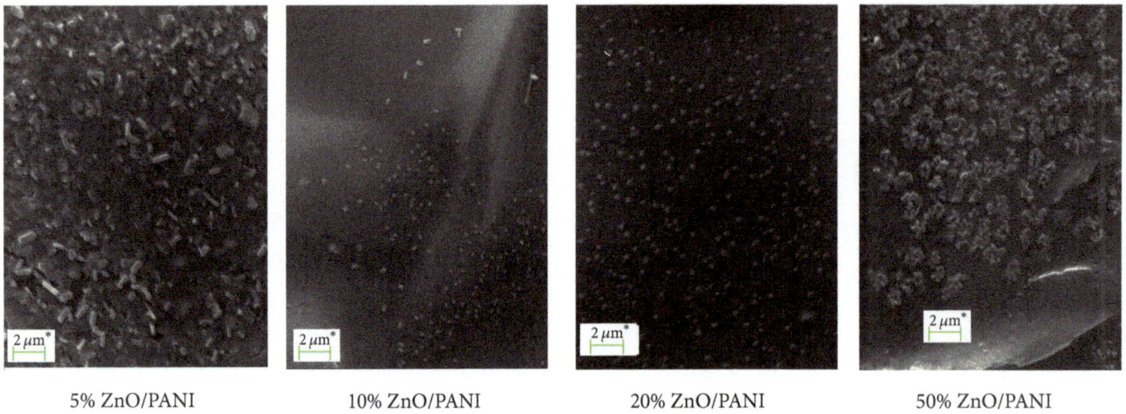

FIGURE 2: SEM image of PANI and PANI/ZnO composites.

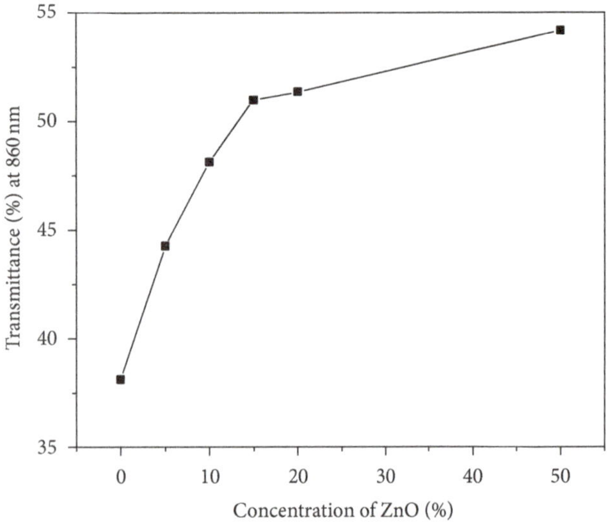

FIGURE 3: Optical transmittance as a function of ZnO concentration.

of the curve. The energy gap for pure PANI is observed to be 2.8 eV. The higher value of optical band gap is due to the reduction in particle size of the synthesized ZnO powder. Composites with 5 wt%, 10 wt%, 15 wt%, 20 wt%, and 50 wt% of ZnO in PANI exhibit the energy gap of 2.63 eV, 2.59 eV, 2.56 eV, 2.51 eV, and 2.47 eV, respectively.

3.3.3. Determination of Refractive Index. Reflectance, transmission, and absorption from any material depend upon the refractive index of that material. Reflectance is given as $R = (n-1)^2/(n+1)^2$; from this relation, the refractive index as a function of wavelength in terms of reflectance has been derived and reported earlier [14] as $n(\lambda) = -(R+1) \pm \sqrt{(-3R^2 + 10R - 3)}/2(R-1)$. The reflectance of the samples was extracted from the transmittance and absorbance data. Figure 5 shows the variation of refractive index with respect to the concentration of ZnO at a wavelength of 800 nm. The refractive index is found to be increasing with respect to ZnO up to 20% of ZnO and again decreases at higher dispersion level of ZnO.

3.3.4. Band Gap: Refractive Index Relationship. The refractive index and energy gap of semiconductors represent two fundamental physical aspects that characterize their optical and electronic properties. Optical band gap of a semiconductor determines the threshold for the absorption of photons, and the refractive index determines the transparency of the material. Therefore, the correlation between these two properties obtains the significant importance. Theoretically energy gap-refractive index relationship was obtained by Ravindra et al. [15] using Penn model, $n(Y) = A + B_1x + B_2x^2 + B_3x^3 + B_4x^4$. The same method has been employed to determine the relation between band gap and refractive index theoretically. Figure 6 shows how the refractive index varies with respect to the optical band gap. The relation is well fit into a fourth order binomial equation. The relation between "x" and "E_g" is determined as $n = -60.57+94.47E_g - 53.99E_g^{\ 2} + 13.71E_g^{\ 3} - 1.31E_g^{\ 4}$.

3.3.5. Fluorescence Spectroscopic Studies. From previous studies, it is concluded that PANI/ZnO composites exhibit good optical properties and conducting properties. In order to check whether the films can emit radiation, fluorescence studies were carried out. PANI/ZnO nanocomposites were subjected to fluorescence spectroscopy to observe the excitation and emission phenomenon of as-synthesized composites. Figure 7 shows the fluorescence spectra of PANI and PANI/ZnO composites which reveal the emission of radiation in the UV region.

Inset of Figure 7 shows the variation of emission wavelength with respect to ZnO composition. It is observed that the emission wavelength increases as ZnO increases up to 20%, and further addition of ZnO leads to the decrease of emission wavelength. Hence by adding different composition of ZnO with PANI the radiation of desired wavelength can be obtained. Further it is also observed that the emission is very strong (highly intense) at 20% of ZnO. Therefore, the PANI

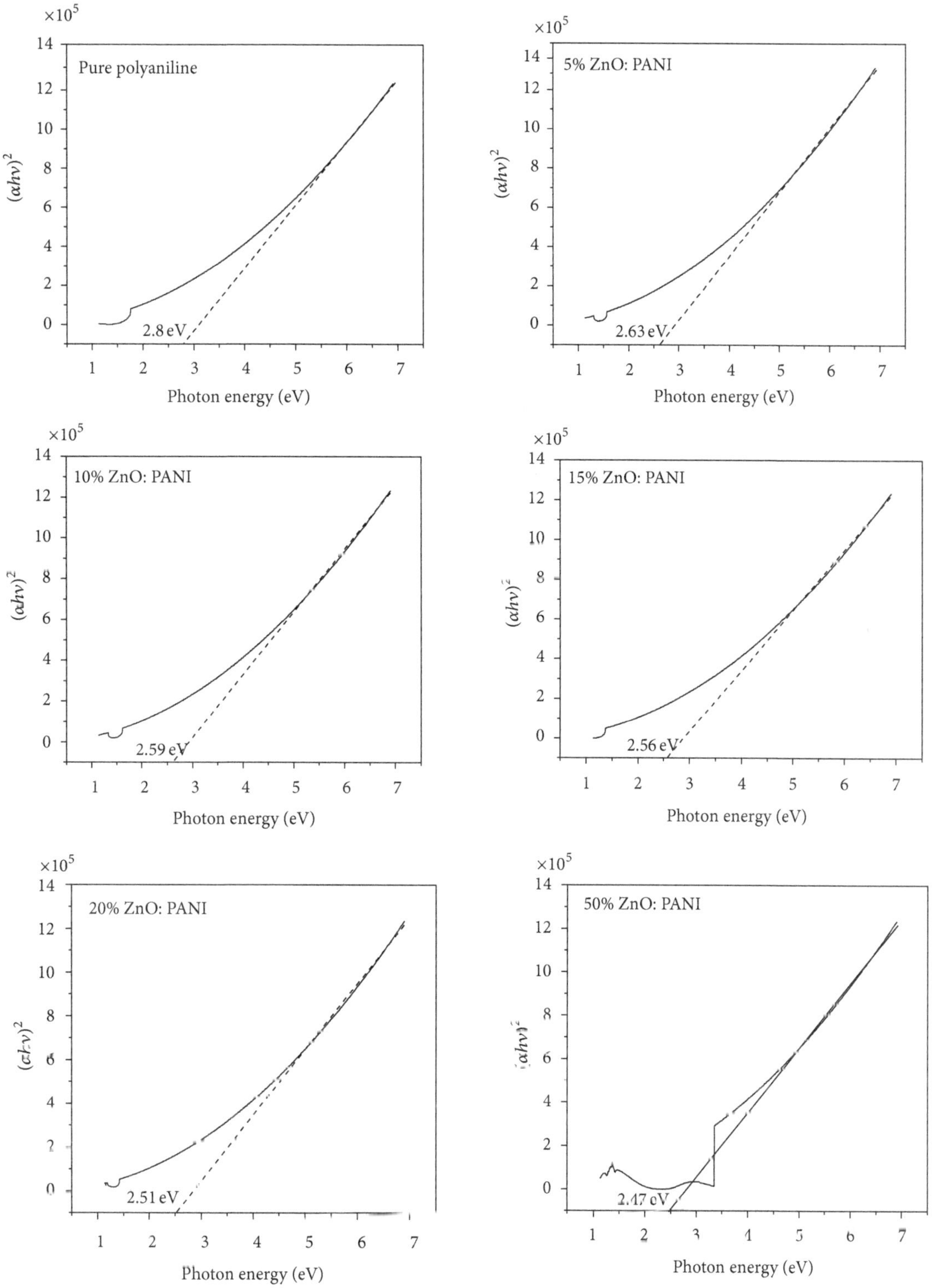

FIGURE 4: Tauc's plot for various compositions of ZnO in PANI.

film with 20% of ZnO emits the radiation of 395 nm with high intensity.

3.4. Electrical Properties. Figure 8 shows the variation of optical band gap and the electrical conductivity with ZnO concentration. The band gap is observed to be decreased by the addition of ZnO nanoparticles. The higher level of incorporation of ZnO reduces the band gap due to the formation of donor level at the bottom of the conduction band which contributes to electrical conductivity. The widening

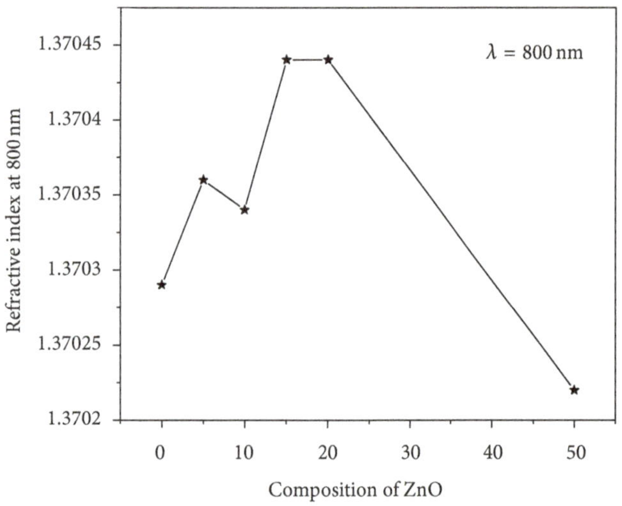

FIGURE 5: Refractive index versus ZnO concentration at $\lambda = 800$ nm.

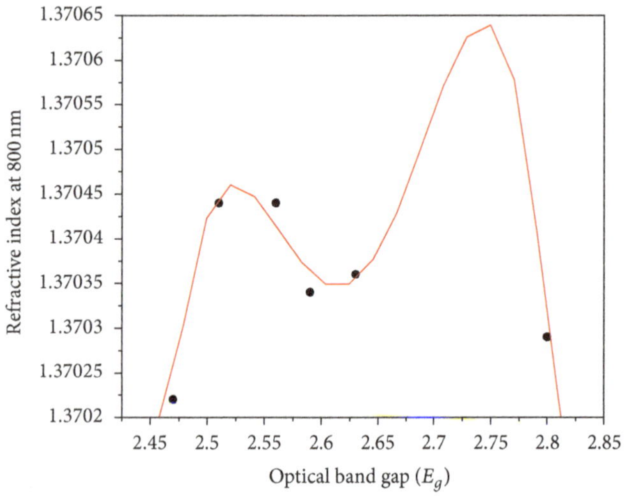

FIGURE 6: Relation between refractive index and band gap.

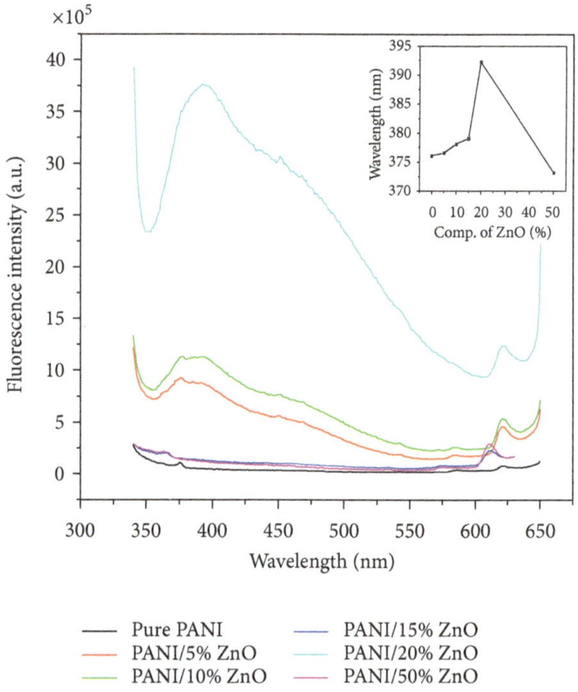

FIGURE 7: Fluorescence spectra of PANI/ZnO films. Inset: Variation of emission wavelength w. r. to ZnO concentration.

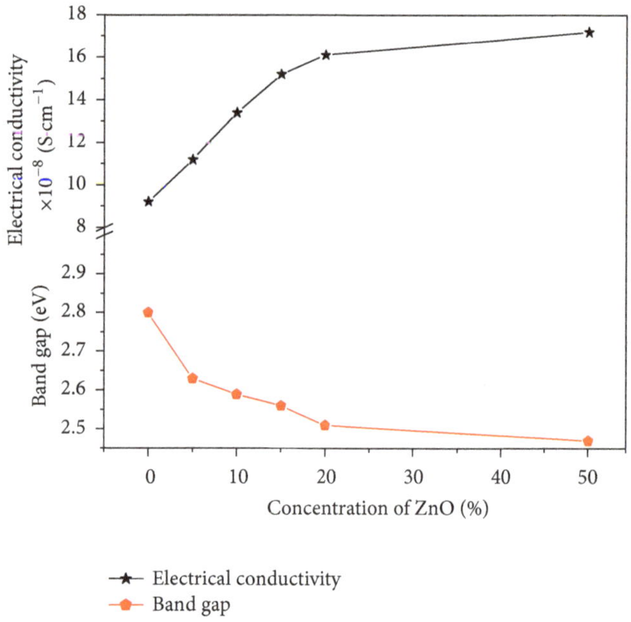

FIGURE 8: Variation of optical band gap and electrical conductivity with respect to ZnO concentration.

of band gap in case of heavily doped semiconductors is due to the blocking of the low-energy transitions by the donor electrons occupying the states at the bottom of the conduction band which is known as the Burstein-Moss effect [16]. According to this effect, the electrical conductivity of the film is expected to be increasing with ZnO concentration. The electrical conductivity of the film was measured using electrometer. The conductivity was found to be increasing with the increase of ZnO concentration and well reflects the Burstein-Moss effect.

4. Conclusion

Thin films of PANI and ZnO nanoparticles dispersed PANI were prepared by PVA assisted method on glass substrates. XRD was carried out to confirm the formation of ZnO nanoparticles and PANI. The film of PANI and ZnO/PANI nanocomposites was subjected to UV-visible spectroscopy and fluorescence spectroscopy. The dispersion of ZnO nanoparticles in the PANI matrix improves the transparency and modifies the optical band gap. The electrical conductivity is found to be enhanced by the dispersion of ZnO. Further, the dispersion of ZnO tunes the emission wavelength and the intensity of the radiation. Hence, it is concluded that

ZnO nanoparticle dispersed PANI is a promising material for emissive layer in polymer light-emitting diodes.

References

[1] A. J. Heeger, "Semiconducting and metallic polymers: the fourth generation of polymeric materials," *Angewandte Chemie International Edition*, vol. 40, no. 14, p. 2591, 2001.

[2] J. Kido and Y. Iizumi, "Fabrication of highly efficient organic electroluminescent devices," *Applied Physics Letters*, vol. 73, no. 19, pp. 2721–2723, 1998.

[3] R. H. Friend, R. W. Gymer, A. B. Holmes et al., "Electroluminescence in conjugated polymers," *Nature*, vol. 397, no. 6715, pp. 121–128, 1999.

[4] H. Hoppe and N. S. Sariciftci, "Organic photovoltaic solar cells: recent advancements in efficiency," *Journal of Materials Research*, vol. 19, pp. 1924–1945, 2004.

[5] N. Stutzmann, R. H. Friend, and H. Sirringhaus, "Self-aligned, vertical-channel, polymer field-effect transistors," *Science*, vol. 299, no. 5614, pp. 1881–1884, 2003.

[6] J. H. Burroughes, D. D. C. Bradley, A. R. Brown et al., "Light-emitting diodes based on conjugated polymers," *Nature*, vol. 347, no. 6293, pp. 539–541, 1990.

[7] C. Winder, A. Andreev, H. Sitter, G. Matt, N. S. Sariciftci, and D. Meissner, "Optoelectronic devices based on para-sexiphenyl films grown by Hot Wall Epitaxy," *Synthetic Metals*, vol. 139, no. 3, pp. 573–576, 2003.

[8] G. Grem, G. Leditzky, B. Ullrich, and G. Leising, "Blue electroluminescent materials for polymer light-emitting diodes," *Advanced Materials*, vol. 4, no. 1, pp. 36–37, 1992.

[9] Y. Ohmori, M. Uchida, K. Muro, and K. Yoshino, "Blue electroluminescent diodes utilizing poly(alkylfluorene)," *Japanese Journal of Applied Physics*, vol. 30, no. 11 B, pp. 19–1941, 1991.

[10] J. Roncali, "Conjugated poly(thiophenes): synthesis, functionalization, and applications," *Chemical Reviews*, vol. 92, no. 4, pp. 711–738, 1992.

[11] K. S. Cho, J. I. L. Hong, and C. I. Chung, "Effects of ZnO nano particles on thermal stabilization of polymers," *Polymer Engineering and Science*, vol. 44, no. 9, pp. 1702–1706, 2004.

[12] A. El-Shaer, A. Dev, J. P. Richters et al., "Hybrid LEDs based on ZnO-nanowire arrays," *Physica Status Solidi B*, vol. 247, no. 6, pp. 1564–1567, 2012.

[13] S. B. Kondawar, S. D. Bompilwar, V. S. Khati, S. R. Thakre, V. A. Tabhane, and D. Burghate, "Characterizations of zinc oxide nanoparticles reinforced conducting PANI composite," *Archives of Applied Science Research*, vol. 2, no. 1, p. 247, 2010.

[14] B. K. Periyasamy, R. S. Jebas, N. Gopalakrishnan, and T. Balasubramanian, "Development of NLO tunable band gap organic devices for optoelectronic applications," *Materials Letters*, vol. 61, no. 21, pp. 4246–4249, 2007.

[15] N. M. Ravindra, P. Ganapathy, and J. Choi, "Energy gap-refractive index relations in semiconductors—an overview," *Infrared Physics and Technology*, vol. 50, no. 1, pp. 21–29, 2007.

[16] E. Burstein, "Anomalous optical absorption limit in InSb," *Physical Review Letters*, vol. 93, no. 3, pp. 632–633, 1954.

22

Microwave Assisted Synthesis of ZnO Nanoparticles: Effect of Precursor Reagents, Temperature, Irradiation Time, and Additives on Nano-ZnO Morphology Development

Gastón P. Barreto, Graciela Morales, and Ma. Luisa López Quintanilla

Centro de Investigación en Química Aplicada, Boulevard Enrique Reyna 140, 25253 Saltillo, COAH, Mexico

Correspondence should be addressed to Gastón P. Barreto; gbarreto@fio.unicen.edu.ar

Academic Editor: Antoni Morawski

The effect of different variables (precursor reagents, temperature, irradiation time, microwave radiation power, and additives addition) on the final morphology of nano-ZnO obtained through the microwave assisted technique has been investigated. The characterization of the samples has been carried out by field emission scanning electron microscopy (FE-SEM) in transmission mode, infrared (FTIR), UV-Vis spectroscopy, and powder X-ray diffraction (XRD). The results showed that all the above-mentioned variables influencedto some extent the shape and/or size of the synthetized nanoparticles. In particular, the addition of an anionic surfactant (sodium di-2-ethylhexyl-sulfosuccinate (AOT)) to the reaction mixture allowed the synthesis of smaller hexagonal prismatic particles (100 nm), which show a significant increase in UV absorption.

1. Introduction

ZnO powder has been widely used into numerous materials and products including paints, plastics, ceramics, and adhesives. It is a semiconductor of the II–VI semiconductor group with several favorable properties such as high electron mobility, wideband gap, and strong room temperature luminescence. These properties make ZnO an attractive compound for different emerging applications.

In the last two decades many methods ranging from gas-phase processes to solution routes have been investigated for the synthesis of ZnO nanoparticles including solution precipitation [1, 2], spray pyrolysis [3, 4], hydrothermal synthesis [5–7], sol-gel processes [8–11], and microemulsion synthesis [12].

In cases where the synthesis has been carried out through a conventional thermostatic system, the walls of the reactor are heated by convection or conduction, the core of the sample needs longer time to achieve the target temperature, and this may result in inhomogeneous temperature profiles. One possible solution to this problem is the use of microwave heating, which has become a very promising method of synthesis for both organic [13, 14], and inorganic [15] chemistry. This technique enables the rapid and homogenous heating of the reaction mixture to the desire temperature, which saves time and energy.

The microwave heating is based on two conversion mechanisms of the electromagnetic radiation into heat energy, namely, dipolar rotation and ionic conduction, which are directly related to the chemistry composition of the reaction mixture. So that, different compounds have different microwave absorbing properties, and this behavior allows a selective heating of compounds in the reaction mixture.

The general advantages of microwave mediated synthesis over conventional ones are (1) reaction rate acceleration as a consequence of high heating rates, (2) wide range of reaction conditions, that is, mild conditions or autoclave conditions, (3) high reaction yields, (4) reaction selectivity due to different microwave absorbing properties, (5) excellent control over reaction conditions, and (6) simple handling, allowing simple and fast optimization of experimental parameters.

Since 2007, the number of publications dealing with the microwave assisted synthesis of ZnO has increased significantly. In this sense there have been many published results concerned with aqueous and nonaqueous solution microwave assisted synthesis where the effect of different experimental variables over ZnO size, shape, and physicochemical behavior have been analyzed [16–18]. This irradiation technique has been used to obtain nanostructures with various morphologies including spherical particles, rods, hexagonal rings, hexagonal columns, Flowers, wires, hexagonal tubes, and sheets.

Some authors [16] demonstrated the possibility of ZnO synthesis with a novel three-dimensional morphology, where microwave radiation as heating source plays an important role in the formation of complex nanostructures. In addition, the absence of metal catalyst, template, or surfactant in this method avoids the subsequent effort for the removal of residual additives.

Other studies [18] dealt with the effect of power of microwave radiation on the shape and size of the synthetized ZnO nanostructures, where the average particle size decreases with decreasing microwave power being the optical properties sensitive to the variation of this parameter.

On the other hand, other authors [19] analyzed the effect of solvent (water, ethanol, and isopropanol) on the synthetized ZnO nanostructures, and an important change in the size and shape of nanoparticles was reported. When distilled water was used as a solvent, nanoparticles presented ellipse shape and larger size (larger axis was about 100 nm and size of the other axis was about 40 nm). With ethanol, the obtained nanostructures had rod form with smaller sizes than those obtained in water, and when isopropanol was used as the reaction medium, spherical particles with radius of 10–12 nm were obtained.

However, most synthetic studies did not show a complete study about different parameters involved during the microwave assisted technique. In this sense, this work deals with a complete analysis of different variables (precursor reagents, temperature, irradiation time, microwave radiation power, and the addition of additives) and their effect on the final morphology of nano-ZnO obtained through the microwave assisted technique.

2. Experimental

2.1. Synthesis of ZnO. In a typical synthesis process, 15 mL (1.6 mol L^{-1}) of a precursor salt ($Zn(NO_3)_2 \cdot 6H_2O$, $Zn(CH_3COO)_2 \cdot 2H_2O$ or $ZnCl_2$) were diluted in 32 mL of deionized water (or a sodium di-2-ethylhexyl-sulfosuccinate aqueous solution) to obtain a Zn^{+2} solution. Afterwards, 4 mL of a base (3.2 mol L^{-1}) (NaOH, KOH, or NH_4OH) was added dropwise (2 min) into the above-described solution with magnetic stirring at room temperature to get a colloid system, which was maintained under stirring for 10 min. Then, the reaction mixture was transferred into a Teflon autoclave and treated at selected temperature (80, 100, 120, or 140°C) for specific time (5, 10 or 20 min) under temperature-controlled mode in a microwave accelerated reaction system (Mars-X) operating at different powers (300, 600, or 1200 W). It must be mentioned that the system was programmed with a first temperature ramp where the target temperature was reached after 10 min in all cases (Table 1). When the reaction was finished and cooled to room temperature, the white precipitate was collected by filtration and washed with deionized water and ethyl alcohol several times. Finally, the product was dried at 65°C in a vacuum oven for 3 h.

2.2. Characterization. To investigate the structural properties of ZnO nanoparticles, field emission scanning electron microscopy (FE-SEM), infrared (FTIR), UV-Vis spectroscopy, and powder X-ray diffraction (XRD) studies were carried out. ZnO powders synthetized were characterized by X-ray diffraction with a Siemens D500 diffractometer. Diffraction patterns were recorded from 10 to 80° 2θ with a step size of 0.06° at 35 kV and 25 mA. The average crystallite sizes D (Table 1) of the ZnO nanostructures were calculated using Scherrer's formula

$$D = \frac{0.9\lambda}{\beta \cos\theta}, \tag{1}$$

where λ is the X-ray wavelength of Cu-Kα radiation source, β is the full width at half maximum intensity of the diffraction peak located at 2θ, and θ is the Bragg angle.

On the other hand, an aliquot of solid state material was placed in a carbon label for analysis by FE-SEM (JSM-7401F). Samples were analyzed using a secondary electron detector. The infrared spectra of ZnO nanoparticles were taken in a Nicolet spectrophotometer model Nexus 470 Nicolet brand in transmittance mode. The sample preparation was made in tablet way by mixing nano-ZnO and KBr in an agate mortar. The solid samples were dispersed in deionized water and ultrasonicated for 10 min. Then UV-Vis spectra of colloidal systems were obtained through a Shimadzu double beam spectrophotometer model UV-2401PC.

Characterization data of all ZnO synthesis carried out is described in Electronic Supplementary Material (ESM) that accompanies this paper (see Supplemetary Material available online at http://dx.doi.org/10.1155/2013/478681).

3. Results and Discussion

3.1. Temperature Effect. The microwave system employed to synthetize the ZnO particles allows to set a constant reaction temperature. In order to analyze the temperature effect on the final morphology of the product obtained, the synthetic reactions were carried out in the temperature interval 80–140°C using $Zn(NO_3)_2$ as the precursor salt (from MW-1 to MW-4, Table 1). XRD studies (Figure 1) show that, when the reaction is carried out for 20 minutes at 80 or 100°C (MW-1 and MW-2, resp.), the signals are not clear, and some peaks attributable to precursor reagents and unknown impurities are observed. On the contrary, at 120 and 140°C (MW-3 and MW-4, resp.) all the diffraction peaks match those of the wurtzite ZnO with lattice constants of $a = 3.250$ Å and $c = 5.207$ Å. The strong diffraction peaks appear at

TABLE 1: Experimental conditions employed during microwave assisted synthesis of ZnO.

Synthesis code	Experimental conditions (irradiation time[a], temperature, and microwave power)	Zn precursor, 15 mL 1.6 mol L^{-1}	Base, 4 mL 3.2 mol L^{-1}	Yield, %	Average crystallite size, nm	Average particle size[h], μm
MW-1	20 min, 80°C, 600 W	$Zn(NO_3)_2\ 6H_2O$	NaOH	N.A	N.A	N.A
MW-2	20 min, 100°C, 600 W	$Zn(NO_3)_2\ 6H_2O$	NaOH	N.A	N.A	0.413 (W[i]) 0.088 (L[j])
MW-3	20 min, 120°C, 600 W	$Zn(NO_3)_2\ 6H_2O$	NaOH	93.51	8.05	1.116 (W) 3.099 (L)
MW-4	20 min, 140°C, 600 W	$Zn(NO_3)_2\ 6H_2O$	NaOH	89.71	8.25	1.927 (W) 6.512 (L)
MW-5[b]	20 min, 140°C, 600 W	$Zn(NO_3)_2\ 6H_2O$	NaOH	99.31	6.41	1.534 (W) 3.577 (L)
MW-6[c]	20 min, 140°C, 600 W	$Zn(NO_3)_2\ 6H_2O$	NaOH	87.02	6.41	0.128 (W) 0.192 (L)
MW-7[d]	20 min, 140°C, 600 W	$Zn(NO_3)_2\ 6H_2O$	NaOH	92.30	6.31	0.123 (W) 0.059 (L)
MW-8[e]	20 min, 140°C, 600 W	$Zn(NO_3)_2\ 6H_2O$	NaOH	82.72	7.53	0.394 (W) 0.311 (L)
MW-9[f]	20 min, 140°C, 600 W	$Zn(NO_3)_2\ 6H_2O$	NaOH	89,73	8.22	1.308 (W) 4.092 (L)
MW-10	0 min, 140°C, 600 W	$Zn(NO_3)_2\ 6H_2O$	NaOH	N.A	8.25	N.A
MW-11	5 min, 140°C, 600 W	$Zn(NO_3)_2\ 6H_2O$	NaOH	91.17	8.22	1.467 (W) 3.256 (L)
MW-12	10 min, 140°C, 600 W	$Zn(NO_3)_2\ 6H_2O$	NaOH	92.23	8.23	1.006 (W) 2.493 (L)
MW-13	20 min, 140°C, 300 W	$Zn(NO_3)_2\ 6H_2O$	NaOH	81.40	8.26	1.487 (W) 3.924 (L)
MW-14	20 min, 140°C, 1200 W	$Zn(NO_3)_2\ 6H_2O$	NaOH	90.35	8.40	1.351 (W) 3.154 (L)
MW-15	20 min, 140°C, 600 W	$Zn(NO_3)_2\ 6H_2O$	KOH	87.00	7.77	0.870 (W) 2.375 (L)
MW-16	20 min, 140°C, 600 W	$Zn(NO_3)_2\ 6H_2O$	KOH[g]	N.A	7.81	0.220 (W) 1.063 (L)
MW-17	20 min, 140°C, 600 W	$Zn(NO_3)_2\ 6H_2O$	NH_4OH	N.A	7.54	1.689 (W) 4.673 (L)
MW-18	20 min, 100°C, 600 W	$Zn(CH_3COO)_2\ 2H_2O$	NaOH	N.A	8.05	3.716 (W) 3.662 (L)
MW-19	20 min, 120°C, 600 W	$Zn(CH_3COO)_2\ 2H_2O$	NaOH	N.A	7.97	3.128 (W) 3.951 (L)
MW-20	20 min, 140°C, 600 W	$Zn(CH_3COO)_2\ 2H_2O$	NaOH	N.A	8.15	1.656 (W) 1.735 (L)
MW-21	20 min, 140°C, 600 W	$ZnCl_2$	NaOH	N.A	N.A	N.A

[a]An initial 10 min period required to reach the target temperature must be considered in all cases.
[b]Solid product was isolated by centrifugation. The product was not washed.
[c]Change in the addition order.
[d]0.24 g of sodium di-2-ethylhexyl-sulfosuccinate dissolved in water were added before base addition.
[e]0.50 g of sodium di-2-ethylhexyl-sulfosuccinate dissolved in water were added before base addition.
[f]1 g of ethylene glycol dissolved in water was added before base addition.
[g]13 mL of base were added instead of 4 mL.
[h]Calculated from SEM images employing ImageJ Software.
[i]Average width of the rod-like particles.
[j]Average length of the rod-like particles.
N.A: not available.

$2\theta = 31.8, 34.3$, and $36.5°$, which correspond to the (100), (002), and (101) planes of wurtzite ZnO, respectively. FE-SEM images show an almost homogeneous morphological distribution at a set reaction temperature of 140°C. In this case most ZnO particles were formed as bihexagonal rods which were probably stacked from sheets with a starting growth from both sides. Additionally, a low population of ZnO particles with a more complex three-dimensional structure can be observed (MW-4, Figure 1) where a radial growth from a center seems to be favored yielding nanodandelions-like morphologies [16]. On the other hand, when the precursors are irradiated at 120°C (MW-3, Figure 1), the FE-SEM images show that, while some particles are formed, there is still a heterogeneous background. A possible explanation for this behavior can be linked to the existence of a greater irradiation to achieve a higher temperature, which increases the polarization of the system by producing a more orderly system that favors the nucleation and growth of the ZnO nanoparticles as was also described by other authors [18].

On the other hand, FTIR studies (Figure 1) show the characteristic vibrations of ZnO in the product obtained at 140°C. In all the cases the FTIR spectrums show absorption bands centered at about 3430 and 2344 cm^{-1}, which can be assigned to the O–H stretching vibrations, and bands centered at 1630 and 1384 cm^{-1} that correspond to the asymmetric and symmetric C=O stretching modes (CO_2 modes). At low wavenumbers there can be observed two peaks located at approximately 420 and 512 cm^{-1}, corresponding to the ZnO rod like-nanostructures [20, 21].

3.2. Precursors Effect. Three different precursors were employed to synthetize ZnO (temperature 140°C; power 600 W): $Zn(NO_3)_2 \cdot 6H_2O$ (MW-4), $Zn(CH_3COO)_2 \cdot 2H_2O$ (MW-20), and $ZnCl_2$ (MW-21). Figure 2 shows the corresponding morphologies as well as XRD analysis for these samples.

As it can be observed from Figure 2, XRD analysis shows that when $Zn(NO_3)_2$ and $Zn(CH_3COO)_2$ are the precursor salts, all the diffraction peaks match those of the wurtzite ZnO as it was described above. The morphological distribution is homogeneous in both cases, being in the first case particles with a bihexagonal rod-like structure of about 6 μm long and 2 μm wide; meanwhile in the second case the morphology is of the hexagonal prismatic type, and the average particle size is of about 3-4 μm long and 1 μ wide. In these both systems the larger particles observed could be grown from small primary particles through an oriented attachment process in which the adjacent nanoparticles are self-assembled by sharing a common crystallographic orientation and docking of these particles at a planar interface.

On the contrary, when $ZnCl_2$ is used as the precursor salt, the corresponding diffractogram shows many signals attributed to precursor reagents and unknown impurities, and a heterogeneous and non-well-defined morphology and a broad particle size distribution can be observed.

Taking into account that the growth of large crystals depends on the counteranion as it was previously described [22], the difference in the morphology of the three analyzed systems can be attributed to the fact that, in the case of the halide anion, it adsorbs more strongly on surfaces than acetate ions and nitrate ions which exhibits very weak surface interaction. As the adsorption on surfaces increases, the growth process decreases conducting particles with lower sizes as it can be observed in Figure 2 where the size of the synthesized ZnO particles decreases in the following order: $NO_3^- > CH_3COO^- > Cl^-$.

This behavior can also be confirmed by the fact that nitrate anion is a noncoordinative ligand, so that the hexagonal wurtzite prefers growing in the c axis (0001 plane); meanwhile in the case of acetate ion it is a chelating ligand that allows a better control on the ZnO particle development yielding particles with lower particle size.

On the other hand and for the case of $Zn(NO_3)_2$, the order of reagents addition during the synthesis of nano-ZnO was analyzed; when a solution of $Zn(NO_3)_2$ precursor was added dropwise over NaOH solution (basic route), different morphologies (short rows, etc.) were obtained in the final product (MW-6) with a wide particle size distribution and arranged in the form of agglomerates (Figure 3). This behavior can be explained attending to the following reaction mechanism. The first step in the synthetic reaction, irrespective of the route employed (acidic or basic), corresponds to the formation of a colloidal system:

$$Zn^{2+}(aq) + 2OH^-(aq) \longrightarrow Zn(OH)_2(s) \downarrow \qquad (2)$$

The zinc cations are known to react with hydroxide anions to form stable $Zn(OH)_4{}^{2-}$ complexes, which could act as the growing unit of ZnO nanostructures [23, 24]. Therefore, the growth mechanism can be considered as follows when the synthesis is carried out through the acidic route:

$$\begin{aligned} Zn(OH)_2(s) + 2H_2O &\xrightarrow{\text{Microwave}} Zn(OH)_4{}^{2-} + 2H^+ \\ &\xrightarrow{\text{Microwave}} ZnO + 3H_2O \end{aligned} \qquad (3)$$

When the addition is made in such a way that OH^- is aggregated over the Zn^{+2} solution, reaction symbolized by (2) takes place, and the formation of growing units during the irradiation stage is favored as it is symbolized in (3). In this sense, $Zn(OH)_4{}^{2-}$ species are generated under microwave effect, and it is well known that growing species can be polarized [18] allowing a controlled and directed growth of the particles.

On the other hand, if Zn^{+2} is added slowly over the OH^- solution as in the case of MW-6, the growing units can be formed rapidly, and the following reaction can occur before irradiation:

$$Zn(OH)_2(s) + 2OH^- \longleftrightarrow Zn(OH)_4{}^{2-} \qquad (4)$$

The $Zn(OH)_4{}^{2-}$ species formed have no polarization by microwave radiation, and the growth of the particles is in a random way.

3.3. Washing Solvent Effect. There has been described in the literature the effect of different solvents (ethanol, water, or

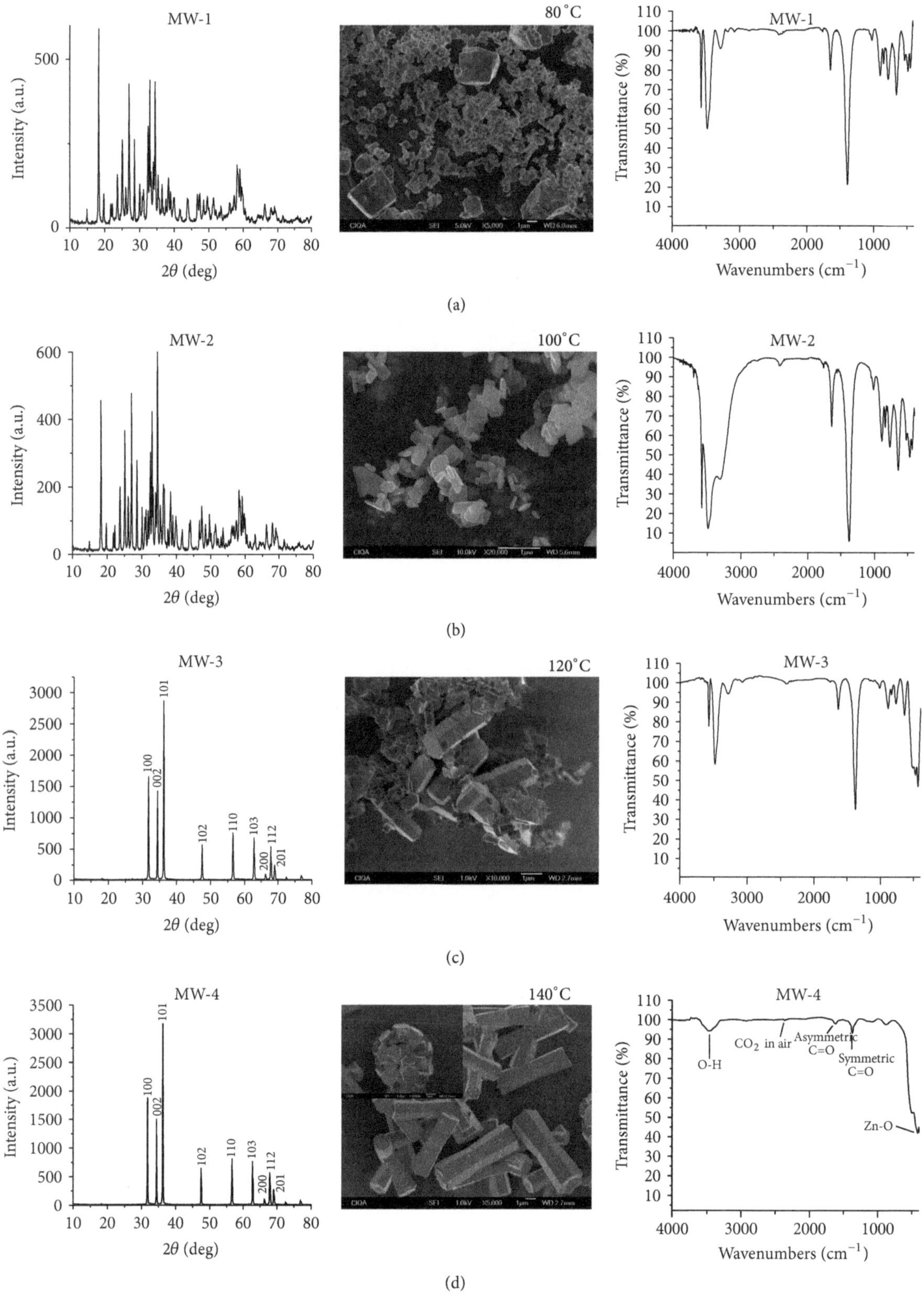

FIGURE 1: Temperature effect during microwave synthesis of ZnO on its morphological development.

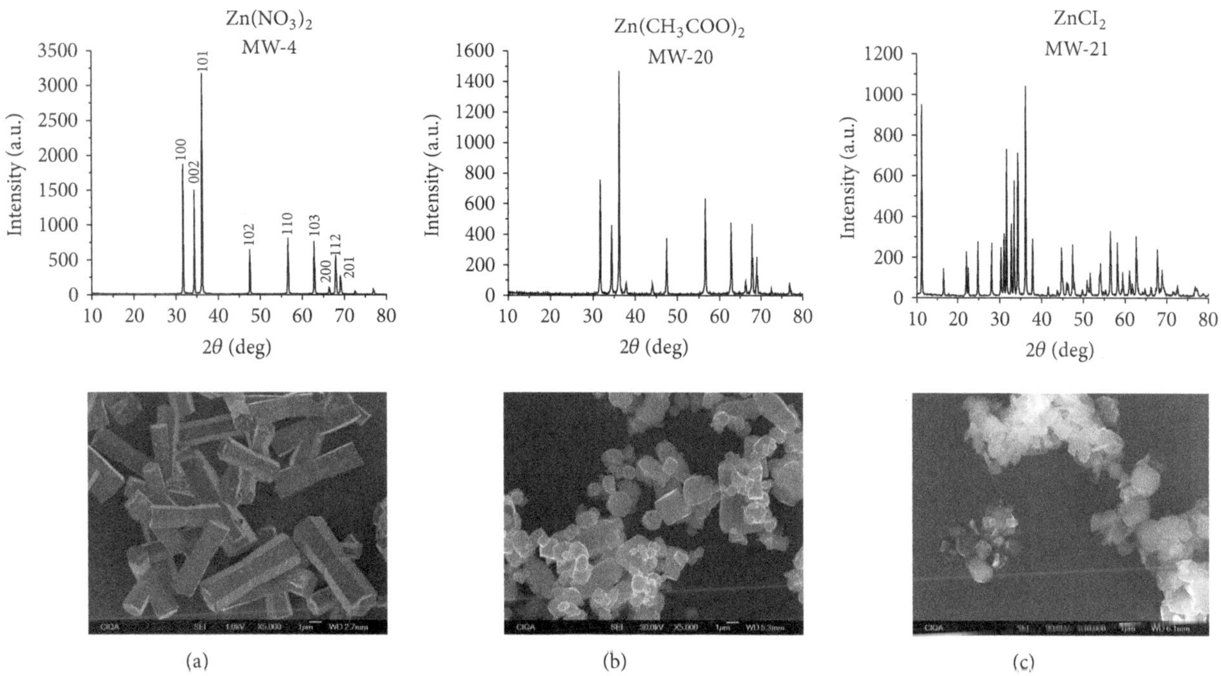

FIGURE 2: Effect of precursor salt during microwave synthesis of ZnO on its morphological development, at 140°C and 600 W.

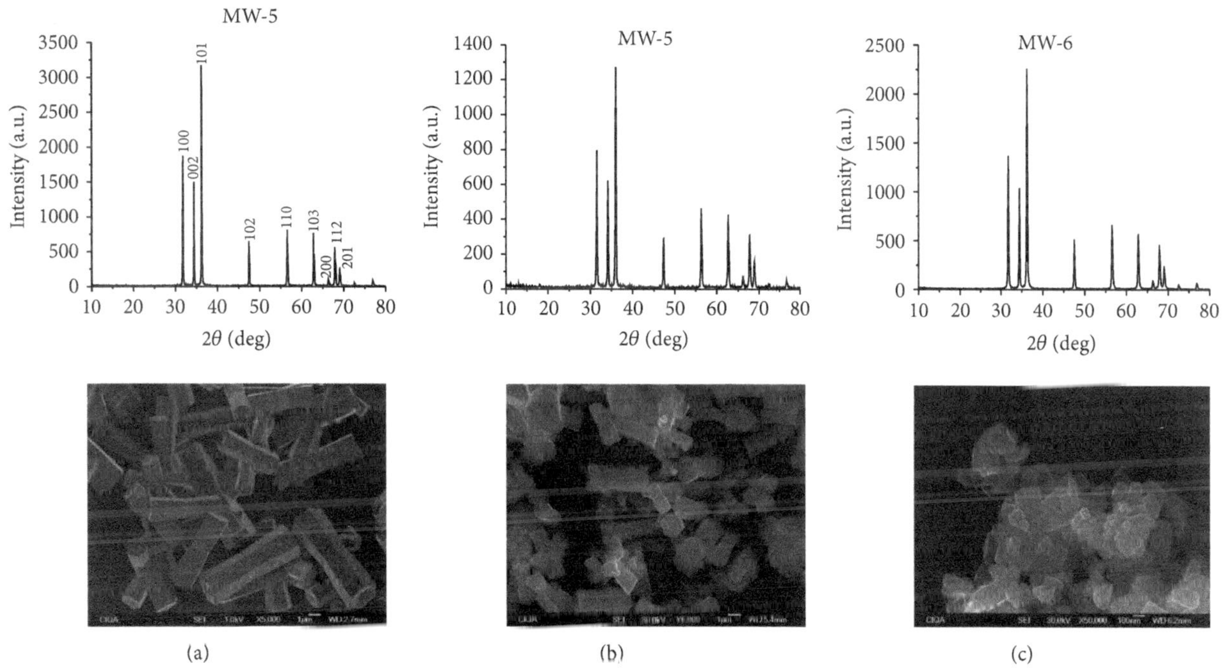

FIGURE 3: Isolation and precursors order addition effects on ZnO final morphology.

both) used during the washing step on the morphological parameters of the nano-ZnO [25], and it was evidenced that they can affect the morphological characteristics of the final product damaging the surface of ZnO, depending on the solvent used and on its polarity and vapor pressure. Attending to these results it was decided to carry out a synthesis where the isolation of nano-ZnO was made by centrifugation (10 min at 10000 rpm; MW-5) without previous washes; instead water is a preferable solvent to be used for ZnO [25]. XRD analysis for the former case, it means centrifugation without further purification, shows the wurtzite hexagonal ZnO characteristics (Figure 3). From the images obtained by

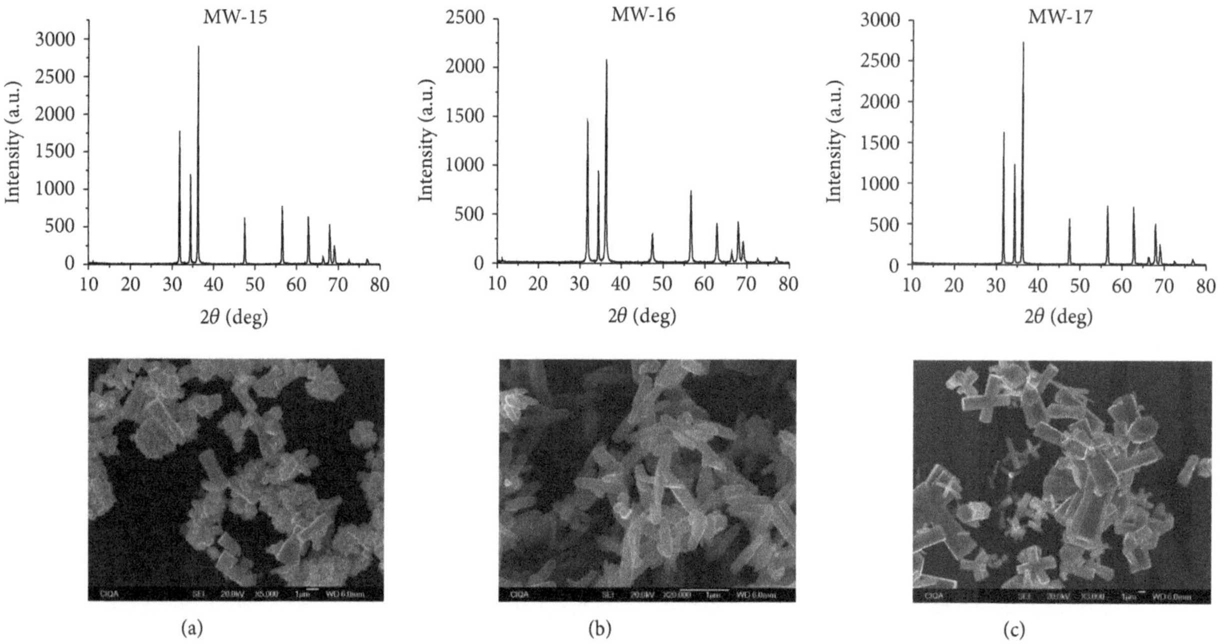

FIGURE 4: Base effect in microwave assisted ZnO synthesis. (a) KOH, (b) KOH 13 mL, and (c) NH_4OH.

FE-SEM a more dispersed system from both, morphological and size distribution aspects (Figure 3), can be observed with a less resolution on the lattices of the bihexagonal rod-like structure.

3.4. Precursor Base Effect. The results presented up to this point have been obtained using NaOH as the precursor base. However, KOH (MW-15) and NH_4OH (MW-17) have also been used under the same experimental conditions in order to analyze their effect on the morphological characteristics of the nanoparticles produced. In the case of using KOH two populations of well-defined morphologies can be observed (Figure 4(a)); one of them is similar to that obtained when NaOH is used as the precursor base (Figure 1, MW-4), in the form of bihexagonal rods, and there is another population of particles in the form of agglomerates which are constituted by small particles. When NH_4OH is used as the base, two morphologies can be clearly observed; the characteristic hexagonal rods previously described in the case of NaOH and KOH and other one where two- or three-dimensional growths are favored (Figure 4(c)). In this last case the growth is favored radially and in different directions yielding three-dimensional complex particles, similar to the dandelions particles obtained by Huang et al. [16]. The changes in morphology are attributed to the different degrees of electrostatic effects of the hydrated ion (Na^+, K^+, or NH_4^+) which can act on the surface of the growing crystals causing different growth on their surfaces. Therefore, the size of hydrated ions ($K^+ \cong NH_4^+ < Na^+$) and their charge are important factors in this approach.

In this sense, when the more bulky Na^+ is present, the Zn^{+2} is preferably adsorbed on the $0001^{+/-}$ planes so that the crystals show oriented growth along these directions. On the contrary, when K^+ is the counterion, there is a competition between K^+ and Zn^{+2}, and K^+ inhibited the adsorption of Zn^{+2} on the $0001^{+/-}$ planes, so Zn^{+2} must be adsorbed not only on the $0001^{+/-}$ planes but also onto the other six planes yielding particles with shorter length and higher diameter, in a similar way as the behavior described by Nejati et al. [26].

3.5. Stoichiometry Effect. The stoichiometry relation between Zn^{+2} and OH^- was also studied, and instead of adding the amount of base described in the synthesis ($Zn^{+2}/OH^- = 1.87/1$) (Section 2.1), that allows carrying out the reaction at pH of 5.8, 13 mL of KOH ($Zn^{+2}/OH^- = 1/1.73$) (MW-16) were added, and a pH of 10.8 was reached. This modification in OH^- concentration has been directly reflected in a morphological change of the particles obtained (Figure 4(b)). This effect is attributed to a different reaction mechanism in the generation of ZnO particles.

If an excess of OH^- is added (like in MW-16 synthesis) to the aqueous solution, the precipitated $Zn(OH)_2(s)$ species formed in (2) would dissolve back into the alkali solution to form the initial ions [27], and, after certain reaction time during the stirring period, the reaction symbolized by (4) can occur before irradiation with the formation of $Zn(OH)_4^{2-}$ complexes. As KOH concentration increases, the $Zn(OH)_4^{2-}$ complexes are surrounded by a large amount of OH^- so that the differences in the observed morphologies (Figures 4(a) and 4(b)) can be attributed to the different stages where growth unit can be formed and to the initial zinc species and their environment in the reactor as it was demonstrated by Xu et al. in the case of ZnO synthesis via a hydrothermal method [28].

3.6. Power and Irradiation Time Effect. In order to analyze the microwave irradiation system controllable variables, such as power and irradiation time on the produced ZnO morphology, some experiments were carried out under the same experimental conditions but with different irradiation times: only initial ramp (10 min) (MW-10) to reach the target temperature (140°C), and 5, 10, and 20 min after reaching the desirable temperature (Table 1; MW-11, MW-12, and MW-4, resp.). By XRD and IR techniques it could be observed that only in the case of using the initial ramp (MW-10) there is evidence of signals indicating the presence of impurities in the final product. In the other three cases (MW-11, MW-12, and MW-4) only wurtzite hexagonal signals and ZnO characteristics vibrations are present. In Figure 5 the morphology evolution of ZnO nanoparticles with irradiation time can be observed where an increase in this parameter generates more defined structures and a final system with a more homogeneous particle size and morphology distribution as a consequence of more time for the 0001 plane to growth.

On the other hand, when the effect of power microwave irradiation was analyzed (samples MW-13, MW-4, and MW-14 for 300, 600, and 1200 W, resp., Table 1) (Figure 6), it was observed that the formation of aggregates is favored as the microwave power increases. As microwave power increases, there is an increase in the mixture aqueous solution temperature enhancing the tendency of the ZnO nuclei to aggregate, in a similar way to that described by Al-Gaashani et al. [18]. Moreover, when the power is set to 1200 W (Figure 6(c)), the generation of agglomerates makes the system not homogeneous and not well dispersed like the one obtained at 600 W (Figure 6(b)).

3.7. Additives Effect. It is well known that the use of additives leads to changes in ZnO crystal growth [29]. In this work the effect of the addition of an anionic surfactant (sodium di-2-ethylhexyl-sulfosuccinate (AOT)) and a dispersant (ethylene glycol) on ZnO particles obtained by microwave synthetic method was evaluated. In these experiments the syntheses were carried out in the same way as it was previously described by dissolving Zn^{+2} in an additive (AOT or ethylene glycol) aqueous solution.

As it can be inferred from Figure 7(a), when 1 g (16 mmol) of ethylene glycol was added, (MW-9) to the reaction media, there were no significant changes neither in the morphology nor in the size of the obtained ZnO particles in comparison with the product obtained in the absence of any additive (MW-4).

On the contrary, when AOT surfactant was added, important changes could be observed. With 0.24 g of AOT (0.5 mmol) (MW-7) the particles show a different morphology (Figure 7(b)) in the sense that there is a lower growth in c axis ([001] direction), so smaller hexagonal prismatic particles are formed with sizes in the interval of 60–80 nm in diameter and length between 90 and 110 nm when compared with MW-4. Similar results were obtained when 0.5 g of AOT (1.1 mmol) were added (MW-8). In this case characteristic vibrations of surfactant could be observed by FTIR due to surfactant traces are still present after washes made with ethanol and deionized water.

The previous results can be attributed to the fact that AOT can be acting as an external factor that can control the growth rate of various faces of the crystals, due to the adsorption of the counterions on the negatively charged surfaces (precursor $Zn(OH)_4^{2-}$) reducing the surface energy of the specific crystal faces. This behavior can lead to the control of the crystal growth while retaining in some way the morphology.

3.8. Optical Properties. Having in mind that the absorption of colloidal system depends on the particle size and shape [30], absorption spectra of some synthetized ZnO nanoparticles were obtained. As it can be observed from Figure 8, particles with size higher than 500 nm do not show important absorption band, and no maximum can be detected in the absorbance curves (MW-4, MW-15, MW-17, and MW-20). However, when surfactant AOT is added to the reaction media and smaller particles are produced (particles size < 100 nm), a significant increase in UV absorption and a well defined maximum at 374 nm can be observed. This absorption is caused by two factors; the first one is due to the excitation of the electrons from the valence band to the conduction band when they are irradiated with UV light, which causes the absorption of UV radiation. The second is light scatter; the particle size of a colloidal solution within the wavelength of UV light results in light scattering, which also causes the absorption.

4. Conclusions

A simple and effective microwave assisted aqueous solution method has been used to synthetize ZnO with high crystallinity and purity when $Zn(NO_3)_2$ is used as precursor reagent. In addition, several parameters have been evaluated in order to analyze their effects on the ZnO nanoparticles obtained. An increase from 80°C to 140°C in set reaction temperature allowed to obtain a system with high purity and homogeneous in size and shape. On the other hand, three different Zn^{+2} precursors were employed to synthetize the nanoparticles: $Zn(NO_3)_2{\cdot}6H_2O$, $Zn(CH_3COO)_2{\cdot}2H_2O$, and $ZnCl_2$. It was observed that, as the adsorption of counteranions on surfaces increases, the synthesized ZnO particles sizes decrease in the following order: $NO_3^- > CH_3COO^- > Cl^-$.

When different OH^- precursors were employed (NaOH, KOH, and NH4OH), some changes were observed in the morphological characteristics of the synthesized ZnO. If NH_4OH is used as the base, two morphologies were found: the hexagonal rods described in the case of NaOH and KOH and other ones where two- or three-dimensional growths are favored. The final reaction pH can be mentioned as another important variable that causes significant changes on the morphology of the final product.

An increase in irradiation time generates more defined structures and a final system with a more homogeneous particle size and morphology distribution as a consequence of more time for the c direction to growth. Furthermore, it

(a) (b) (c)

Figure 5: FE-SEM images of ZnO obtained at different irradiation times: (a) only initial temperature ramp, (b) 5 min at 140°C, and (c) 10 min at 140°C.

(a) (b) (c)

Figure 6: FE-SEM images of ZnO obtained at different irradiation powers at 140°C: (a) 300 W, (b) 600 W, and (c) 1200 W.

was observed that the formation of aggregates is favored as the microwave power increases.

The addition of an anionic surfactant (AOT) to the reaction media allowed the synthesis of smaller particles and a significant increase in UV absorption, and a well-defined maximum at 374 nm was observed.

As final conclusions it can be enounced that to obtain a very pure phase of ZnO of high density, employing microwave as energy source, it is recommended to use $Zn(NO_3)_2$ as precursor.

Acknowledgments

The authors would like to thank Miriam Lozano and Jesús Angel Cepeda Garza for their contribution on the characterization of the ZnO nanoparticles by FE-SEM. Also, the authors wish to thank the Mexican National Council for Science and Technology (CONACYT) for its financial support for the realization of this study through a postdoctoral fellowship for Gastón P. Barreto, Ph.D. (CB-2008-01 Project no. 101934).

References

[1] L. Wang and M. Muhammed, "Synthesis of zinc oxide nanoparticles with controlled morphology," *Journal of Materials Chemistry*, vol. 9, no. 11, pp. 2871–2878, 1999.

[2] R. Hong, T. Pan, J. Qian, and H. Li, "Synthesis and surface modification of ZnO nanoparticles," *Chemical Engineering Journal*, vol. 119, no. 2-3, pp. 71–81, 2006.

[3] F. Paraguay, W. Estrada, D. R. Acosta, E. Andrade, and M. Miki-Yoshida, "Growth, structure and optical characterization of high quality ZnO thin films obtained by spray pyrolysis," *Thin Solid Films*, vol. 350, no. 1, pp. 192–202, 1999.

[4] T. Tani, L. Mädler, and S. E. Pratsinis, "Homogeneous ZnO nanoparticles by flame spray pyrolysis," *Journal of Nanoparticle Research*, vol. 4, no. 4, pp. 337–343, 2002.

[5] J. Wang and L. Gao, "Wet chemical synthesis of ultralong and straight single-crystalline ZnO nanowires and their excellent UV emission properties," *Journal of Materials Chemistry*, vol. 13, no. 10, pp. 2551–2554, 2003.

[6] B. Liu and H. C. Zeng, "Hydrothermal synthesis of ZnO nanorods in the diameter regime of 50 nm," *Journal of the American Chemical Society*, vol. 125, no. 15, pp. 4430–4431, 2003.

[7] U. Pal and P. Santiago, "Controlling the morphology of ZnO nanostructures in a low-temperature hydrothermal process," *Journal of Physical Chemistry B*, vol. 109, no. 32, pp. 15317–15321, 2005.

[8] L. Spanhel and M. A. Anderson, "Semiconductor clusters in the sol-gel process: quantized aggregation, gelation, and crystal growth in concentrated ZnO colloids," *Journal of the American Chemical Society*, vol. 113, no. 8, pp. 2826–2833, 1991.

[9] M. Ristić, S. Musić, M. Ivanda, and S. Popović, "Sol-gel synthesis and characterization of nanocrystalline ZnO powders," *Journal of Alloys and Compounds*, vol. 397, no. 1-2, pp. L1–L4, 2005.

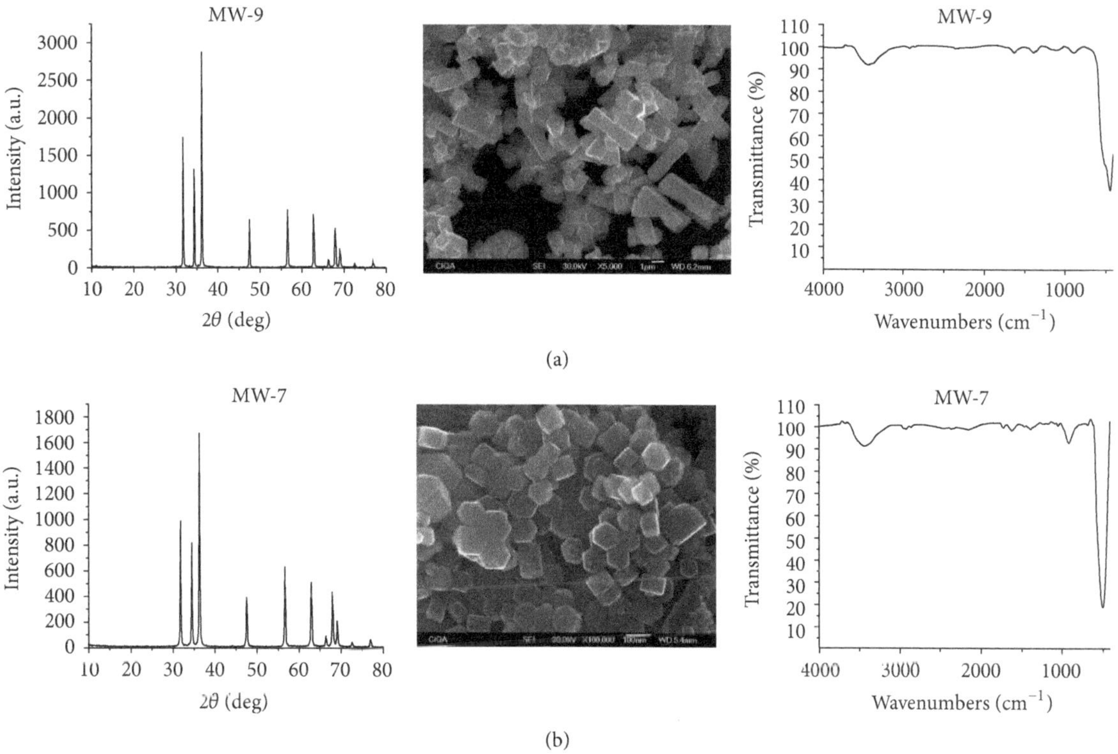

FIGURE 7: XRD, FE-SEM, and FTIR characterization of ZnO synthetized with different additives: (a) 1 g ethylene glycol, (b) 0.24 g AOT, and (c) 0.5 g AOT (experimental conditions: 140°C, 20 min, and 600 W).

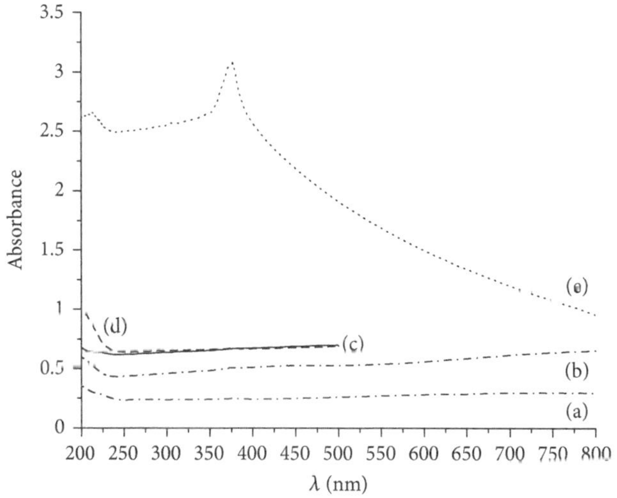

FIGURE 8: UV-Vis spectra of ZnO nanoparticles dispersed in water. (a) MW-17, (b) MW-15, (c) MW-20, (d) MW-4, and (e) MW-7.

[10] H. M. Cheng, H. C. Hsu, S. L. Chen et al., "Efficient UV photoluminescence from monodispersed secondary ZnO colloidal spheres synthesized by sol-gel method," *Journal of Crystal Growth*, vol. 277, no. 1–4, pp. 192–199, 2005.

[11] Y. Y. Tay, S. Li, F. Boey, Y. H. Cheng, and M. H. Liang, "Growth mechanism of spherical ZnO nanostructures synthesized via colloid chemistry," *Physica B*, vol. 394, no. 2, pp. 372–376, 2007.

[12] Ö. A. Yıldırım and C. Durucan, "Synthesis of zinc oxide nanoparticles elaborated by microemulsion method," *Journal of Alloys and Compounds*, vol. 506, no. 2, pp. 944–949, 2010.

[13] V. Polshettiwar and R. S. Varma, "Microwave-assisted organic synthesis and transformations using benign reaction media," *Accounts of Chemical Research*, vol. 41, no. 5, pp. 629–639, 2008.

[14] C. O. Kappe, "Controlled microwave heating in modern organic synthesis," *Angewandte Chemie—International Edition*, vol. 43, no. 46, pp. 6250–6284, 2004.

[15] I. Bilecka and M. Niederberger, "Microwave chemistry for inorganic nanomaterials synthesis," *Nanoscale*, vol. 2, no. 8, pp. 1358–1374, 2010.

[16] J. Huang, C. Xia, L. Cao, and X. Zeng, "Facile microwave hydrothermal synthesis of zinc oxide one-dimensional nanostructure with three-dimensional morphology," *Materials Science and Engineering B*, vol. 150, no. 3, pp. 187–193, 2008.

[17] D. Sharma, S. Sharma, B. S. Kaith, J. Rajput, and M. Kaur, "Synthesis of ZnO nanoparticles using surfactant free in-air and microwave method," *Applied Surface Science*, vol. 257, no. 22, pp. 9661–9672, 2011.

[18] R. Al-Gaashani, S. Radiman, N. Tabet, and A. R. Daud, "Effect of microwave power on the morphology and optical property of zinc oxide nano-structures prepared via a microwave-assisted aqueous solution method," *Materials Chemistry and Physics*, vol. 125, no. 3, pp. 846–852, 2011.

[19] T. D. Canh, N. V. Tuyen, and N. N. Long, "Influence of solvents on the growth of zinc oxide nanoparticles fabricated by microwave irradiation," *VNU Journal of Science, Mathematics-Physics*, vol. 25, pp. 71–76, 2009.

[20] L. Wu, Y. Wu, Y. Shi, and H. Wei, "Synthesis of ZnO nanorods and their optical absorption in visible-light region," *Rare Metals*, vol. 25, no. 1, pp. 68–73, 2006.

[21] H. Kleinwechter, C. Janzen, J. Knipping, H. Wiggers, and P. Roth, "Formation and properties of ZnO nano-particles from gas phase synthesis processes," *Journal of Materials Science*, vol. 37, no. 20, pp. 4349–4360, 2002.

[22] Z. Hu, G. Oskam, R. L. Penn, N. Pesika, and P. C. Searson, "The influence of anion on the coarsening kinetics of ZnO nanoparticles," *Journal of Physical Chemistry B*, vol. 107, no. 14, pp. 3124–3130, 2003.

[23] W. Li, E. Shi, W. Zhong, and Z.-W. Yin, "Growth mechanism and growth habit of oxide crystals," *Journal of Crystal Growth*, vol. 203, no. 1-2, pp. 186–196, 1999.

[24] S. Yamabi and H. Imai, "Growth conditions for wurtzite zinc oxide films in aqueous solutions," *Journal of Materials Chemistry*, vol. 12, no. 12, pp. 3773–3778, 2002.

[25] Y. C. Chen and S. L. Lo, "Effects of operational conditions of microwave-assisted synthesis on morphology and photocatalytic capability of zinc oxide," *Chemical Engineering Journal*, vol. 170, no. 2-3, pp. 411–418, 2011.

[26] K. Nejati, Z. Rezvani, and R. Pakizevand, "Synthesis of ZnO nanoparticles and investigation of the ionic template effect on their size and shape," *International Nano Letters*, vol. 1, no. 2, pp. 75–81, 2011.

[27] S. Erten-Ela, S. Cogal, and S. Icli, "Conventional and microwave-assisted synthesis of ZnO nanorods and effects of PEG400 as a surfactant on the morphology," *Inorganica Chimica Acta*, vol. 362, no. 6, pp. 1855–1858, 2009.

[28] H. Xu, H. Wang, Y. Zhang et al., "Hydrothermal synthesis of zinc oxide powders with controllable morphology," *Ceramics International*, vol. 30, no. 1, pp. 93–97, 2004.

[29] J. H. Park and S. G. Oh, "Preparation of CaO as OLED getter material through control of crystal growth of $CaCO_3$ by block copolymers in aqueous solution," *Materials Research Bulletin*, vol. 44, no. 1, pp. 110–118, 2009.

[30] E. Koushki, M. H. Majles Ara, S. H. Mousavi, and H. Haratizadeh, "Temperature effect on optical properties of colloidal ZnO nanoparticles," *Current Applied Physics*, vol. 11, no. 5, pp. 1164–1167, 2011.

23

Corrosion Inhibition of Aluminium in HCl by Amine Modified Epoxy Resin

Makanjuola Oki,[1,2] Kayode Oki,[3] John Otaigbe,[4] and Samuel Otikor[4]

[1] *Institute of Petroleum Studies, University of Port Harcourt, PMB 2, Choba, Port Harcourt 500102, Rivers State, Nigeria*
[2] *Department of Chemical Sciences, Oduduwa University, Ipetumodu, Ile-Ife 220371, State of Osun, Nigeria*
[3] *Technical Department, Greenfield-Oaks Limited, London SE25 4HL, UK*
[4] *Department of Pure and Industrial Chemistry, University of Port Harcourt, PMB 5323, Port Harcourt 500102, Rivers State, Nigeria*

Correspondence should be addressed to Makanjuola Oki; m.oki11@yahoo.com

Academic Editor: Wolfgang Maser

N-Butyl amine has been grafted unto Dow epoxy resin. The product was evaluated as an acid inhibitor by gravimetric techniques. The reduction in corrosion of aluminium in 1 M HCl solution was 92% at 30°C and 83% at 60°C with an inhibitor concentration of 100 ppm. The corrosion rate at this concentration was 3.0×10^{-2} and 1.5×10^{-1} mm/yr at 30° and 60°C. For the specimens in HCl without inhibitor at 10-hour immersion period, the corrosion rate was 3.8×10^{-1} and 8.6×10^{-1} mm/yr, respectively, at 30° and 60°C. Data obtained correlate with Langmuir isotherm.

1. Introduction

Aluminium and its alloys find extensive application in the construction of heat exchangers, radiators, and related components in water cooling/treatment facilities. It is often observed that such facilities may suffer corrosion damage from ingress of chlorides and other extraneous anions as a result of inadequate provision for corrosion mitigation such as faulty design features and nonintroduction of suitable corrosion inhibitors into the system. Pitting corrosion at crevices and under sediments is often encountered in such industrial situations.

The use of organic inhibitors is the most economical and practical method of protecting metals against aqueous corrosion [1–3]. These inhibitors contain heteroatoms such as Sulphur, Oxygen, and Nitrogen in their structures in addition to multiple bonds and may contain cyclic hydrocarbons as well [3–6]. The mechanism of inhibition of these organic compounds is thought to be by adsorption which may depend on the number of heteroatoms and the surface area of the corroding metal/alloy [7, 8].

Various investigators [8–13] have researched on polymers as corrosion inhibitors because of their large surface areas, cost effectiveness, and inherent stability. These polymers have been reported to be very efficient at low concentrations in very aggressive media although efficiencies vary widely as a result of differences in composition. Chemically modified polyurethane [14] gave about 95% inhibition efficiency for mild steel corroding in acidic medium. On the other hand, polyacrylamide [15] achieved an efficiency of about 63% for aluminium corroding freely in 0.5 M HCl solution.

In order to contribute towards the search for suitable corrosion inhibitors of polymeric origin, in the present investigation, a Dow epoxy resin was chemically modified with n-butyl amine at suitable temperatures and examined as corrosion inhibitor for aluminium in 1 M HCl solution.

2. Experimental

2.1. Materials. 99.99% aluminium with nominal impurities of 0.004 wt% Fe, 0.002 wt% Cu, and 0.003 wt% Si was made into electrodes measuring 10 mm × 10 mm × 1 mm. The epoxy resin, Resin 828, is a product of Dow Chemicals, USA, while all other chemicals employed in this investigation are laboratory grade reagents from BDH Chemicals, UK.

2.2. Methods

2.2.1. Inhibitor Formulation. The inhibitor employed in this investigation was synthesised by mixing in a 500 mL beaker, equal moles of epoxy resin which had amber colouration, and n-butyl amine in a water bath, set at a controlled temperature of 100°C. A consistent brown coloured product, Amine Modified Epoxy Resin (AMER), was obtained after 30 mins of constant stirring.

The various concentrations of inhibitor were obtained by serially diluting 1 g/litre solution of the inhibitor with the corrodents of interest, 1 M HCl solution.

2.2.2. Corrosion Measurements. The aluminium electrodes were etched in 10% NaOH solution for 3 mins, rinsed in running doubly distilled water and thereafter desmutted in 50% HNO_3 for 2 mins, rinsed in doubly distilled water and in acetone, dried in air for 10 minutes prior to being stored in desiccators. Triplicate experiments, at 30°C and 60°C were performed by separately immersing 10 preweighed electrodes, with exposed area of 2.3 cm^2, in 200 mL of 1 M HCl contained in 250 mL beakers for times ranging from 1 hr to 10 hrs. After each period of immersion, the electrodes were washed in running distilled water for 3 mins while scrubbing the surfaces with a soft brush and later in acetone for 30 sec. After drying in a stream of cool air, the electrodes were reweighed. The mean of the weight changes were taken as the weight change for each period of immersion. Similar procedure was performed in 1 M HCl with different concentrations of AMER as inhibitor of interest.

3. Results and Discussions

3.1. Corrosion Rates. For the calculation of corrosion rates, the relationship in (1) as described by Jones [1] and other authors [15–17] was employed

$$\mathrm{CR} = \frac{87.6\Delta \mathrm{wt}}{\rho_{\mathrm{A}} t}, \quad (1)$$

where CR is the corrosion rate in mm/yr, Δwt is the weight loss, ρ_A is the density of Aluminium in g/cm^3, and t is the time of exposure.

The efficiency of the inhibitor, %Inh., was obtained from the following equation

$$\%\mathrm{Inh.} = \left[\frac{\mathrm{CR_O} - \mathrm{CR_1}}{\mathrm{CR_O}}\right] \times \frac{100}{1}, \quad (2)$$

where CR_O and CR_1 are Corrosion rates without and in the presence of inhibitor, respectively.

θ, the surface coverage, is given by the relationship, $\theta = [(\mathrm{CR_O} - \mathrm{CR_1})/\mathrm{CR_O}]$, where all the symbols have their usual meanings.

The weight loss and corrosion rates of aluminium specimens in hydrochloric acid at both 30°C and 60°C are displayed in Figures 1 and 2, respectively. The rate of change in weight at 30°C increased rapidly between 1 and 2 hours with a rate that decreased with time between 2 and 10 hours of immersion in HCl. The initial rapid increase in weight loss is consistent with the rapid dissolution of the amphoteric air formed oxide skin on aluminium in acidic media which exposes freshly prepared aluminium surface to rapid dissolution of the substrate metal.

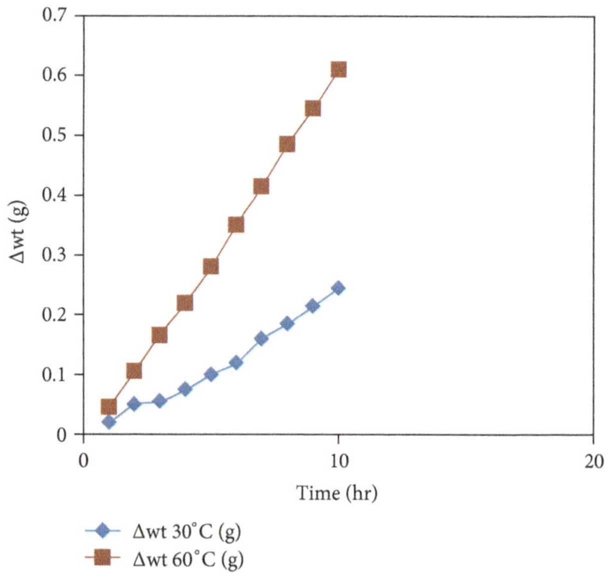

Figure 1: Weight loss of aluminium with time of immersion in 1 M HCl at 30°C and 60°C.

However, the decrease in rate of dissolution may be attributed to the formation of corrosion products at cathodic sites on the aluminium electrode. Such corrosion products will retard the corrosion process for a while before they are subsequently dissolved by the acidic medium to reveal further areas of the specimen and corrosion processes will proceed.

At 60°C, the weight loss is linear with time of immersion in HCl solution as all corrosion products formed are dissolved rapidly in the warm corrodent.

The gravimetric method for corrosion studies is a time tested and reliable technique. Several authors have reported good correlation between this technique and gasometric, impedance, and polarisation techniques [16–20]. The corrosion rate of aluminium in HCl at 60°C (Figure 2) increased at a rate which decreased with exposure time and was maximum at 8.6×10^{-2} mm/yr at 10 hours. On the other hand, the corrosion rate at 30°C was at its peak, 3.5×10^{-2} mm/yr, after 2 hours of exposure in the corrodent which corroborates the weight loss measurement described earlier. Thereafter, the corrosion rate decreased to about 2.2×10^{-2} mm/year at 3 hours of exposure, beyond which steady increase in corrosion rate was observed and was about 3.5×10^{-2} mm/yr at 10 hours in HCl solution.

From Figure 3, the corrosion rates decreased exponentially as the concentration of the inhibitor was increased. Thus, it can be safely inferred that the inhibitor reduced the corrosion rates of aluminium at both 30°C and 60°C.

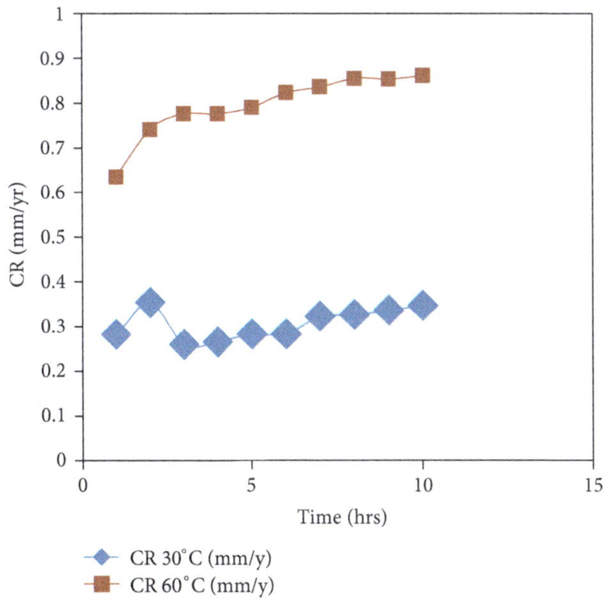

FIGURE 2: Corrosion rates of aluminium in 1 M HCl at 30°C and 60°C.

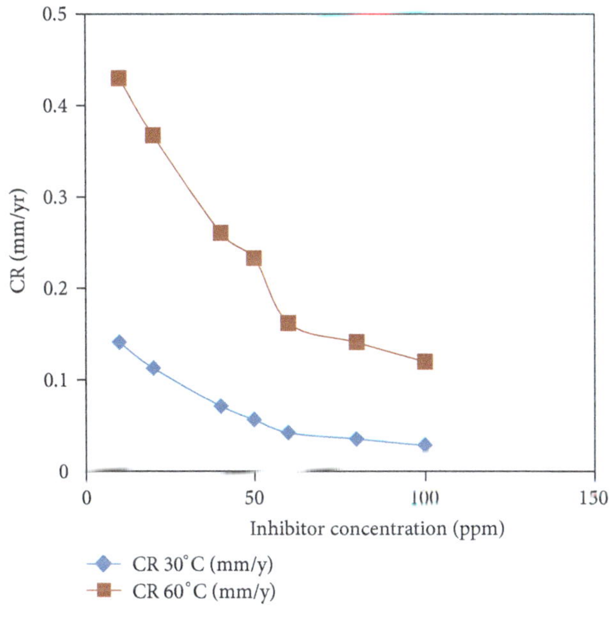

FIGURE 3: Corrosion rates of aluminium in 1 M HCl solution in the presence of various concentrations of inhibitor (AMER).

Displayed in Figure 4 is the percent inhibition which increases with increase in the concentration of the inhibitor. Inhibitors of organic origin inhibit corrosion by adsorbing on metals in corroding systems. The increase in % inhibitor efficiency was more pronounced at concentration levels between 40 and 60 ppm probably as the surface of the corroding aluminium specimens became saturated with adsorbed inhibitor species. The efficiency of the inhibitor

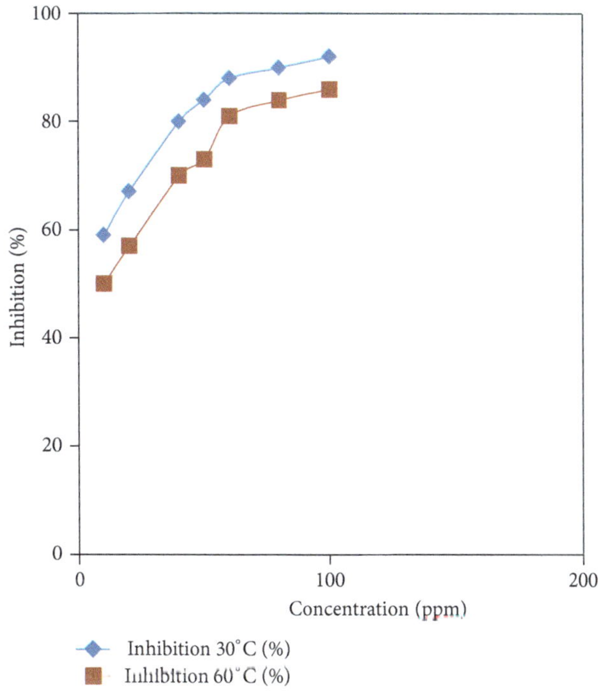

FIGURE 4: Variation of % inhibition with concentration of inhibitor (AMER) at 30°C and 60°C.

was better at lower temperatures and this may be attributed to desorption of inhibitor molecules from the corroding surface at high temperatures as described by El-Etre [18] and Abdallah [19]. However, AMER was more effective at 30°C with an efficiency of about 92% as against 86% at 60°C, with an inhibitor concentration of 100 ppm.

Furthermore, the energy of activation, E_a, for the corrosion reactions in the presence and absence of corrosion inhibitor obtained from the condensed Arrhenius equation (3) is 40.11 kJ/mol and 28.54 kJ/mol, respectively. The increase in E_a in the presence of inhibitor further demonstrates that there was a hindrance placed on the electrochemical reaction path on the substrate by some adsorbed inhibitor species. With a positive E_a value, the mechanism of inhibition is likely to be by physical adsorption [21–23] of the inhibitor species

$$\log\left[\frac{CR_{60}}{CR_{30}}\right] = \frac{E_a}{2.303R}\left[\frac{1}{T_1} - \frac{1}{T_2}\right]. \tag{3}$$

The interpretation of adsorption isotherms has been employed by several authors [24–27] to describe the mode(s) of interaction of inhibitor molecules with corroding surfaces in various media in which, the values for surface coverage of inhibitors, θ, were plotted against concentration to fit Frumkin [28] or, C versus θ/C to establish a Langmuir [29] type of adsorption isotherm. Displayed in Figure 5 are curves for Langmuir adsorption.

Displayed in Figure 5 are curves of Langmuir adsorption for AMER at both 60° and 30°C where it can be observed that the inhibitor obeys this adsorption isotherm. From observed data, it is obvious that the adsorption isotherm curves are

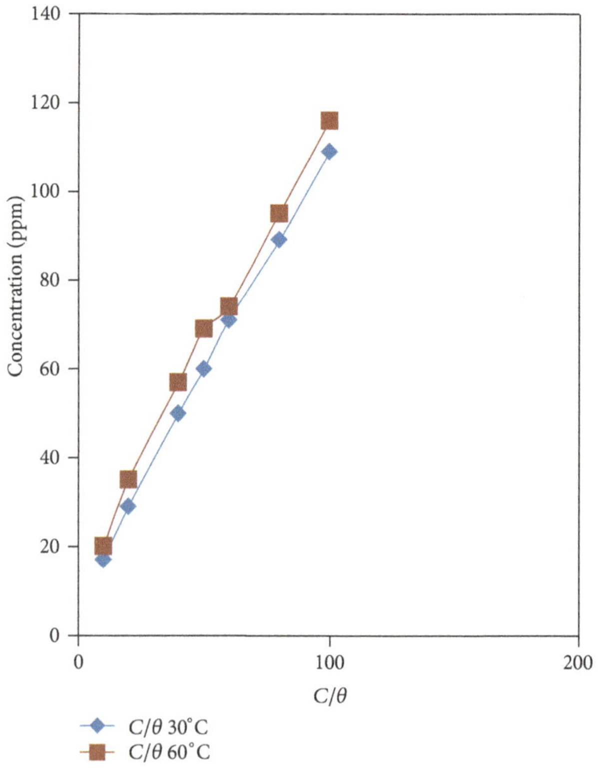

FIGURE 5: Test of variables to fit Langmuir adsorption isotherm.

neither in agreement with Temkin nor Frumkin adsorption isotherms, respectively. Thus there are no interactions among the adsorbed inhibitor molecules on the corroding surfaces [28].

The heat of adsorption, Q_{ads}, can be obtained from the relationship

$$Q_{ads} = 2.303R\left[\log\frac{\theta_2}{1-\theta_2} - \log\left(\frac{\theta_1}{1-\theta_1}\right)\right]\left[\frac{T_1T_2}{T_2} - T_1\right], \quad (4)$$

where T_1 is the temperature of the corroding system at 30°C and T_2 the corresponding temperature at 60°C. θ_1 is the surface coverage of the inhibitor at 30°C while θ_2 is the surface coverage at 60°C.

The calculated approximate value of Q_{ads} is −17.52 kJ/mol. The negative value of Q_{ads} is consistent and agrees with the findings of other researchers [20, 21] that for physio-sorption, the heat of adsorption is generally negative and positive values indicate mechanism by chemisorption [20].

4. Conclusions

Amine modified epoxy resin (AMER) effectively reduced the corrosion rate of aluminium in hydrochloric acid by physically adhering to the corroding metallic surface.

The data obtained from the gravimetric studies correlate with the Langmuir adsorption isotherm which indicate that there are no interactions among the adsorbed inhibitor species. However, the reactions at the electrolyte/aluminium interface was sufficiently interrupted by the inhibitor species to lower the corrosion rates by 92% and 86% at 30° and 60°C, respectively.

The corrosion rates of aluminium in 1 M hydrochloric acid in the presence of 100 ppm of AMER at 10 hours of exposure are 3.0×10^{-2} and 1.5×10^{-1} mm/yr, respectively at 60° and 30°C. On the other hand, for specimens exposed in hydrochloric acid without AMER for the same period, the corrosion rates are 3.8×10^{-1} and 8.6×10^{-1} mm/yr at 30° and 60°C.

The mechanism of inhibition is proposed by physical adsorption of the inhibitor species on the corroding aluminium surface.

Acknowledgments

Dr. M. Oki acknowledges the National Universities Commission, Nigeria, for the award of one-year Linkages with Academics and Experts in Diaspora Scheme which was utilised at the Institute of Petroleum Studies/Department of Petroleum and Gas Engineering, University of Port Harcourt, Nigeria. S. Otikor acknowledges the Head of Department of Chemistry and Industrial Chemistry for the use of laboratory facilities while T. k. Oki is grateful to the management of Greenfield-Oaks Ltd., UK, for the use of time share facilities.

References

[1] D. A. Jones, *Principles and Prevention of Corrosion*, Prentice Hall, Upper Saddle River, NJ, USA, 2nd edition, 1996.

[2] G. Y. Elewady, I. A. El-Said, and A. S. Fouda, "Effect of anions on the corrosion inhibition of aluminum in HCl using ethyl trimethyl ammonium bromide as cationic inhibitor," *International Journal of Electrochemical Science*, vol. 3, pp. 644–655, 2008.

[3] E. A. Noor, "The inhibition of mild steel corrosion in phosphoric acid solutions by some N-heterocyclic compounds in the salt form," *Corrosion Science*, vol. 47, no. 1, pp. 33–55, 2005.

[4] K. Babic-Samardzija, C. Lupu, N. Hackerman, and A. R. Barron, "Inhibitive properties, adsorption and surface study of butyn-1-ol and pentyn-1-ol alcohols as corrosion inhibitors for iron in HCl," *Journal of Materials Chemistry*, vol. 15, no. 19, pp. 1908–1916, 2005.

[5] D. Turcio-Ortega, T. Pandiyan, J. Cruz, and E. Garcia-Ochoa, "Interaction of imidazoline compounds with Fe_n (n = 1-4 Atoms) as a model for corrosion inhibition: DFT and electrochemical studies," *The Journal of Physical Chemistry C*, vol. 111, no. 27, pp. 9853–9866, 2007.

[6] E. E. Oguzie, S. G. Wang, Y. Li, and F. H. Wang, "Influence of iron microstructure on corrosion inhibitor performance in acidic media," *The Journal of Physical Chemistry C*, vol. 113, no. 19, pp. 8420–8429, 2009.

[7] R. Solmaz, G. Kardaş, M. Çulha, B. Yazici, and M. Erbil, "Investigation of adsorption and inhibitive effect of 2-mercaptothiazoline on corrosion of mild steel in hydrochloric acid media," *Electrochimica Acta*, vol. 53, no. 20, pp. 5941–5952, 2008.

[8] P. Manivel, S. Sathiyanarayanan, and G. Venkatachari, "Synthesis of poly(p-phenylene diamine) and its corrosion inhibition effect on iron in 1M HCl," *Journal of Applied Polymer Science*, vol. 110, no. 5, pp. 2807–2814, 2008.

[9] S. Banerjee, A. Mishra, M. M. Singh, B. Maiti, B. Ray, and P. Maiti, "Highly efficient polyurethane ionomer corrosion inhibitor: the effect of chain structure," *RSC Advances*, vol. 1, pp. 199–210, 2011.

[10] M. A. Amin, S. S. A. EI-Rehim, E. E. F. El-Sherbini, O. A. Hazzazi, and M. N. Abbas, "Polyacrylic acid as a corrosion inhibitor for aluminium in weakly alkaline solutions. Part I: weight loss, polarization, impedance EFM and EDX studies," *Corrosion Science*, vol. 51, no. 3, pp. 658–667, 2009.

[11] A. Y. Musa, A. A. H. Kadhum, A. B. Mohamad, M. S. Takrif, and E. P. Chee, "Inhibition of aluminum corrosion by phthalazinone and synergistic effect of halide ion in 1.0M HCl," *Current Applied Physics*, vol. 12, no. 1, pp. 323–330, 2012.

[12] C. Jeyaprabha, S. Sathiyanarayanan, and G. Venkatachari, "Co-adsorption effect of polyaniline and halide ions on the corrosion of iron in 0.5 M H_2SO_4 solutions," *Journal of Electroanalytical Chemistry*, vol. 583, no. 2, pp. 232–240, 2005.

[13] V. Srivastava, S. Banerjee, and M. M. Singh, "Inhibitive effect of polyacrylamide grafted with fenugreek mucilage on corrosion of mild steel in 0.5 M H_2SO_4 at 35°C," *Journal of Applied Polymer Science*, vol. 116, no. 2, pp. 810–816, 2010.

[14] S. Banerjee, A. Mishra, M. M. Singh, and P. Maiti, "Effects of naniclay and polyurethanes on inhibition of mild steel corrosion," *Journal of Nanoscience and Nanotechnology*, vol. 11, no. 2, pp. 966–978, 2011.

[15] S. A. Umoren and M. M. Solomon, "Effect of halide ions additives on the corrosion inhibition of aluminum in HCL by polyacrylamide," *The Arabian Journal of Science and Engineering*, vol. 35, no. 2A, pp. 115–130, 2009.

[16] A. A. El-Maghraby, "Corrosion inhibition of aluminium in HCl using potassium iodate," *The Open Corrosion Journal*, vol. 2, pp. 189–196, 2009.

[17] P. C. Okafor, U. J. Ekpe, E. E. Ebenso, E. E. Oguzie, N. S. Umo, and A. R. Etor, "Extract of Allium cepa and Allium sativum as corrosion inhibitors of mild steel in HCl solution," *Transactions of the SAEST*, vol. 41, no. 2, pp. 82–87, 2006.

[18] A. Y. El-Etre, "Inhibition of aluminum corrosion using Opuntia extract," *Corrosion Science*, vol. 45, no. 11, pp. 2485–2495, 2003.

[19] M. Abdallah, "Antibacterial drugs as corrosion inhibitors for corrosion of aluminium in hydrochloric solution," *Corrosion Science*, vol. 46, no. 8, pp. 1981–1996, 2004.

[20] S. A. Umoren, I. B. Obot, E. E. Ebenso, P. C. Okafor, O. Ogbobe, and E. E. Oguzie, "Gum arabic as a potential corrosion inhibitor for aluminium in alkaline medium and its adsorption characteristics," *Anti-Corrosion Methods and Materials*, vol. 53, no. 5, pp. 277–282, 2006.

[21] P. C. Okafor, E. E. Ebenso, and U. J. Ekpe, "Äzadirachta indica extracts as corrosion inhibitor for mild steel in acid medium," *International Journal of Electrochemical Science*, vol. 5, no. 7, pp. 978–993, 2010.

[22] L. A. Nnanna, I. U. Anozie, A. G. I. Avoaja, C. S. Akoma, and E. P. Eti, "Comparative study of corrosion inhibition of Aluminium alloy type AA 3003 in acidic and alkaline media by Euphorbia hirta extract," *African Journal of Pure & Applied Chemistry*, vol. 5, no. 8, pp. 265–271, 2011.

[23] I. B. Obot and N. O. Obi-Egbedi, "An interesting and efficient green corrosion inhibitor for Aluminium from extracts of Chlomolaena odorata L. in acidic solution," *Journal of Applied Electrochemistry*, vol. 40, pp. 1977–1984, 2010.

[24] M. Oki, E. Charles, C. Alaka, and T. K. Oki, "Corrosion inhibition of mild steel in HCl by tannins from Rhizophora racemosa," *Materials Science and Applications*, vol. 2, pp. 592–595, 2011.

[25] K. F. Khaled and N. Hackerman, "Investigation of the inhibitive effect of ortho-substituted anilines on corrosion of iron in 1 M HCl solutions," *Electrochimica Acta*, vol. 48, no. 19, pp. 2715–2723, 2003.

[26] E. E. Oguzie, "Corrosion inhibition of aluminium in acidic and alkaline media by Sansevieria trifasciata extract," *Corrosion Science*, vol. 49, no. 3, pp. 1527–1539, 2007.

[27] K. F. Khaled, K. Babić-Samardžija, and N. Hackerman, "Cobalt(III) complexes of macrocyclic-bidentate type as a new group of corrosion inhibitors for iron in perchloric acid," *Corrosion Science*, vol. 48, no. 10, pp. 3014–3034, 2006.

[28] A. N. Frumkin, "Electrocapillary curve of higher aliphatic acids and the state equation of the surface layer," *Zeitschrift für Physikalische Chemie*, vol. 116, pp. 466–470, 1925.

[29] I. Langmuir, "The adsorption of gases on plane surfaces of glass, mica and platinum," *The Journal of the American Chemical Society*, vol. 40, no. 9, pp. 1361 1403, 1918.

24

Photocatalytic Performance of TiO_2 Nanofibers as a Function of Fiber Diameter Using $TiCl_4$ as a Precursor

Ali Akbar Ashkarran,[1] Sima Eshghi,[2] and Mohammad Reza Nourani[2]

[1] *Department of Physics, Faculty of Basic Sciences, University of Mazandaran, P.O. Box 47416-1467, Babolsar, Iran*
[2] *Department of Material Engineering, Science and Research Branch, Islamic Azad University, Tehran, Iran*

Correspondence should be addressed to Ali Akbar Ashkarran; ashkarran@gmail.com

Academic Editor: Tong Lin

We focus on the production of TiO_2 nanofibers with controllable diameters using a facile electrospinning technique at room temperature. The resulting nanofibers were characterized using X-ray diffraction (XRD), scanning electron microscopy (SEM), and ultraviolet-visible spectroscopy (UV-Vis). The most important electrospinning parameters including potential difference (kV), flow rate (mL/h), and the separation distance between electrodes (cm) were found to have significant influence on the diameter of the produced nanofibers. The photocatalytic performance of TiO_2 nanofibers was successfully demonstrated for decolorization of Rhodamine B (Rh.B) under UV light irradiation. It was found that fiber diameter has a crucial influence on the photocatalytic performance of TiO_2 nanofibers.

1. Introduction

Among several oxide semiconductors, titanium dioxide has a more helpful role in our environmental purification due to its nontoxicity, photocatalytic activity, photoinduced superhydrophilicity, selfcleaning, and antifogging effect [1, 2]. Many researches on titanium dioxide are going on since Fujishima and Honda discovered the photocatalytic activity of titanium dioxide in 1969 [3]. They found out that titanium dioxide decomposes water to hydrogen and oxygen by photoillumination. In this regard, much attention has been drawn towards their photocatalytic properties because of its application in environmental purification and decomposition of toxic and organic compounds into environmentally harmless components [4]. A key requirement for improving the photocatalytic activity is to increase the specific surface area and enhance the crystallinity [5]. Therefore, significant progress has been achieved in the development of one-dimensional (1D) nanomaterials such as nanofibers due to their intrinsically high porosities and large surface areas [6–10].

Among the most successful methods for producing nanofibers such as chemical reduction, layer-by-layer self-assembly, and template-based techniques, electrospinning has attracted much attention since it is one of the most simple, versatile, and low-cost methods for generating nanofibers from a wide range of materials including polymers, composites, ceramics, and semiconductors [7, 11–14]. It provides the ability to produce long continuous fibers with typical diameters ranging from 50 nm up to few micrometers and length of up to several centimeters [15, 16].

In this work, we have prepared TiO_2 nanofibers by electrospinning technique using $TiCl_4$ as a precursor and studied the effect of electrospinning parameters such as applied electric field, distance between nozzle and collector and flow rate on structure, morphology and fiber diameter. We have also evaluated the diameter dependency of electrospun TiO_2 nanofiber on photocatalytic performance. In the majority of published papers, at least 5 to 6 precursors are used for preparation of primary sol, but in this method we have just used $TiCl_4$ and ethanol for formation of TiO_2 and also PVP as an appropriate polymer. To the best of author knowledge there were no published report, on electrospun TiO_2 fibers using $TiCl_4$ and ethanol which is a rather fast and one-step approach.

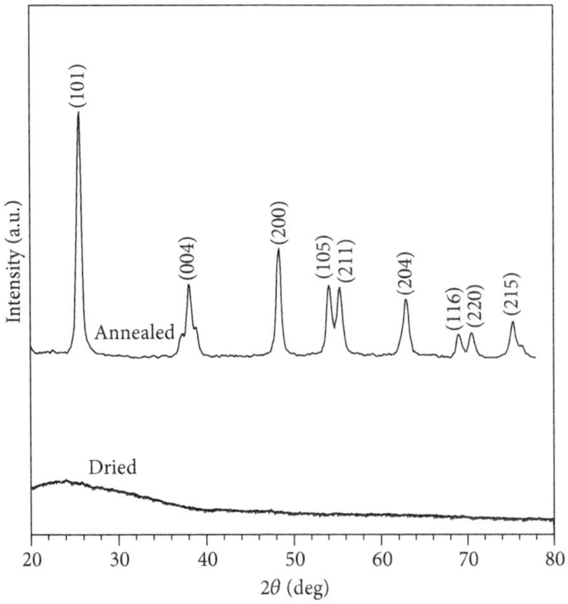

FIGURE 1: XRD pattern of dried at 100°C and heat-treated TiO_2 nanofibers at 550°C for 2 hours.

2. Materials and Methods

2.1. Synthesis of TiO_2 Nanofibers. The experimental setup of synthesis process which was arranged vertically, consists of a homemade DC high-voltage power source (30 kV, 5 mA), a double-syringe pump (Ascor AP 22), positive and negative electrodes, grounded conductive collector, and 20 mL syringes with syringe needles. The electrospinning was conducted in an open environment at the room temperature of approximately 25°C. An aluminum plate covered with aluminum foil was used as the collector screen and connected to the negative pole while positive pole was connected to the syringe needle. For preparation of electrospinning solution, 0.45 g PVP (Mw = 1300000) was dissolved in 10 mL ethanol (Merck, 99.8%). 1 mL $TiCl_4$ (Merck, 99%) was slowly added dropwise to the previous solution under vigorous stirring at room temperature and stirred for another 30 min at room temperature to get a homogenous solution. The final solution was then loaded into the syringe pump for electrospinning. In this work, a 2.5 to 15 kV voltage was applied in the distance range of 5 to 15 cm and flow rate of 0.3 to 10 mL/h. The as-spun nanofibers were dried at 100°C for 1 hour and then heat treated at 550°C for 1 h in order to remove PVP and crystallize TiO_2.

2.2. Characterization. Analysis of the crystalline structures was performed by XRD diffractometer (X'pert Philips) with wavelength of Cu $K\alpha$ radiation in 2θ range from 20° to 80° by 0.04° sec^{-1} steps. UV-Vis spectroscopy of the samples was taken out by a Lambda 950 spectrophotometer (Perkin Elmer) from 200 nm to 1100 nm wavelengths. SEM analysis was taken out by a SEM instrument (Philips XL30) at 5–20 keV accelerating energy, and TEM analysis was performed by a LEO 912 AB instrument at 200 keV accelerating energy by deposition of TiO_2 nanofibers onto the copper grid at room temperature.

2.3. Photocatalytic Study. Photocatalytic activity of the nanofibers was measured by photodegradation of (Rh.B), with the initial concentration and volume of 10^{-5} M and 30 mL, respectively, at the presence of 30 mL 0.01 wt% TiO_2 nanofibers. Actually 0.01 g of final residue which was a white fibrous substance of TiO_2 nanofibers dispersed in DI water to obtain a 0.01 wt% nanofibrous TiO_2 solution. Then, the mixed (Rh.B) and TiO_2 solution was stirred in dark for 30 min to equilibrate the adsorption/desorption between dye molecules and TiO_2 photocatalysts. Then, it was irradiated at room temperature by a 15 W low-pressure mercury lamp, 254 nm strongest peak wavelength. The degradation rate was measured by UV-Vis spectrophotometer at the maximum absorption wavelength of (Rh.B).

3. Results and Discussion

3.1. Crystal Structure. XRD analysis was taken out to find the crystalline phase and structure of the produced structures. The results for dried and heat-treated sample is depicted in Figure 1. As it is clear from the XRD pattern, the dried fibers at 100°C are amorphous and heating them at 550°C for 1 hour leads to formation of a white residue which is the crystalline TiO_2 nanofibers in anatase phase. The obtained spectrum after heat treatment at 550°C in air for 1 hour has the TiO_2 anatase phase peaks at $2\theta = 25.3, 37.7, 47.8, 54,$ 62.7, 68.6, 70.3, and 75.2 degrees with no impurity which is in agreement with 21-1272 standard cards from JCPDS. Heating the as-spun TiO_2 nanofibers results in PVP removing and crystallization of TiO_2. These results are in good agreement with previous reports [6, 10].

3.2. Microscopic Study. Applied electric field is the most important parameter in the electrospinning process due to its direct impact on the dynamics of the fluid flow and as a result the structure and morphology of produced fibers. The

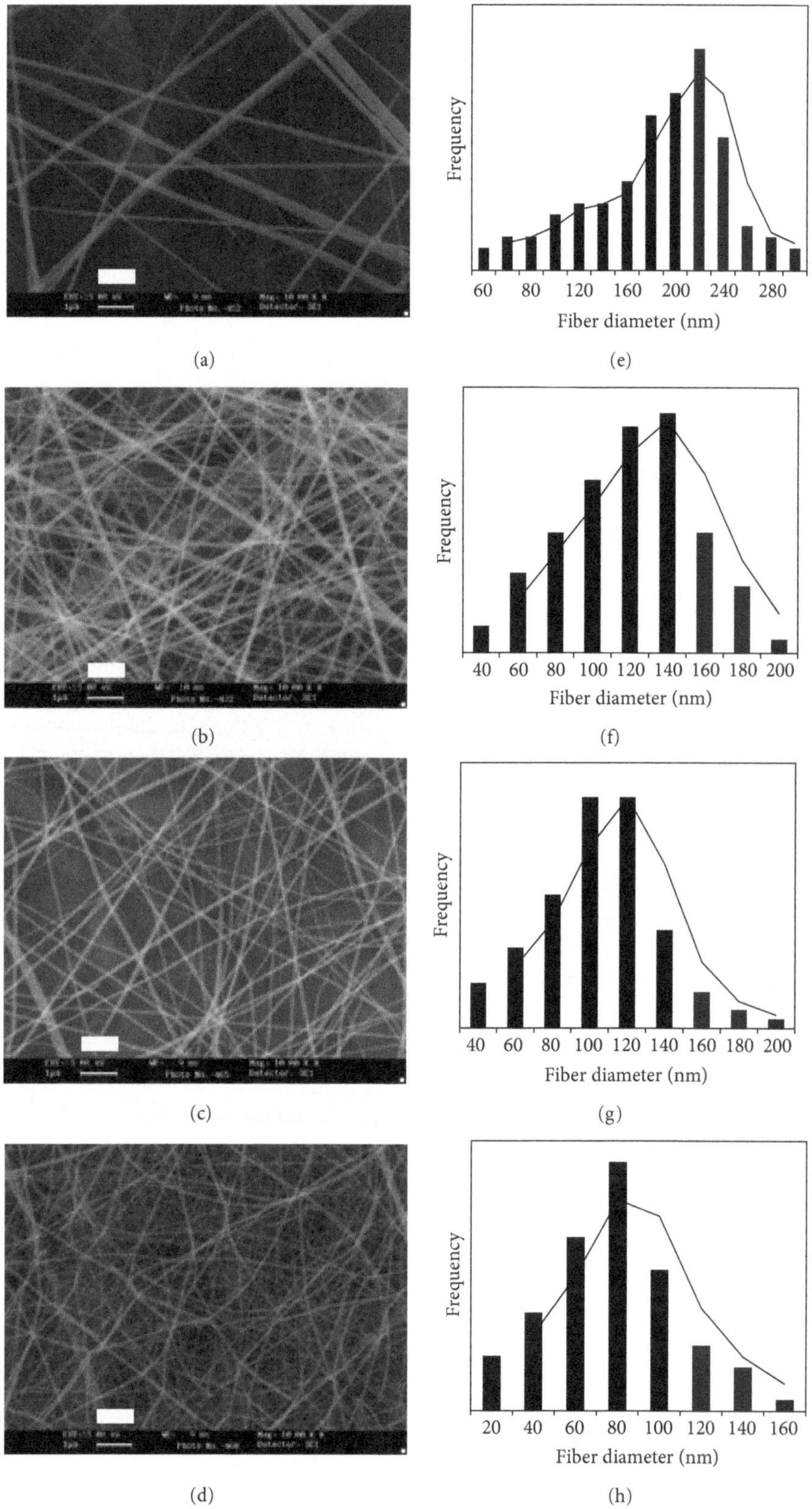

FIGURE 2: SEM images and the corresponding measured average fiber diameter of TiO_2 nanofibers synthesized at a fixed 5 cm distance and (a) 2.5 kV, (b) 5 kV, (c) 7.5 kV, and (d) 10 kV potential difference (the white scale bars in all SEM images are 1 μm).

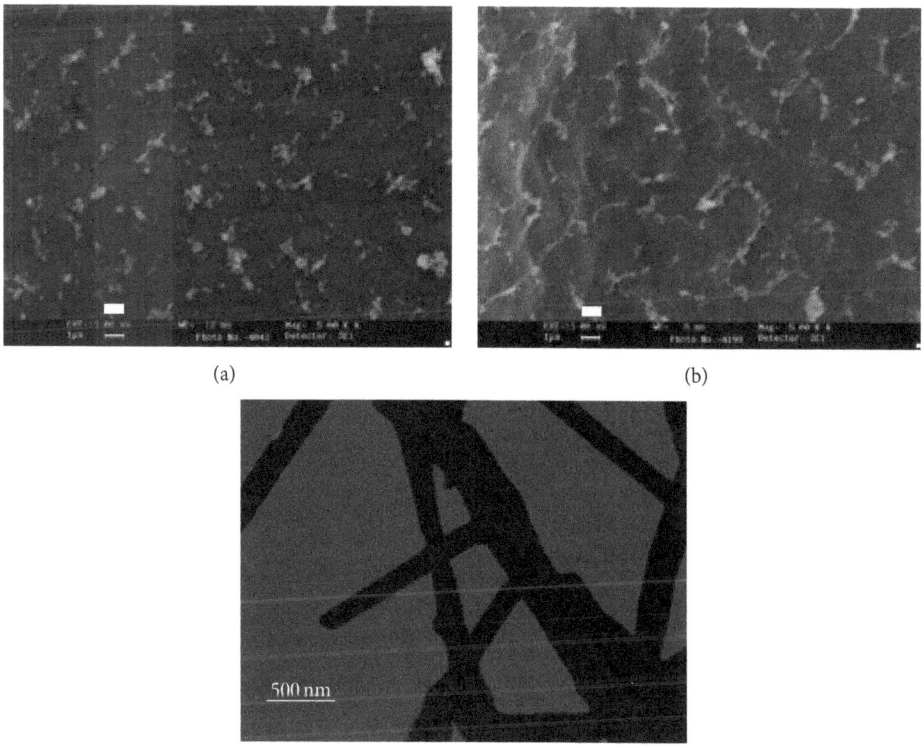

(a) (b) (c)

FIGURE 3: SEM images of final products prepared at lower than threshold. (a) electric field, (b) flow rate, and (c) typical TEM image of TiO_2 nanofiber prepared at 5 cm distance and 5 kV potential difference.

changes in the applied voltage will be reflected on the shape of the pendant droplet at the needle tip, its surface charge, dripping rate, velocity of the flowing fluid, and hence on the fiber structure and morphology [17]. Figures 2(a)–2(d) show SEM images of TiO_2 nanofibers synthesized at a fixed 5 cm distance and different electric fields. As is clear from SEM images, the mean diameter of fibers decreases by increasing applied voltage (scale bars in all SEM images are 1 μm). The applied voltage imposed an electrical polarization stress on the fluid drop at the needle tip and causes the elongation of the droplet. As the charged droplet ejects from the needle tip, it accelerates toward the collector and during this process the charged fluid encounters bending instability due to columbic repulsion between charged sections within the fluid jet. This bending instability is responsible for the stretching of the strands as they change from liquid into a solid form. Consequently, an inverse relationship exists between the diameter of the nanofibers and the applied voltage between the electrodes, and by increasing potential difference, the fiber diameter decreases [17]. By more increase of the applied voltage in a fixed 5 cm distance, a visible arc happens between the needle and collector. The minimum fiber diameter was found about 80 ± 16.6 nm which was obtained at 10 kV. Table 1 summarizes all the experiments and the resulting calculated average fiber diameter. Fibers were not formed at potential differences lower than 2.5 kV and also flow rates lower than 0.3 mL/h in a fixed 5 cm separation distance (see Figures 3(a) and 3(b)). Figure 3(c) shows the typical TEM image of TiO_2 nanofibers at 5 cm distance and 5 kV applied voltage.

TABLE 1: Summary of experimental conditions.

No.	Applied voltage (kV)	Separation distance (cm)	Flow rate (mL/h)	Average fiber diameter (nm)
1	2.5	5	0.3	220 ± 43.6
2	5	5	0.3	140 ± 21.4
3	7.5	5	0.3	110 ± 18.3
4	10	5	0.3	80 ± 16.6
5	10	10	0.3	60 ± 8.2
6	10	15	0.3	40 ± 6.3
7	10	5	1	100 ± 17.6
8	10	5	3	120 ± 20.8
9	10	5	10	160 ± 27.6

Similarly, the needle tip to collector distance is also a major factor in determining the fiber diameter. The effects of varying voltage and needle collector distance were studied individually by keeping other parameters constant. Figure 4 illustrates SEM images of TiO_2 nanofibers prepared at a fixed 10 kV applied voltage and different needle tip to collector distances. By increasing distance in a constant electric field, the average fiber diameter decreases. Increasing the separation distance between the needle and collector results in longer pass time for the charged fluid jet ejected from the needle. Therefore, as the distance increases, the fibers tend to become more elongated and smaller in diameter. Thus,

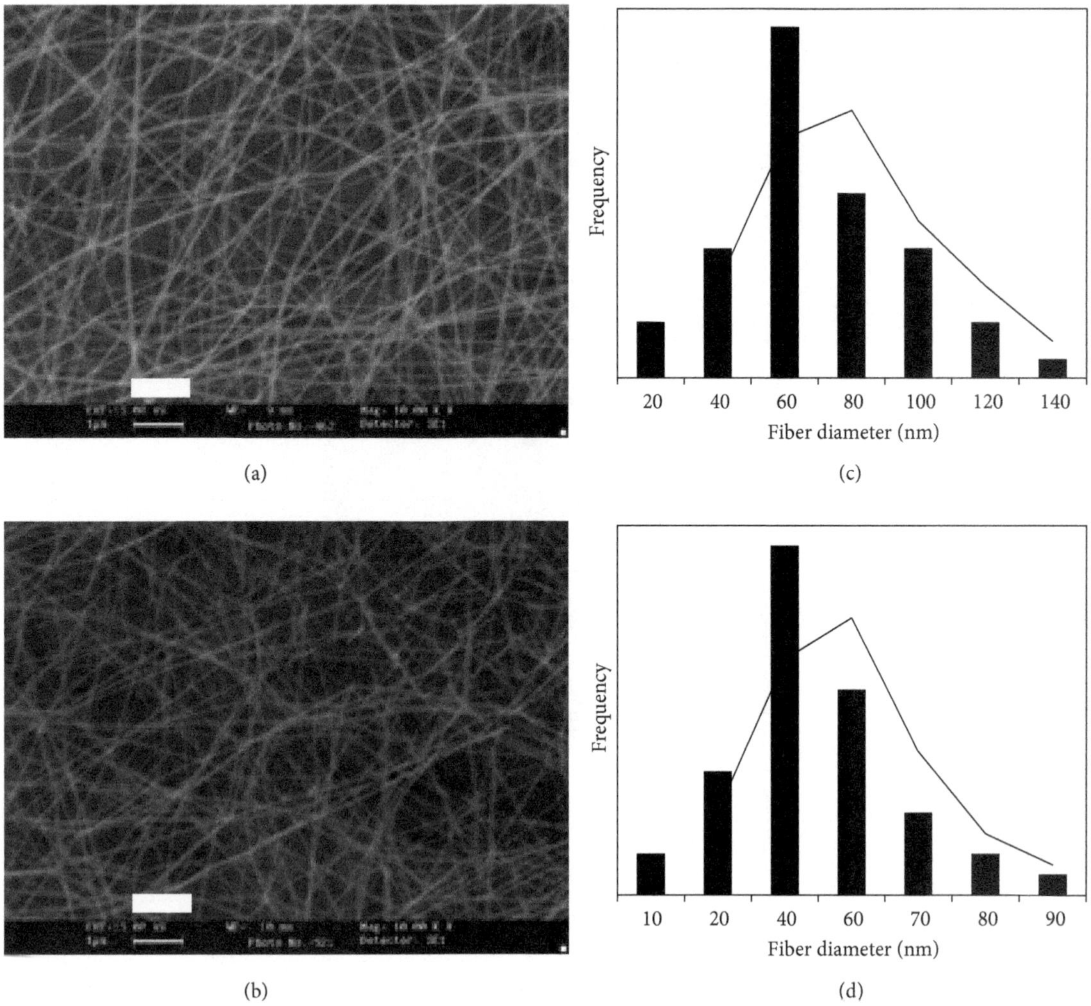

Figure 4: SEM images and the corresponding measured average fiber diameter of TiO_2 nanofibers prepared at a fixed 10 kV applied voltage and (a) 10 cm and (b) 15 cm separation distance.

an inverse relationship was observed between the separation distance and the TiO_2 nanofiber diameter. Accordingly, smaller diameter nanofibers were produced as the separation distance increased [17].

SEM images of TiO_2 nanofibers prepared in fixed 5 cm distance, 10 kV applied voltage, and different flow rates are demonstrated in Figure 5. It is clear from SEM images that by increasing the flow rate from 0.3 mL/h to 10 mL/h and keeping the potential difference and separation distance constant, the mean fiber diameter increases. In fact, increasing the mass throughput at the needle tip in a fixed electrical field and separation distance results in a higher rate of TiO_2 deposition onto the collector surface. Thus, nanofibers with bigger diameters were formed as the flow rate increases.

Recently, Ray and Lalman [17] both theoretically and experimentally showed that three preliminary experimental factors including potential difference, flow rate, and separation distance have crucial effect on the fiber diameter. They have found that by increasing the applied voltage and separation distance the mean fiber diameter decreases, which is in close agreement with our results.

3.3. Photocatalytic Activity Measurements. Changes in the absorption spectrum of Rh.B as a standard dye under UV illumination at different irradiation times for TiO_2 nanofibers prepared at different experimental conditions and one Rh.B sample as reference are illustrated in Figure 6. No remarkable changes in the concentration of the Rh.B solution were observed in the absence of TiO_2 nanofibers. Therefore, decomposition of Rh.B only depends on the photoexcitation of TiO_2 nanofibers. It was observed that by increasing the irradiation time the maximum absorption peak decreases. This indicates that the concentration of Rh.B is decreasing at the presence of TiO_2 nanofibers and UV light illumination. Figures 6(a)–6(c) demonstrate the changes in concentration, of Rh.B as a function of time under UV irradiation for TiO_2 nanofibers with different diameters. C and C_0 are the reactant concentrations at time $t = t$ and $t = 0$, respectively. As is clear from the obtained results, the photocatalytic performance

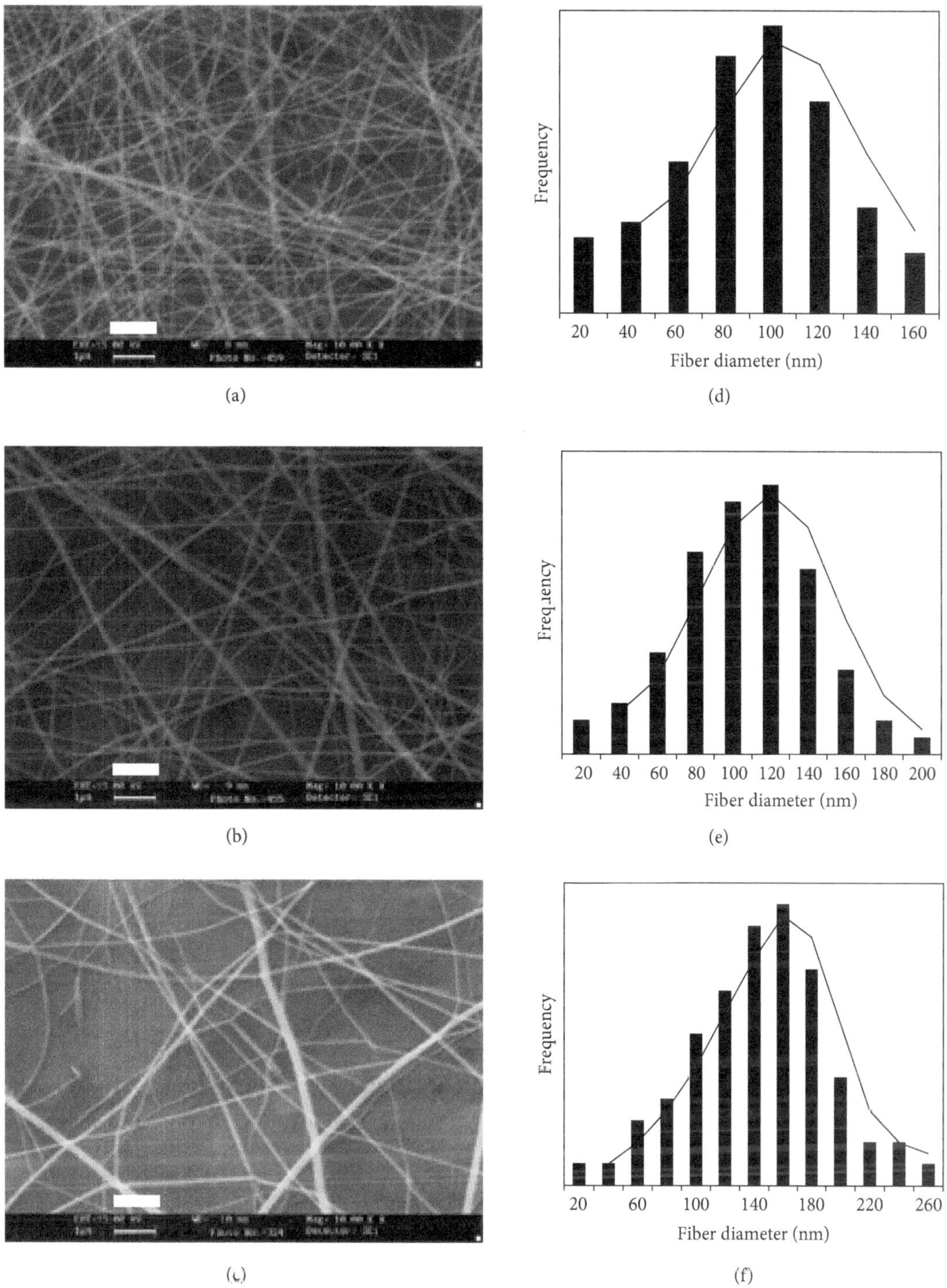

FIGURE 5: SEM images and the corresponding measured average fiber diameter of TiO_2 nanofibers prepared in 5 cm distance, 10 kV applied voltage and (a) 1 mL/h, (b) 3 mL/h and (c) 10 mL/h flow rates.

is stronger in all smaller diameter fibers rather than larger-diameter fibers.

4. Conclusion

In summary, TiO_2 nanofibers were prepared using the simple, facile, and inexpensive electrospinning technique at room temperature. Fiber diameter was successfully controlled by controlling the applied voltage, flow rate, and spinning distance. We have found that by increasing the potential difference and needle to collector distance the mean fiber diameter decreases. Also, we have observed that by increasing the flow rate and keeping the potential difference and separation distance constant the average fiber diameter increases.

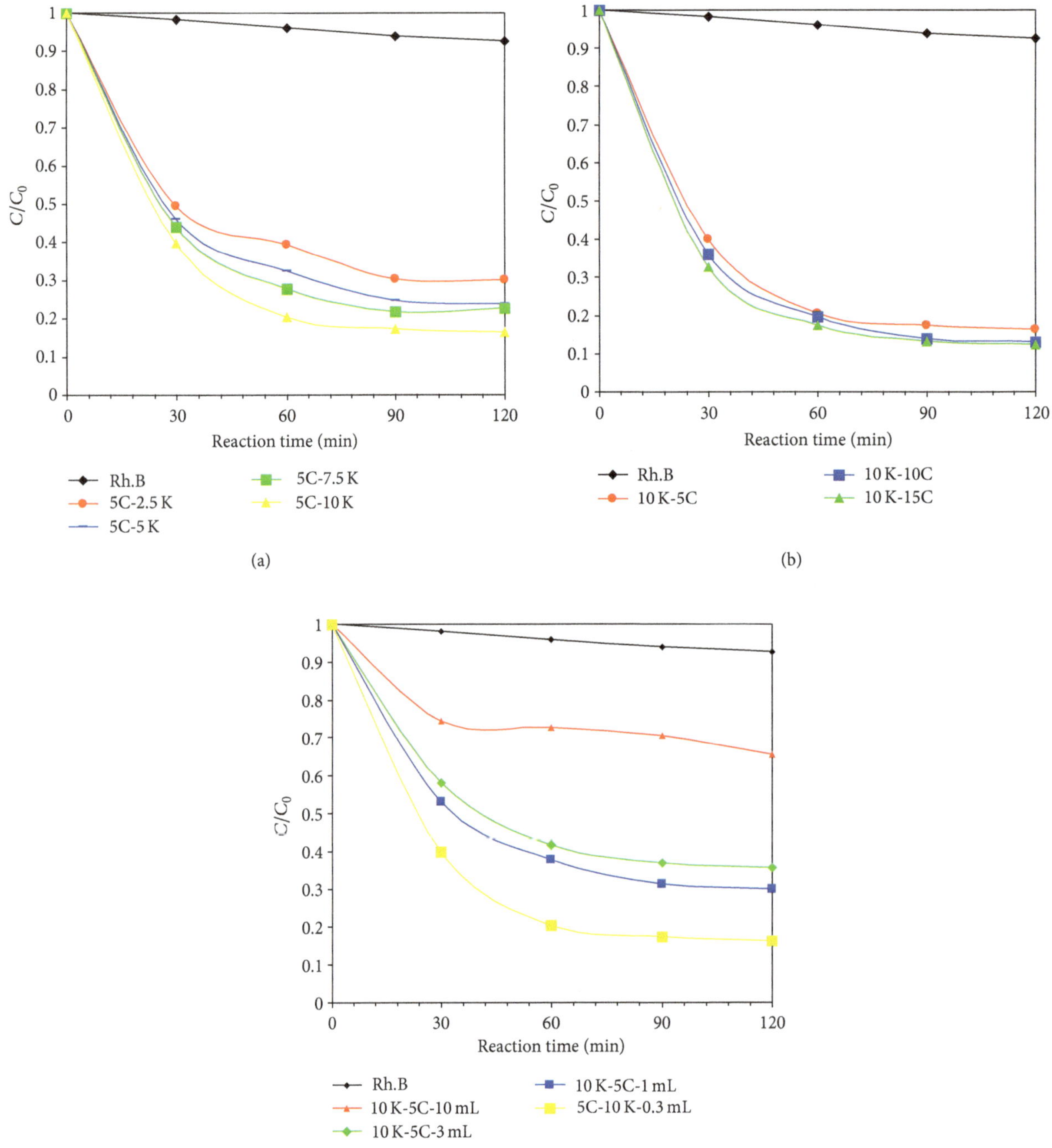

Figure 6: Evaluation of photocatalytic activity of TiO_2 nanofibers prepared at (a) fixed 5 cm distance and different electric fields, (b) fixed 10 kV applied voltage and different separation distances, and (c) fixed 5 cm separation distance, 10 kV applied voltage and different flow rates by photodegradation of Rh.B. under UV light irradiation.

Photocatalytic performance of prepared TiO_2 nanofibers was measured by photodegradation of Rh.B under UV illumination as a model dye. It was found that photocatalytic activity of TiO_2 nanofibers depends on electrospinning parameters and consequently the fiber diameter.

Conflict of Interests

The authors confirm that they do not have any financial relation with the commercial identities mentioned in their paper.

Acknowledgement

This work was supported by Iran national science foundation (INSF).

References

[1] A. A. Ashkarran, M. Kavianipour, S. M. Aghigh, S. A. A. Afshar, S. Saviz, and A. I. Zad, "On the formation of TiO_2 nanoparticles via submerged arc discharge technique: synthesis, characterization and photocatalytic properties," *Journal of Cluster Science*, vol. 21, no. 4, pp. 753–766, 2010.

[2] A. A. Ashkarran and M. R. Mohammadizadeh, "Superhydrophilicity of TiO_2 thin films using $TiCl_4$ as a precursor," *Materials Research Bulletin*, vol. 43, no. 3, pp. 522–530, 2008.

[3] A. A. Ashkarran and M. R. Mohammadizadeh, "The effect of heat treatment on superhydrophilicity of TiO_2 nano thin films," *EPJ Applied Physics*, vol. 40, no. 2, pp. 155–162, 2007.

[4] M. K. Aminian, N. Taghavinia, A. Iraji-Zad, S. M. Mahdavi, M. Chavoshi, and S. Ahmadian, "Highly porous TiO_2 nanofibres with a fractal structure," *Nanotechnology*, vol. 17, no. 2, pp. 520–525, 2006.

[5] A. A. Ashkarran, "Metal and metal oxide nanostructures prepared by electrical arc discharge method in liquids," *Journal of Cluster Science*, vol. 22, no. 2, pp. 233–266, 2011.

[6] S. J. Doh, C. Kim, S. G. Lee, S. J. Lee, and H. Kim, "Development of photocatalytic TiO_2 nanofibers by electrospinning and its application to degradation of dye pollutants," *Journal of Hazardous Materials*, vol. 154, no. 1–3, pp. 118–127, 2008.

[7] A. Greiner and J. H. Wendorff, "Electrospinning: a fascinating method for the preparation of ultrathin fibers," *Angewandte Chemie International Edition*, vol. 46, no. 30, pp. 5670–5703, 2007.

[8] Q. P. Pham, U. Sharma, and A. G. Mikos, "Electrospinning of polymeric nanofibers for tissue engineering applications: a review," *Tissue Engineering*, vol. 12, no. 5, pp. 1197–1211, 2006.

[9] D. H. Reneker and A. L. Yarin, "Electrospinning jets and polymer nanofibers," *Polymer*, vol. 49, no. 10, pp. 2387–2425, 2008.

[10] X. Zhang, S. Xu, and G. Han, "Fabrication and photocatalytic activity of TiO_2 nanofiber membrane," *Materials Letters*, vol. 63, no. 21, pp. 1761–1763, 2009.

[11] D. W. Hutmacher and P. D. Dalton, "Melt electrospinning," *Chemistry: An Asian Journal*, vol. 6, no. 1, pp. 44–56, 2011.

[12] D. Li and Y. Xia, "Electrospinning of nanofibers: reinventing the wheel?" *Advanced Materials*, vol. 16, no. 14, pp. 1151–1170, 2004.

[13] C. Wang, Y. Tong, Z. Sun, Y. Xin, E. Yan, and Z. Huang, "Preparation of one-dimensional TiO_2 nanoparticles within polymer fiber matrices by electrospinning," *Materials Letters*, vol. 61, no. 29, pp. 5125–5128, 2007.

[14] F. C. Wang, C. H. Liu, C. W. Liu, J. H. Chao, and C. H. Lin, "Effect of pt loading order on photocatalytic activity of Pt/TiO_2 nanofiber in generation of H2from neat ethanol," *Journal of Physical Chemistry C*, vol. 113, no. 31, pp. 13832–13840, 2009.

[15] S. Agarwal, J. H. Wendorff, and A. Greiner, "Use of electrospinning technique for biomedical applications," *Polymer*, vol. 49, no. 26, pp. 5603–5621, 2008.

[16] S. A. Sell, M. J. McClure, K. Garg, P. S. Wolfe, and G. L. Bowlin, "Electrospinning of collagen/biopolymers for regenerative medicine and cardiovascular tissue engineering," *Advanced Drug Delivery Reviews*, vol. 61, no. 12, pp. 1007–1019, 2009.

[17] S. Ray and J. A. Lalman, "Using the Box-Benkhen design (BBD) to minimize the diameter of electrospun titanium dioxide nanofibers," *Chemical Engineering Journal*, vol. 169, no. 1–3, pp. 116–125, 2011.

25

Bulk Metallic Glasses and Their Composites: A Brief History of Diverging Fields

Douglas C. Hofmann

Jet Propulsion Laboratory, California Institute of Technology, MS 18-105, 4800 Oak Grove Drive, Pasadena, CA 91109, USA

Correspondence should be addressed to Douglas C. Hofmann; dch@jpl.nasa.gov

Academic Editor: Ram Gupta

Bulk metallic glasses (BMGs) and their derivative metal matrix composites (BMGMCs) are emerging high-performance engineering materials that are on the precipice of widespread commercialization. This review article discusses the origin of these materials and how their applications and research focus have divided into two distinct fields, one primarily focused on the plastic-like processability of BMGs and the other on the enhanced fracture mechanics of BMGMCs. Although the materials are of similar composition and origin, it is argued that their implementation will be widely varying due to their different processing requirements and intended uses. BMGs will likely find use as plastic-replacement components in cosmetic applications (e.g., watches, cell phones, biomedical implants) while BMGMCs will be used in structural applications (e.g., golf clubs, hardware for defense, energy absorbing structures).

Since 1960, with the first publication of a metallic glass system in AuSi [1], amorphous metals and their composites have produced widespread scientific and commercial interest [2–4]. Initially, the focus of research in this area was centered on rapidly quenched thin foils and ribbons of metallic glass and, in particular, the scientific curiosity of how deeply undercooled liquids could avoid nucleation and growth of crystals when cooled below their glass transition temperature [5, 6]. Commercial applications soon followed with the identification of unique mechanical and magnetic properties obtained in thin layers of the materials. These included transformer coils produced by winding ribbons of Ni-, Co-, and Fe-based metallic glasses and corrosion resistant spray coatings of glass-forming alloys on drilling pipes, for example [7]. Two decades of further research led to the development of multicomponent alloy compositions with deep eutectics, large atomic size mismatch between constituents and sluggish crystallization kinetics such that critical cooling rates to prevent crystallization could be reduced by orders of magnitude (from 10^6 in quenched ribbons to 0.7 K/s in Pd-Cu-Ni-P). These new alloys, identified as *bulk* metallic glasses (BMGs), exhibited robust glass-forming ability and could be fabricated at thicknesses greater than 1 mm, thus creating a potential for structural hardware [8, 9]. The development of BMGs in practical (low cost) alloy compositions, such as Cu-Zr-Ni-Al and Zr-Ti-Cu-Ni-Be, opened the door for widespread commercialization into applications such as cell phone cases and golf clubs. The surge in commercial interest, starting in the mid-1990s, identified a wide range of applications for BMGs but also identified a number of flaws in the material when used as structural hardware, particularly the low fracture toughness, low fatigue limit and complete lack of ductility [10, 11]. Some research shifted towards toughening strategies for BMGs and in the early 2000s, progress was made by recognizing that the brittle failure of BMGs could be mitigated with the addition of crystalline phases into a BMG matrix [12]. This started a parallel field of research in BMG matrix composites (BMGMCs), with the aim of producing centimeter or greater thickness amorphous alloys for structural (load bearing) applications. Vigorous funding and increased world-wide research interest on BMGs and BMGMCs led to a resurgence in scientific progress on the materials by the end of the decade. Further understanding of the intricacies of processing amorphous metals has now brought us to the brink of global commercialization for these novel alloys and to a place where future activities in

BMGs and their composites, while equally bright, diverge. BMGs are poised to make an impact in low-thickness, high-precision cosmetic applications usually reserved for polymers, while the high-strength, toughness, and hardness of BMGMCs are being investigated for high-performance load-bearing applications, such as spacecraft shielding and panels for military vehicles. Metallic glass coatings, in contrast, have already broken into the commercial marketplace with successful ventures achieved primarily in the oil and gas industry (see, e.g., websites for Armacor, Scoperta and the NanoSteel Company).

With access to infinite cooling rates, all metals and metal alloys can be undercooled into an amorphous state because crystal nucleation and growth are time dependent phenomena. Practically, the highest laboratory cooling rates that can be achieved to form a free-standing metal structure are obtained through splat or ribbon quenching, reaching initial cooling rates of $\sim 10^7$ K/s. Unfortunately, for all metallic elements (and most metal alloys), crystal nucleation and growth intervene on a shorter timescale than is achievable in the laboratory. The scientific breakthrough in this area was the identification of alloys with deep eutectics that could be undercooled to below their glass transition temperature at cooling rates of less than 10^7 K/s [5]. Successful fabrication of amorphous AuSi foils were obtained by splat quenching, a technique that involves levitating a metal droplet (0.25–1 g) using a conical induction coil, turning off the power to allow the droplet to fall through a laser beam, and splatting the droplet between two copper paddles to achieve amorphous foils 10–100 μm in thickness, see Figure 1(a) [1]. Owing to the high laboratory cooling rates achieved through splat-quenching, many alloy compositions were discovered that could be formed into thin foils, including alloys using low-cost elements such as Fe-Ni-B, Fe-Ni-P-B, and Ni-P [5]. The development of ribbon quenching techniques allowed these new alloys to be fabricated in hundreds-of-meters long strips, which are still commercialized today, see Figure 1(b). The ribbon quenching technique is a continuous casting process that involves spraying molten alloy onto a spinning copper wheel. The alloys cools from one side by the wheel and the other side is free-cooled, leading to similar cooling rates as the splat quenching technique ($\sim 10^6$ K/s) but with the added improvement of continuous casting.

By the 1980s, some research focus in metallic glass research turned towards developing alloys that could be formed into bulk dimensions, typically greater than 1 mm in thickness or diameter [5, 6]. Forming BMGs presented a major practical challenge because of the three orders of magnitude increase in the critical cooling rate needed to form bulk parts (e.g., 10^7 K/s in splat quenching to 10^3 K/s in copper mold casting). Early success in the fabrication of BMGs involved Mg or La-based alloys until practical Zr-Cu-Al-Ni alloys were developed in 1990, followed by Zr-Ti-Cu-Ni-Be alloys in 1993. In the new Zr-based BMGs, critical cooling rates could be as low as 1 K/s to vitrify the alloys, which opened the door for die-casting-type processes for fabrication of net-shaped hardware. Alloy rods of up to 85 mm diameter have been demonstrated in the best glass-forming system yet known, Pd-Cu-Ni-P, which exhibits the lowest known critical cooling rate (0.067 K/s) [5–7]. In the last three decades, dozens of alloys systems have been discovered that can be formed into BMGs, including ferromagnetic alloy systems (Fe-, Ni-, and Co-based), precious metals systems (Au-, Ag-, Pd-, Pt-based), and practical systems (Ti-, Cu-, and Zr-based). Commercial applications for BMGs, starting in the late 1990s by Liquidmetal Technologies, in California, and several Japanese ventures, demonstrated the promise of the new materials [6, 7]. Applications ranging from golf clubs, springs, optical devices, cell phone cases, biomedical implants, sporting equipment, and so forth were quickly developed through a variety of die-casting methods. The unique chemistry of BMGs made them ideal materials for applications typically held by polymers. BMGs exhibit low melting temperatures (which means that simple Cu or steel molds can be used repeatedly with little wear), low viscosities in the melt (which allows for easy mold filling and complex net-shaped parts), low critical cooling rates (which means parts up to several centimeters thick can be made amorphous), and excellent mechanical properties when cooled into the amorphous state (high strength, hardness and surface finish). Figure 2 shows several BMG parts sitting on a $\sim 1\ m^2$ plate of die-cast Vitreloy 1 (parts fabricated by Liquidmetal Technologies and Howmet). The parts shown include a golf club driver, an iron, a USB hard drive case, a camera body, an optical device, eight ingots of feedstock, and a 12 mm diameter rod, all from the alloy Vitreloy 1 (trade name LM1). Figure 3 shows die-cast BMG components used to fabricate a watch.

Despite the initial promise of replacing plastics, as was the early objective, the widespread commercialization of BMGs did not happen because of several unforeseen drawbacks. It is now known that the mechanical properties of BMGs (particularly fracture toughness and fatigue limit) are highly dependent on material quality, oxygen content and the presence of unwanted phases or partial crystallization [10, 11]. This presents a major challenge to commercial manufacturing because lower quality material is typically used to make products cost competitive. Moreover, large-scale die-casting generally involves melting crucibles and low-vacuum systems, which add contaminants and oxygen to BMGs. High volume parts made from BMGs can become brittle if low-quality material is used and unwanted phases can tarnish the cosmetic finish usually desired from consumer hardware. Despite these shortcomings, BMGs (of any quality or composition) are inherently brittle, which limits any application where a part may be subjected to unconfined (tensile) loads. Owing to a lack of microstructure to stop crack formation, no monolithic metallic glass exhibits significant ductility in tension, despite some alloys that have large fracture toughness. Currently, there are now well-established correlations between thickness, fracture toughness, plastic zone size and bending ductility, which demonstrates that in plates or rods greater than ~3 mm in thickness, BMGs fail in a brittle manner [13–16]. Although these problems have been addressed recently by creating ultratough BMGs and designing applications for BMGs to be under 3 mm, the solution in the early 2000s was to develop BMG composites,

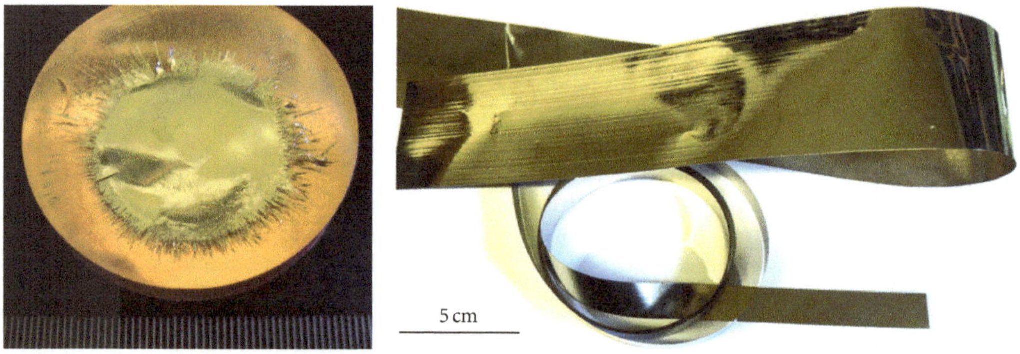

(a) (b)

Figure 1: (a) A splat quenched foil of the Fe-based metallic glass. (b) Commercially available ribbons of the metallic glass Metglas 2826 in different thicknesses and widths. Samples courtesy of Allied Metglas.

Figure 2: Examples of bulk metallic glass hardware, including golf clubs, electronic casings, optical hardware, ingots, 12 mm diameter rod, and large plate (under the samples). Hardware courtesy of Liquidmetal Technologies.

Figure 3: A net-shaped cast of a watch case directly out of the mold with no finishing or polishing. Hardware courtesy of Liquidmetal Technologies.

where additional phases could be selectively added to BMGs to arrest crack propagation.

The ideas of a partially crystalline metallic glass or a combination of metallic glass with some inclusion have been around nearly as long as metallic glasses. Early annealing experiments to study the glassy structure resulted in partial crystallization while other composites were stumbled upon through alloy development. In the late 1990s, attempts were made at developing "ex situ" composites, where BMGs were infiltrated over wires or powders of various crystalline materials to create two-phase mixtures [17, 18]. Compression tests from two ex situ BMG composites reinforced with different volume fractions of soft Zr-Nb particles is shown in Figure 4(a). Despite shear band formation and plasticity in bending, Figure 4(b), tensile ductility is absent. The addition of wires and particles was intended to improve the plastic strain associated with failure in the BMG but also to increase the density for possible kinetic energy penetrator (KEP) applications, when W or Ta wires were used. These ex situ BMG composites showed some improvement in compressive plasticity over monolithic BMGs, but ductility in tension was still absent unless the wires spanned the length of the rod, creating an iso-strain loading configuration and a "rule-of-mixtures" ductility. The major breakthrough in the formation of BMG composites was the development of "in situ" matrix composites (BMGMCs), where a second-phase dendritic structure could grown in equilibrium with a glass forming alloy [12]. This was accomplished by utilizing chemical decomposition in Zr-Ti-Cu-Ni-Be, recognizing that Cu, Ni, and Be have little or no solubility in the body centered cubic (bcc) phase of Zr-Ti. By increasing the concentration of Zr and Ti, while reducing Cu, Ni, and Be, a thermodynamic equilibrium is created between bcc dendrites and a eutectic liquid. When quenched, the eutectic vitrifies as a glass matrix with the dendrites evenly spaced within. The first toughened in situ BMGMCs were developed by Hays and Kim at Caltech in 1999 (later published in 2000), by recognizing that the dendritic phase could be softened relative to the BMG matrix by adding a beta stabilizer, Nb. The six component BMGMC $Zr_{56.2}Ti_{13.8}Nb_5Cu_{6.9}Ni_{5.6}Be_{12.5}$ (60% BMG, 40% bcc) was the first alloy of metallic glass to exhibit both the high tensile strength of BMGs, but also significant necking in tension (~3%) [19]. This discovery started the parallel field of BMG composites, for those researchers interested in improving the mechanical properties of amorphous metals for structural applications.

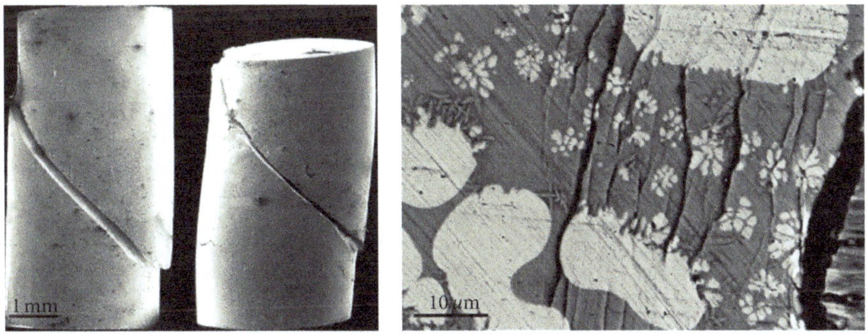

(a) (b)

FIGURE 4: (a) Ex situ metallic glass matrix composites exhibiting different volume fractions of soft particles and therefore, different levels of compressive plasticity. (b) SEM micrograph of a particle-reinforced ex situ metallic glass matrix composite exhibiting shear band arresting. Smaller dendrites and needle crystal phase are also present.

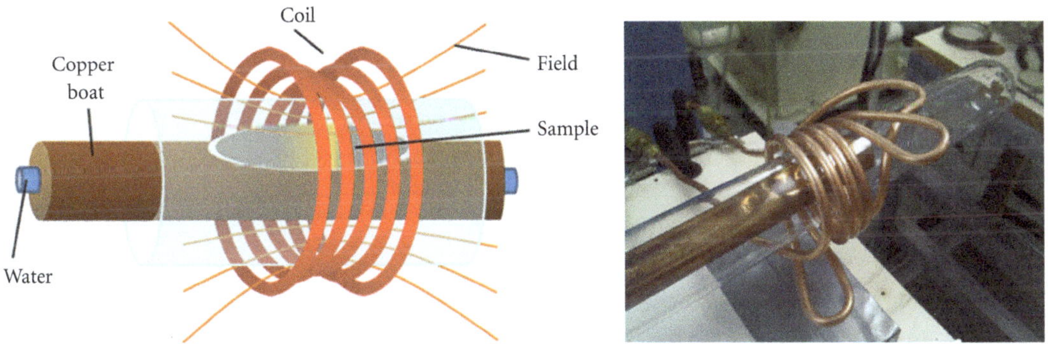

(a) (b)

FIGURE 5: (a) Diagram of a water-cooled copper boat used for semisolid processing of in situ metallic glass matrix composites. (b) Example of a custom coil on a water-cooled copper boat.

During the early 2000s, research in amorphous metals increased dramatically, with numerous countries supporting large research efforts and university professors (USA, UK, Germany, China, Korea, etc.). However, despite early promise, commercial applications for BMGs did not become widespread due to problems with material cost, quality, production volume, and toughness, among others. Progress in BMGMCs was also hampered by a number of scientific issues which prevented significant improvements over the $Zr_{56.2}Ti_{13.8}Nb_5Cu_{6.9}Ni_{5.6}Be_{12.5}$ composite published in 2001. Although many BMGMCs were discovered in a variety of systems, these alloys did not outperform the original alloy in terms of mechanical properties such as strength and ductility [20–25]. The primary reasons for this included the unique chemistry of Be-bearing composites, the composition of the dendrites, the length scale of the dendrites, and the toughness of the glass matrix. By 2007, progress was made in the emerging parallel field of BMGMCs by developing a series of guidelines for creating toughened composites [26]. (1) The composite system needs a highly processable monolithic metallic glass with sluggish crystallization. Zr-Be-based composites are optimal for BMGMC formation due to the lack of stable compounds between the constituents and Be, which prevents heterogeneous nucleation triggered by the dendrites during quenching. (2) The composite system needs a stable crystal that does not cause heterogeneous nucleation of the matrix. These dendrites can be formed in equilibrium two-phase systems, or by partially crystallizing the BMG during cooling. (3) The shear modulus (G) of the dendrite needs to be lower than the matrix. The inclusion needs to be soft relative to the matrix such that cracks are "attracted" to them, rather than propagating through the matrix alone. (4) The microstructure should be coarsened to match the inclusion size and spacing with the plastic zone size (critical shear band length) of the BMG matrix. This rule was developed by observing a cooling rate dependence on the scale of the dendrites in a BMGMC. When cooled from the liquidus, time for nucleation and growth is limited, creating a microstructure that varies throughout the quenched sample depending on the local cooling rate. The major breakthrough was the understanding that toughness was related to the size of these dendrites, and that regions where they were coarser produced higher toughness. This led to the development of semisolid processing, an intermediate quenching step where the BMGMC is held just above the solidus temperature in the two-phase region and the dendrites are allowed to thermodynamically coarsen prior to final quenching, see Figure 5. This procedure proved extremely successful at creating large

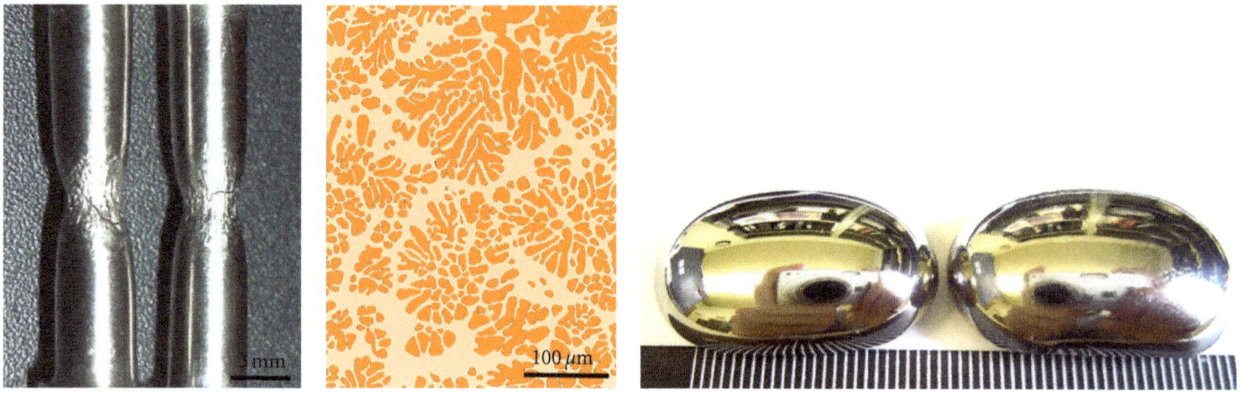

(a) (b) (c)

Figure 6: (a) Necking in tension from two different Zr-Ti-Nb-Cu-Be bulk metallic glass matrix composites. (b) Contrast adjusted SEM micrograph of the dendrite structure in a bulk metallic glass matrix composite. (c) Comparison in surface appearance between a monolithic metallic glass (left) and an in situ composite. Both alloys were cooled from the liquidus state using an arc melter.

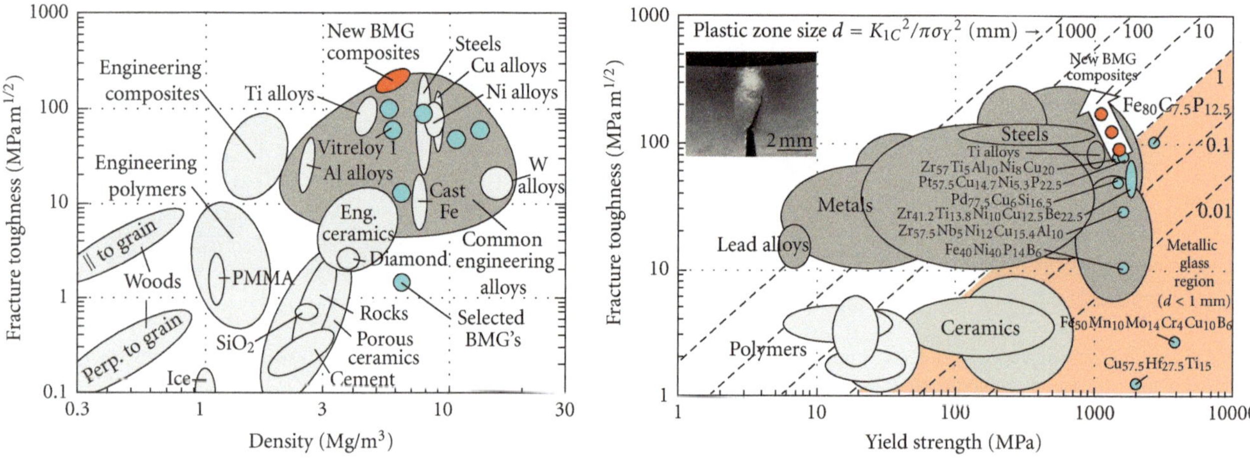

(a) (b)

Figure 7: (a) Ashby map of fracture toughness versus density showing the location of select bulk metallic glasses and Zr-Ti-Nb-Cu-Be bulk metallic glass matrix composites. The composite alloys have among the highest fracture toughness for their density. (b) Ashby map of fracture toughness versus yield strength showing select metallic glasses and composites. Contour lines indicate plastic zone size and the shaded region signifies the $d < 1$ mm region that typically confines monolithic metallic glasses. The plastic zone size of the toughened composites is shown in the inset.

coarsened microstructures and ductility in BMGMCs was increased to >10% while fracture toughness increased to >160 MPa $m^{1/2}$ [27]. (5) There also must be a suitably high fracture toughness interface between the inclusion and the BMG matrix such that crack growth does not proceed along the boundary of the inclusions. This rule was developed to explain the lack of ductility observed in powder-reinforced ex situ BMGMCs, even those where the size and composition of the dendrite matches the in situ composite. Oxide layers or porosity at the interface between the inclusions and the matrix generally leads to cracks propagating along the boundaries. Figure 6 shows images from a BMGMC, termed DH1 ($Zr_{36.6}Ti_{31.4}Nb_7Cu_{5.9}Be_{19.1}$) with significant necking in tension (Figure 6(a)) due to the distribution of soft dendrites, shown in Figure 6(b) using a color adjusted scanning electron microscope image. BMGMCs, which generally contain at least 40% crystal phase, exhibit a rough surface when cooled in an ingot form, compared with a monolithic BMG which exhibits a perfect mirror finish, see Figure 6(c).

Recent progress in both monolithic BMGs and BMGMCs, particularly in the area of processing, would seem to indicate a bright future for these materials. The field of BMGs is poised to have widespread applications in the areas of consumer electronics, jewelry, fuel cells, coatings, and nano/microtechnology due to a much greater understanding of both the material properties and their processing potential. New toughened BMGMCs are now being investigated for high-performance structural applications in the areas of defense, aerospace, aeronautics, automotive, and sporting equipment. These include energy

FIGURE 8: (a) Thermoplastic processing in a Zr-Ti-Cu-Be bulk metallic glass. (b) Blow molding demonstrated in a Zr-Ti-Cu-Be bulk metallic glass. (c) Capacitive discharge forming of a metallic glass showing the part and the mold in which it was cast in 40 milliseconds. These are all demonstrations of thermoplastic processing. Images from G. Duan and M.D. Demetriou.

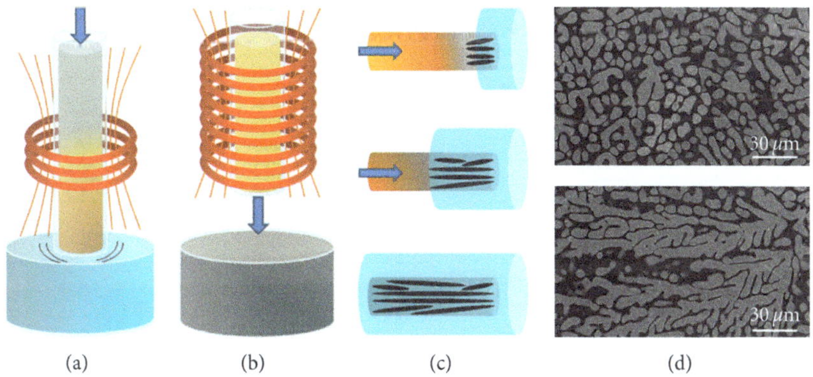

FIGURE 9: (a-b) Schematics of directional solidification of bulk metallic glass composites into water and a liquid metal bath. (c) Schematic of dendrite growth under directional solidification. (d) SEM micrographs from a bulk metallic glass matrix composite with no texture (top) and directional texture (bottom).

absorbing cellular structures, spacecraft debris shielding, golf clubs, car door panels, gears, armor, and penetrators. Ashby maps for fracture toughness versus density and fracture toughness versus yield strength are shown in Figures 7(a) and 7(b). Several BMGs are shown along with the new BMGMCs based in Zr-Ti-Cu-Nb-Be (termed DH1-3) [27]. These maps demonstrate how amorphous alloys and composites compare with all other known engineering materials in terms of toughness, density, and strength. In the case of the toughened BMG composites, the maps show that their properties lie at the boundaries with all available materials, including plastic zone size, d, which has limited BMG use in thicknesses greater than 1 mm.

The processing possibilities of monolithic BMGs have been the focus of considerable research effort in recent years. New capabilities for casting samples from the molten state, including tilt casting [28], squeeze casting, injection casting, and so forth, have resulted in high-quality, tough parts with excellent cosmetic finishes. These have included cell phone cases, hollow tubes, sensors, and plates [6]. Enormous progress has also been made in the area of thermoplastic processing, which involves heating a BMG to the supercooled liquid region between the glass transition and the crystallization temperatures and forming it like a plastic, see Figure 8(a) [29–41]. Thermoplastic forming has been used in a variety of novel ways recently, including blow molding, shown in Figure 8(b) [30]. This technique has been used to make BMG bottles, holograms, spheres, and hollow watch cases [35, 39, 41]. Recently, rapid discharge forming (RDF) was demonstrated as a new technique to heat and thermoplastically process a BMG in milliseconds, which takes advantage of the near constant resistivity as function of temperature in metallic glasses to create uniform volumetric heating, Figure 8(c) [29]. Thermoplastic forming has also been used recently to develop micro and nano applications for metallic glasses [39]. Since BMGs are liquids when formed, there is no limit to the scale of features that can be fabricated. Successful demonstrations have included nanowires, microgears, MEMS devices, ultra-smooth surfaces, and large area fuel cells [29–41]. A large number of worldwide commercial and government ventures to create hardware from BMGs demonstrates the continued interest in turning amorphous metals into everyday objects.

New processing techniques and alloy compositions have also driven progress in BMGMCs. A recent breakthrough was obtained by creating a composite with a ductile shape memory alloy phase in Cu-Zr-Al-Co [42, 43]. This alloy, along with another demonstration in Ti-Zr-Ni-Ta-Be, illustrate that transformation induced plasticity (TRIP) is also possible in BMGMCs. Numerous research studies have been undertaken to understand semisolid processing strategies in BMGMCs [44–49]. These have included directional solidification, Bridgman solidification, semisolid forging, squeeze casting, and electrostatic levitation, among others, see Figure 9 [44–49]. Continued government funding in the USA, from agencies such as NASA, the Office of Naval Research, and the Office of Air Force Scientific Research, demonstrates the promise for BMGMCs while widespread commercial interest may soon transform these materials into everyday hardware.

Acknowledgments

This work was supported by the Jet Propulsion Laboratory, California Institute of Technology, NASA's Exploration Systems Mission Directorate, under Contract no. NNH10ZTT001N and by the Air Force Office of Scientific Research under Grant no. RF01152700/PO60020925.

References

[1] W. Klement, R. H. Willens, and P. Duwez, "Non-crystalline structure in solidified Gold-Silicon alloys," *Nature*, vol. 187, no. 4740, pp. 869–870, 1960.

[2] M. Telford, "The case for bulk metallic glass," *Materials Today*, vol. 7, no. 3, pp. 36–43, 2004.

[3] Y. H. Liu, G. Wang, R. J. Wang, D. Q. Zhao, M. X. Pan, and W. H. Wang, "Super plastic bulk metallic glasses at room temperature," *Science*, vol. 315, no. 5817, pp. 1385–1388, 2007.

[4] H. Guo, P. F. Yan, Y. B. Wang et al., "Tensile ductility and necking of metallic glass," *Nature Materials*, vol. 6, no. 10, pp. 735–739, 2007.

[5] A. Inoue, "Stabilization of metallic supercooled liquid and bulk amorphous alloys," *Acta Materialia*, vol. 48, no. 1, pp. 279–306, 2000.

[6] A. Inoue and A. Takeuchi, "Recent development and application products of bulk glassy alloys," *Acta Materialia*, vol. 59, no. 6, pp. 2243–2267, 2011.

[7] N. Nishiyama, K. Amiya, and A. Inoue, "Novel applications of bulk metallic glass for industrial products," *Journal of Non-Crystalline Solids*, vol. 353, no. 32-40, pp. 3615–3621, 2007.

[8] A. Peker and W. L. Johnson, "A highly processable metallic glass: $Zr_{41.2}Ti_{13.8}Cu_{12.5}Ni_{10.0}Be_{22.5}$," *Applied Physics Letters*, vol. 63, no. 17, pp. 2342–2344, 1993.

[9] H. B. Lou, X. D. Wang, F. Xu et al., "73 mm-diameter bulk metallic glass rod by copper mould casting," *Applied Physics Letters*, vol. 99, Article ID 051910, 2011.

[10] C. A. Schuh, T. C. Hufnagel, and U. Ramamurty, "Mechanical behavior of amorphous alloys," *Acta Materialia*, vol. 55, no. 12, pp. 4067–4109, 2007.

[11] P. Lowhaphandu and J. J. Lewandowski, "Fracture toughness and notched toughness of bulk amorphous alloy: Zr-Ti-Ni-Cu-Be," *Scripta Materialia*, vol. 38, no. 12, pp. 1811–1817, 1998.

[12] C. C. Hays, C. P. Kim, and W. L. Johnson, "Microstructure controlled shear band pattern formation and enhanced plasticity of bulk metallic glasses containing in situ formed ductile phase dendrite dispersions," *Physical Review Letters*, vol. 84, no. 13, article no. 2901, 2000.

[13] M. D. Demetriou, G. Kaltenboeck, J. Y. Suh et al., "Glassy steel optimized for glass-forming ability and toughness," *Applied Physics Letters*, vol. 95, no. 4, Article ID 041907, 2009.

[14] R. D. Conner, Y. Li, W. D. Nix, and W. L. Johnson, "Shear band spacing under bending of Zr-based metallic glass plates," *Acta Materialia*, vol. 52, no. 8, pp. 2429–2434, 2004.

[15] R. D. Conner, W. L. Johnson, N. E. Paton, and W. D. Nix, "Shear bands and cracking of metallic glass plates in bending," *Journal of Applied Physics*, vol. 94, no. 2, pp. 904–911, 2003.

[16] M. D. Demetriou, M. E. Launey, G. Garrett et al., "A damage-tolerant glass," *Nature Materials*, vol. 10, no. 2, pp. 123–128, 2011.

[17] H. Choi-Yim, R. D. Conner, F. Szuecs, and W. L. Johnson, "Processing, microstructure and properties of ductile metal particulate reinforced $Zr_{57}Nb_5Al_{10}Cu_{15.4}Ni_{12.6}$ bulk metallic glass composites," *Acta Materialia*, vol. 50, no. 10, pp. 2737–2745, 2002.

[18] H. Choi-Yim, R. Busch, and W. L. Johnson, "The effect of silicon on the glass forming ability of the $Cu_{47}Ti_{34}Zr_{11}Ni_8$ bulk metallic glass forming alloy during processing of composites," *Journal of Applied Physics*, vol. 83, no. 12, pp. 7993–7997, 1998.

[19] F. Szuecs, C. P. Kim, and W. L. Johnson, "Mechanical properties of $Zr_{56.2}Ti_{13.8}Nb_{5.0}Cu_{6.9}Ni_{5.6}Be_{12.5}$ ductile phase reinforced bulk metallic glass composite," *Acta Materialia*, vol. 49, no. 9, pp. 1507–1513, 2001.

[20] Z. Zhu, H. Zhang, Z. Hu, W. Zhang, and A. Inoue, "Ta-particulate reinforced Zr-based bulk metallic glass matrix composite with tensile plasticity," *Scripta Materialia*, vol. 62, no. 5, pp. 278–281, 2010.

[21] M. L. Lee, Y. Li, and C. A. Schuh, "Effect of a controlled volume fraction of dendritic phases on tensile and compressive ductility in La-based metallic glass matrix composites," *Acta Materialia*, vol. 52, no. 14, pp. 4121–4131, 2004.

[22] H. Tan, Y. Zhang, and Y. Li, "Synthesis of La-based in-situ bulk metallic glass matrix composite," *Intermetallics*, vol. 10, no. 11-12, pp. 1203–1205, 2002.

[23] U. Kuhn, J. Eckert, N. Mattern, and L. Schultz, "Microstructure and mechanical properties of slowly cooled Zr-Nb-Cu-Ni-Al composites with ductile bcc phase," *Materials Science and Engineering A*, vol. 375–377, pp. 322–326, 2004.

[24] C. Fan, R. T. Ott, and T. C. Hufnagel, "Metallic glass matrix composite with precipitated ductile reinforcement," *Applied Physics Letters*, vol. 81, no. 6, pp. 1020–1022, 2002.

[25] U. Kühn, J. Eckert, N. Mattern, and L. Schultz, "ZrNbCuNiAl bulk metallic glass matrix composites containing dendritic bcc phase precipitates," *Applied Physics Letters*, vol. 80, no. 14, pp. 2478–2480, 2002.

[26] D. C. Hofmann and W. L. Johnson, "Improving ductility in nanostructured materials and metallic glasses: "three laws"," *Materials Science Forum*, vol. 633-634, pp. 657–663, 2010.

[27] D. C. Hofmann, J. Y. Suh, A. Wiest et al., "Designing metallic glass matrix composites with high toughness and tensile ductility," *Nature*, vol. 451, no. 7182, pp. 1085–1089, 2008.

[28] E. Soinila, T. Pihlajamäki, S. Bossuyt, and H. Hänninen, "A combined arc-melting and tilt-casting furnace for the manufacture of high-purity bulk metallic glass materials," *Review of Scientific Instruments*, vol. 82, no. 7, Article ID 073901, 2011.

[29] W. L. Johnson, G. Kaltenboeck, M. D. Demetriou et al., "Beating crystallization in glass-forming metals by millisecond heating and processing," *Science*, vol. 332, no. 6031, pp. 828–833, 2011.

[30] J. Schroers, "Processing of bulk metallic glass," *Advanced Materials*, vol. 22, no. 14, pp. 1566–1597, 2010.

[31] A. Wiest, G. Duan, M. D. Demetriou et al., "Zr-Ti-based Be-bearing glasses optimized for high thermal stability and thermoplastic formability," *Acta Materialia*, vol. 56, no. 11, pp. 2625–2630, 2008.

[32] G. Duan, A. Wiest, M. L. Lind, A. Kahl, and W. L. Johnson, "Lightweight Ti-based bulk metallic glasses excluding late transition metals," *Scripta Materialia*, vol. 58, no. 6, pp. 465–468, 2008.

[33] G. Duan, A. Wiest, M. L. Lind, J. Li, W. K. Rhim, and W. L. Johnson, "Bulk metallic glass with benchmark thermoplastic processability," *Advanced Materials*, vol. 19, no. 23, pp. 4272–4275, 2007.

[34] D. C. Hofmann, J. Y. Suh, A. Wiest, and W. Johnson, "New processing possibilities for highly toughened metallic glass matrix composites with tensile ductility," *Scripta Materialia*, vol. 59, no. 7, pp. 684–687, 2008.

[35] G. Kumar, A. Desai, and J. Schroers, "Bulk metallic glass: the smaller the better," *Advanced Materials*, vol. 23, no. 4, pp. 461–476, 2011.

[36] M. Carmo, R. C. Sekol, S. Ding, G. Kumar, J. Schroers, and A. D. Taylor, "Bulk metallic glass nanowire architecture for electrochemical applications," *ACS Nano*, vol. 5, no. 4, pp. 2979–2983, 2011.

[37] R. Martinez, G. Kumar, and J. Schroers, "Hot rolling of bulk metallic glass in its supercooled liquid region," *Scripta Materialia*, vol. 59, no. 2, pp. 187–190, 2008.

[38] G. Kumar, P. A. Staffier, J. Blawzdziewicz, U. D. Schwarz, and J. Schroers, "Atomically smooth surfaces through thermoplastic forming of metallic glass," *Applied Physics Letters*, vol. 97, no. 10, Article ID 101907, 2010.

[39] G. Kumar, H. X. Tang, and J. Schroers, "Nanomoulding with amorphous metals," *Nature*, vol. 457, no. 7231, pp. 868–872, 2009.

[40] B. Zhang, D. Q. Zhao, M. X. Pan, W. H. Wang, and A. L. Greer, "Amorphous metallic plastic," *Physical Review Letters*, vol. 94, no. 20, Article ID 205502, 2005.

[41] J. Schroers, Q. Pham, and A. Desai, "Thermoplastic forming of bulk metallic glass—a technology for MEMS and microstructure fabrication," *Journal of Microelectromechanical Systems*, vol. 16, no. 2, pp. 240–247, 2007.

[42] Y. Wu, Y. Xiao, G. Chen, C. T. Liu, and Z. Lu, "Bulk metallic glass composites with transformationMediated work-hardening and ductility," *Advanced Materials*, vol. 22, no. 25, pp. 2770–2773, 2010.

[43] S. Pauly, G. Liu, G. Wang et al., "Modeling deformation behavior of Cu-Zr-Al bulk metallic glass matrix composites," *Applied Physics Letters*, vol. 95, Article ID 101906, 2009.

[44] D. C. Hofmann, H. Kozachkov, H. E. Khalifa et al., "Semi-solid induction forging of metallic glass matrix composites," *Journal of the Minerals, Metals and Materials Society*, vol. 61, no. 12, pp. 11–17, 2009.

[45] D. C. Hofmann, J. Y. Suh, A. Wiest, M. L. Lind, M. D. Demetriou, and W. L. Johnson, "Development of tough, low-density titanium-based bulk metallic glass matrix composites with tensile ductility," *Proceedings of the National Academy of Sciences of the United States of America*, vol. 105, no. 51, pp. 20136–20140, 2008.

[46] J. L. Cheng, G. Chen, F. Xu, Y. L. Du, Y. S. Li, and C. T. Liu, "Correlation of the microstructure and mechanical properties of Zr-based in-situ bulk metallic glass matrix composites," *Intermetallics*, vol. 18, no. 12, pp. 2425–2430, 2010.

[47] J. W. Qiao, J. T. Zhang, F. Jiang et al., "Development of plastic Ti-based bulk-metallic-glass-matrix composites by controlling the microstructures," *Materials Science & Engineering A*, vol. 527, no. 29-30, pp. 7752–7756, 2010.

[48] J. W. Qiao, S. Wang, Y. Zhang, P. K. Liaw, and G. L. Chen, "Large plasticity and tensile necking of Zr-based bulk-metallic-glass-matrix composites synthesized by the Bridgman solidification," *Applied Physics Letters*, vol. 94, no. 15, Article ID 151905, 2009.

[49] J. L. Cheng, G. Chen, F. Xu, Y. L. Du, Y. S. Li, and C. T. Liu, "Correlation of the microstructure and mechanical properties of Zr-based in-situ bulk metallic glass matrix composites," *Intermetallics*, vol. 18, no. 12, pp. 2425–2430, 2010.

26

Studies on Structural, Spectral, and Optical Properties of Organic Nonlinear Optical Single Crystal: 2-Amino-4,6-dimethylpyrimidinium p-Hydroxybenzoate

S. Sudhahar,[1] M. Krishna Kumar,[1] A. Silambarasan,[1] R. Muralidharan,[2] and R. Mohan Kumar[1]

[1] *Department of Physics, Presidency College, Chennai 600 005, India*
[2] *Department of Science and Humanities, Veltech High Tech Dr. R. Rangarajan Dr. Sakunthala Engineering College, Chennai 600 062, India*

Correspondence should be addressed to R. Mohan Kumar; mohan66@hotmail.com

Academic Editor: Matjaz Valant

2-Amino-4,6-dimethylpyrimidinium p-hydroxybenzoate (ADPHB), a novel organic material, was synthesized, and crystals were grown by slow evaporation solution growth method. The cell parameters and crystalline perfection of the grown crystal were studied by single crystal and powder X-ray diffraction analyses. The presence of various functional groups of the grown crystal was confirmed using Fourier Transform infrared spectral analysis. UV-Visible spectrum shows that ADPHB crystals have high transmittance in the range 305–900 nm. The refractive index and transient photoluminescence properties of the grown crystal were analyzed. The laser induced surface damage threshold, optical birefringence, and second harmonic generation efficiency of the grown crystal were studied.

1. Introduction

The research on new organic nonlinear optical (NLO) materials has attracted more research groups to them for their advantages over the inorganic nonlinear optical materials. The organic nonlinear optical molecules generally have a larger second-order nonlinear optical coefficient, and hence they are widely used in many applications such as optical communication, information storage, and optical switching [1–5]. The continuous research on organic NLO materials prompted to find many new nonlinear optical complexes engineered by organic molecules. For instance, nonlinear optical complexes formed from pyridine and carboxylic acids have been crystallized, and the structural, optical, and thermal properties of these crystals have been investigated due to their significant impact on laser technology, optical communication, and optical data storage technology [6]. Organic nonlinear optical compounds have large macroscopic second-order NLO susceptibilities $\chi^{(2)}$, low dielectric dispersion, and ultrafast response to external electric fields [7, 8]. Organic nonlinear optical materials are formed based on NLO chromophores, which should be oriented in noncentrosymmetric way in order to exhibit macroscopic second-order nonlinear optical response. There are several different strategies resulting in macroscopic $\chi^{(2)}$ organic materials, including ferroelectric liquid crystals [9], self-assembled films [10], Langmuir-Blodgett films [11], and poled polymer films [12], as well as single crystals [13]. Among them, polar organic crystals are of special interest, since large NLO effect can be achieved due to the high density of chromophores in the crystals, as well as the most stable packing of chromophores in macroscopic materials, resulting in a superior photochemical and thermal stability as compared with soft category materials [14]. Furthermore, organic NLO crystals can be produced in a bulk form and are therefore not limited to thin-film applications [13–16]. This is particularly important for frequency conversion applications such as terahertz-wave generation, where large interaction lengths

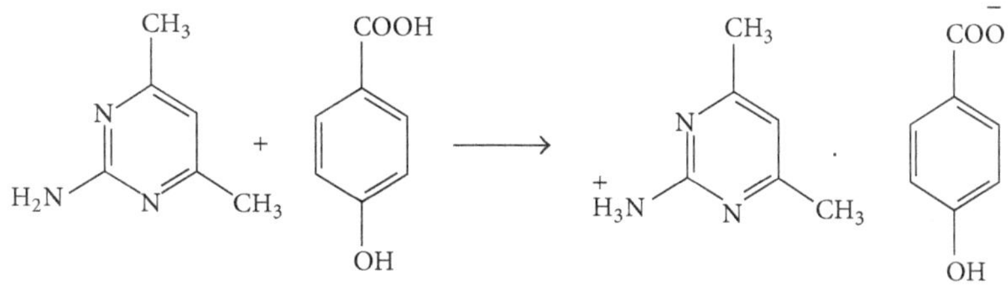

FIGURE 1: Material synthesis scheme of ADPHB.

are required [17]. The asymmetric unit of 2-amino-4,6-dimethylpyrimidinium p-hydroxybenzoate (ADPHB) contains a 2-amino-4,6-dimethylpyrimidine (AMPY) molecule and a 4-hydroxybenzoic acid (4-HBZ) molecule. Both species are neutral; thus ADPHB is an adduct rather than a molecular salt. The atoms O_2 and $-N_2H_2$ group act as hydrogen-bond donors to atoms N_1 and O_3, respectively [18]. In the present investigation, single crystal of the title compound has been grown by slow evaporation method. The grown crystals were subjected to single crystal and powder X-ray diffraction, FTIR, thermal, UV-Visible transmission, refractive index, photoluminescence (PL), laser induced surface damage threshold, birefringence, and second harmonic generation studies.

2. Experimental

2.1. Material Synthesis and Crystal Growth. Commercially available starting materials, 2-amino-4,6-dimethylpyrimidine ($C_6H_9N_3$) (Sigma Aldrich, 99%) and p-hydroxybenzoic acid ($C_7H_6O_3$) (Loba Chemie, 99%), were dissolved in methanol. The clear ADPHB solution was obtained after a continuous stirring for about six hours to achieve homogeneous mixture of the solution. Then, the growth solution was allowed for slow evaporation which yielded the spontaneously nucleated crystals of the synthesized compound. The synthesized compound was further purified by successive recrystallization process in methanol, and it was utilized for the crystal growth. The chemical synthesis scheme of the ADPHB compound is shown in Figure 1.

Single crystal of the ADPHB was grown by slow evaporation method in methanol solution. The saturated solution was prepared by dissolving the purified ADPHB salt in methanol using solubility data. It was stirred well for about 6 h to obtain a homogeneous mixture of the growth solution and kept in the constant temperature bath at 38°C with an accuracy of ±0.01°C. The well-defined ADPHB single crystals with good transparency were collected within the period of 15–18 days with an average size of 5 mm × 5 mm × 1.5 mm as shown in Figure 2.

2.2. Characterization Techniques. The grown single crystal of ADPHB was characterized by different instrumental methods. Single crystal and powder X-ray diffraction analyses were carried out to estimate the lattice parameters and crystalline quality of the grown crystal using Bruker Nonius CAD-4/MACH 3 single X-ray diffractometer and Bruker AXS CAD4 diffractometer, respectively. FT-IR spectrum was recorded for ADPHB using JASCO FTIR 410 spectrometer, and it was effectively used to identify the functional groups of the synthesized compound. TG-DTA analyses were carried out using SII TG/DTA 6300 EXSTAR apparatus with a heating rate of 10°C/min under N_2 atmosphere. The optical transmission spectrum of ADPHB crystal was recorded with PerkinElmer Lambda 35 Spectrophotometer in the range of 190–1100 nm. The photoluminescence property of grown crystal was studied using RF-5301 Spectrometer. The laser damage threshold was recorded for ADPHB using 1064 nm wavelength of Nd : YAG laser radiation. The optical birefringence was observed using modified channeled spectrum method, and the relative second harmonic generation efficiency was evaluated by Kurtz-Perry powder technique using Nd : YAG laser source.

3. Results and Discussion

3.1. X-Ray Diffraction Analyses. The cell parameters of ADPHB crystal were obtained from single crystal XRD data collected at room temperature. The ADPHB compound crystallizes in monoclinic crystal system with noncentrosymmetric space group Cc. The estimated cell parameters are $a = 9.025$ Å, $b = 11.469$ Å, $c = 12.580$ Å, $\alpha = \gamma = 90°$, $\beta = 102.804°$, and $v = 1270$ Å^3 and are in good agreement with the reported data [18]. The ADPHB crystal was crushed to a uniform fine powder and was subjected to powder X ray diffraction study to reveal crystalline perfection of the compound. The observed sharp and intense peaks reveal the high crystallinity of the grown crystal and the corresponding peaks were indexed. The recorded powder X-ray diffraction pattern of ADPHB crystal is shown in Figure 3.

3.2. FT-IR Spectral Analysis. The FTIR spectrum was recorded to understand the chemical bonding, and it provides useful information regarding the molecular structure of the compound. FTIR spectrum was taken for the powdered sample in the wavelength range 4000–400 cm^{-1} as shown in Figure 4. The KBr pellet technique was used to analyze the sample. The powdered ADPHB sample with KBr in the ratio of 1 : 100 was used to make the pellet. The peak at 3476 cm^{-1} is due to phenolic O–H stretching. The peak at 3292 is due to NH stretching vibration of primary amine group in 2-amino-4,6-dimethylpyrimidine. The absorption at 3065 cm^{-1} is due to C–H stretching. The peaks at 2969 and

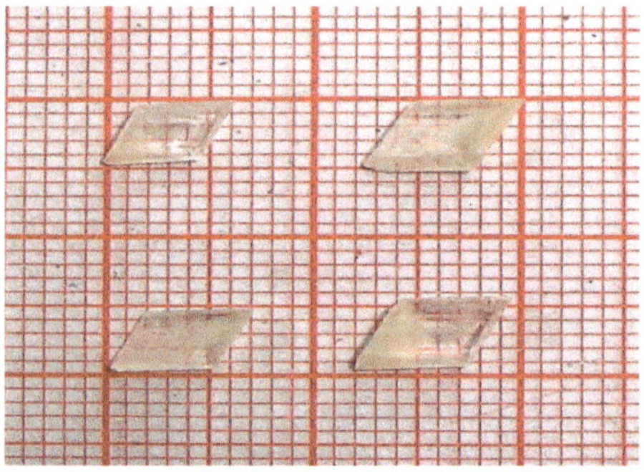

FIGURE 2: Photograph of as-grown ADPHB crystals.

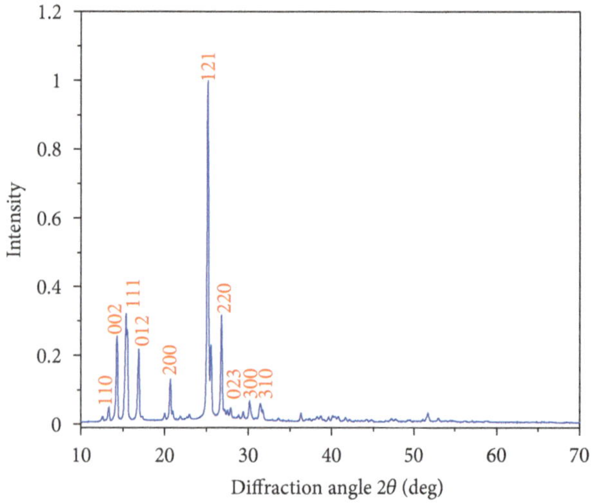

FIGURE 3: Powder XRD pattern of ADPHB crystal.

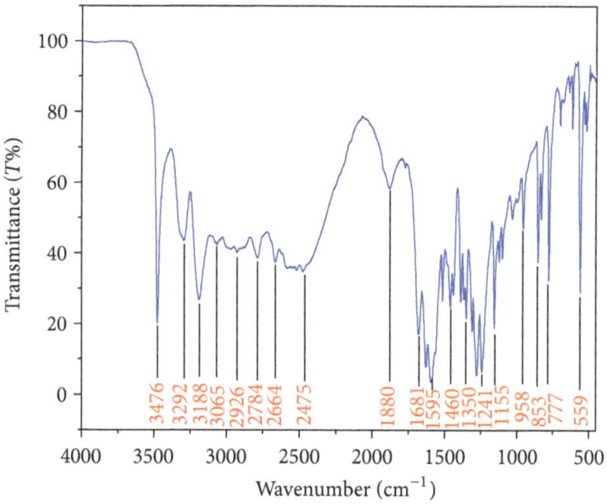

FIGURE 4: FTIR spectrum of ADPHB.

TABLE 1: FTIR frequency assignments of ADPHB.

Functional group assignments	ADPHB sample
O–H phenolic stretching	3476
N–H primary amines	3292
C–H stretching	3065
O–H carboxylic stretching	2969 and 2926
C=N stretching	1628
C=O carbonyl stretching	1681
N–H bend primary amines	1595
C–N stretch in aromatic amines	1366
C–H bending	853, 830, and 777

2926 cm^{-1} are due to O–H stretch of carboxylic acid group present in the compound. The characteristic C=O carbonyl stretching vibration appears as a sharp peak at 1681 cm^{-1}. The C=N stretching vibration appears at 1628 cm^{-1}. The N–H bend primary amine produces absorption peak at 1595 cm^{-1}. The absorption peaks at 1366 cm^{-1} are due to C–N stretch in aromatic amines. The aromatic C–H in-plane bending vibration appears at 853, 830, and 777 cm^{-1}. The assigned vibrational frequencies of functional groups of title compound are listed in Table 1.

3.3. Thermal Analysis. The thermal stability of powdered ADPHB sample was studied by thermogravimetric analysis (TGA) and differential thermal analysis (DTA) in the nitrogen atmosphere with a heating rate of 10°C/min. The ADPHB sample weighing 1.577 mg was used to record TGA and DTA thermogram which is shown in Figure 5. In the TGA curve, there is no weight loss peak that occurs before 130.4°C and it reveals that there is no lattice water molecule in the compound. TG curve shows single stage weight loss patterns it is observed from the temperature 130.4°C up to the 231.9°C and it is approximately 97.1% percentage loss of the compound. Thus, the TGA curve indicates that the sample is stable up to the temperature of 130.4°C for nonlinear optical applications. In the DTA curve, there are two endothermic steps available. The observed first endothermic peak at 183.4°C and a second endothermic peak at 219°C are due to the removal of the hydroxyl and amino groups expelled from the compound respectively [19, 20]. Hence, the recorded thermogram of the compound confirms ADPHB compound, and it is confirmed that the compound is thermally stable up to 130.4°C.

3.4. Optical Studies

3.4.1. UV-Visible Transmission Studies. UV-Visible spectral analysis gives useful information about electronic transitions in the compound [21]. The optical behavior of ADPHB crystal specimen was analyzed using UV-Vis spectrophotometer in the wavelength range of 190–900 nm. For this study, an optically polished single crystal of thickness 1.5 mm was used and the recorded transmittance spectrum is shown in Figure 6. The cut-off wavelength of the crystal is observed at 305 nm, and the crystals show good transmittance percentage in the range of 330–900 nm. This spectral study would be assisted in the understanding of electronic structure and the optical band gap of the crystal. From the spectrum, it is observed that ADPHB crystal has high transmittance in the

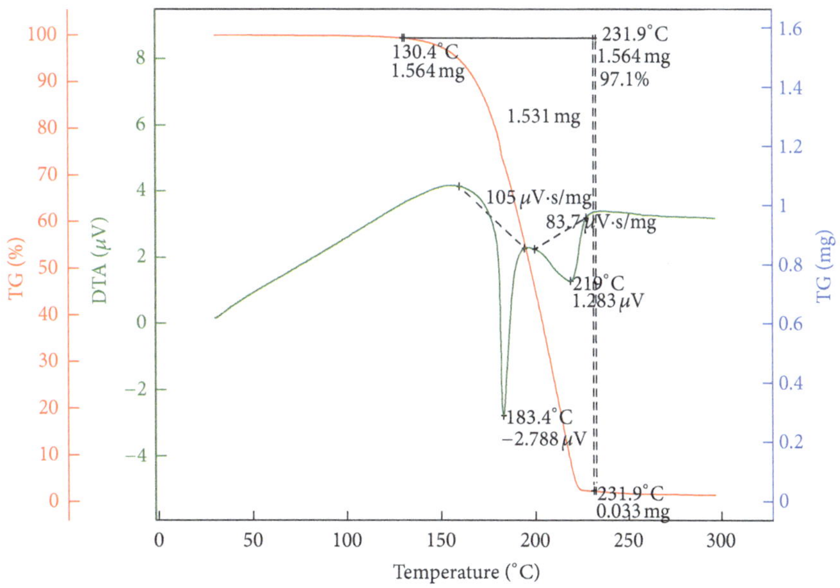

Figure 5: TG/DTA trace of ADPHB.

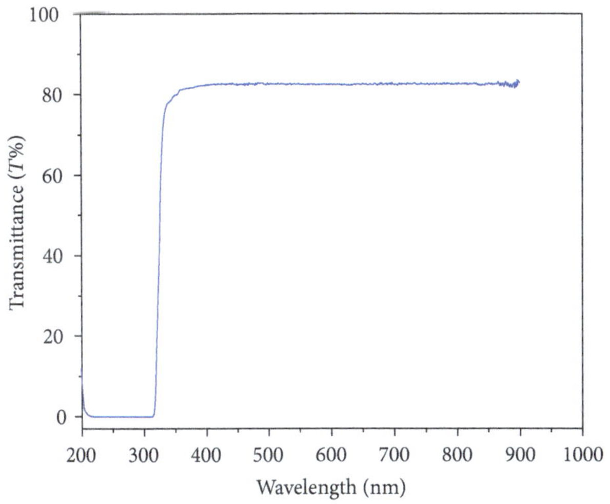

Figure 6: UV-visible transmittance spectrum of ADPHB single crystal.

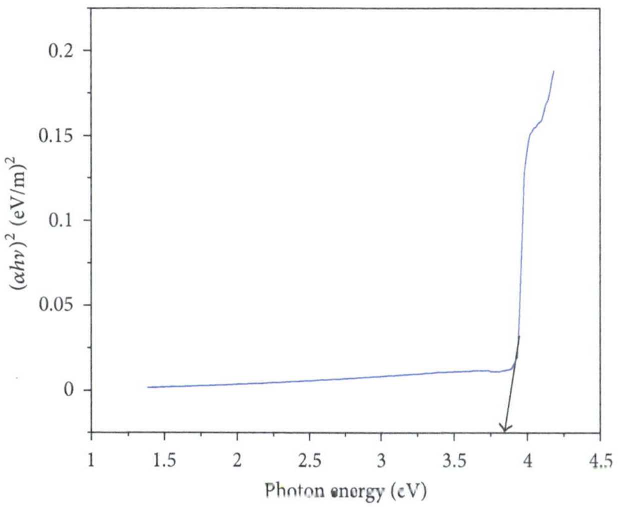

Figure 7: Plot of photon energy versus $(\alpha h\nu)^2$ of ADPHB crystal.

entire visible, near infrared region and this property enables the material holds good for optoelectronic applications.

3.4.2. Optical Band Gap Energy (E_g) Calculation. For optical device fabrication, the crystal should have high transparency in a considerable range of wavelength [22]. The UV cut-off wavelength of the grown ADPHB crystal was found to be 305 nm, and it is useful to make them potential material for optical device fabrication. The optical absorption coefficient (α) was calculated using the relation,

$$\alpha = \left(\frac{1}{d}\right)\log\left(\frac{1}{T}\right), \qquad (1)$$

where d is the thickness of the crystal and T is the transmittance. Owing to the direct band gap, the crystal under study has an absorption coefficient (α) obeying the following relation for high photon energies ($h\nu$),

$$h\nu\alpha = A\left(h\nu - E_g\right)^{1/2}, \qquad (2)$$

where A is a constant, E_g is the optical band gap, h is the Planck's constant, and ν is the frequency of the incident photons [23]. The band gap of grown ADPHB crystal was estimated by plotting $(\alpha h\nu)^2$ versus $h\nu$ and it is shown in Figure 7. The band gap energy of grown ADPHB crystal was found to be 3.84 eV. This indicates the consequences of wide band gap and large transmittance of the grown crystal.

3.4.3. Determination of Optical Constants. The optical behaviour of materials is important to determine their usage in optoelectronic devices [24]. The study of optical constants of a material such as refractive index and extinction coefficient is quite essential to examine the material's potential optoelectronic applications [25]. Further, the optical properties may also be closely related to the material's atomic structure, electronic band structure and electrical properties. The optical constants (n, k) were determined from the transmission (T) and reflection (R) spectrum based on the following relations [26]:

$$T = \frac{(1-R)^2 \exp(-\alpha t)}{1 - R^2 \exp(-2\alpha t)}, \quad (3)$$

where t is the thickness and α is related to extinction coefficient.

k value was determined by

$$k = \frac{\alpha\lambda}{4\pi}. \quad (4)$$

The refractive index (n) can be determined from the reflectance (R) [27]

$$R = \frac{(n-1)^2 + k^2}{(n+1)^2 + k^2}. \quad (5)$$

The reflectance in terms of absorption coefficient was determined using

$$R = 1 \pm \frac{\sqrt{1 - \exp(-\alpha t) + \exp(\alpha t)}}{1 + \exp(-\alpha t)}, \quad (6)$$

and from the above equation, the refractive index n can be derived as

$$n = \frac{-(R+1) \pm \sqrt{-3R^2 + 10R - 3}}{2(R-1)}. \quad (7)$$

The calculated refractive index value using the above equations for the grown ADPHB crystal is 2.562. The other calculated values of extinction coefficient (k) and reflectance (R) are 0.00310 and 0.49, respectively.

3.4.4. Photoluminescence (PL) Spectral Studies. Photoluminescence spectroscopy is a nondestructive method of probing the electronic structure of materials. The photoluminescence spectrum was recorded for the as-grown inclusion free single crystal of ADPHB at room temperature with an excitation wavelength of 316 nm as shown in Figure 8. A very strong intense emission peak was observed at 354 nm (3.502 eV), which corresponds to near band-edge excitons of as-grown crystals. Moreover, absence of visible emission bands has indicated the high crystal quality and perfect crystallinity of as-grown crystals. Therefore, the as-grown ADPHB crystals might be suitable for UV filters and optoelectronic laser devices.

3.4.5. Surface Laser Damage Threshold (LDT) Measurement. One of the essential criteria for the NLO crystal to perform as a device is that it needs high laser damage resistance because high optical intensities are involved in nonlinear process [28]. In the present experiment, an actively Q-switched Nd : YAG laser source with 6 ns pulse width and 10 Hz repetition rate was used. The output intensity of the laser was controlled with variable attenuator and delivered to the test sample located at the near focusing of the converging lens. The energy density of the input laser beam was recorded using power meter till the crystal got damaged. The surface damage threshold of the crystal was calculated using the formula

$$\text{Power density } P_{(d)} = \frac{E}{\tau A}, \quad (8)$$

where E is the energy (mJ), τ is the pulse width (ns) and A is the area of the circular spot size. Multiple shot LDT measurements were made on the well-polished plane of the grown ADPHB crystal. The beam spot size on the sample was 1 mm, and input energy was measured as 40 mJ. The calculated laser damage threshold is 4.73 GW/cm^2. The laser damage property is nearly 23.65 times higher than other standard NLO materials of potassium dihydrogen phosphate (KDP) (0.20 GW/cm^2) and 3.15 times higher than that of urea (1.5 GW/cm^2). The laser damage threshold value of the present sample is compared with other NLO crystals [29] and is given in Table 2. The higher value of LDT will be useful for making laser based devices.

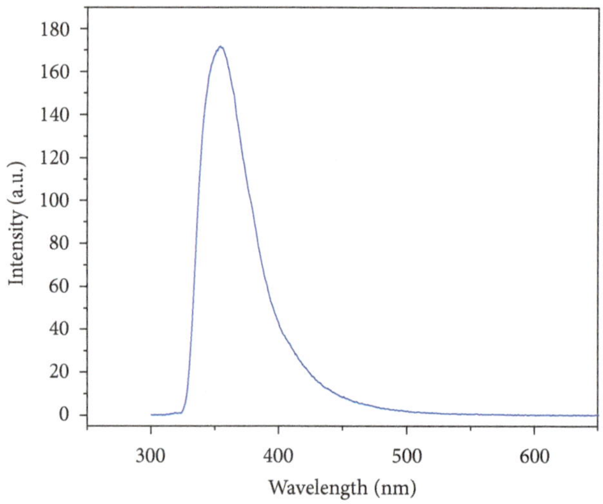

FIGURE 8: Photoluminescence spectrum of ADPHB crystal.

TABLE 2: Comparison of laser damage threshold value of ADPHB crystal with other crystals.

Crystal	Laser damage threshold (GW/cm^2)
KDP	0.2
Urea	1.5
Beta-barium borate	5.0
L-arginine phosphate	10.0
ADPHB	4.73

3.4.6. Birefringence Studies. The experimental setup of modified channel spectrum (MCS) method was employed to reveal

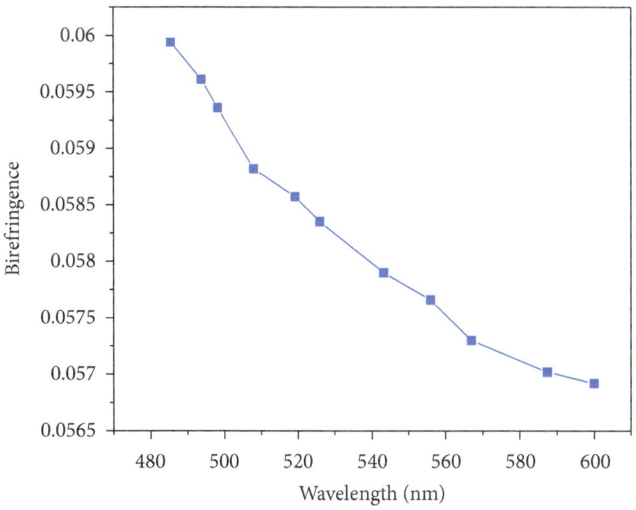

FIGURE 9: Birefringence spectrum of ADPHB crystal.

the birefringence behavior of the specimen [30, 31]. The constant deviation spectrometer was first calibrated using the standard procedure and 1000 W tungsten halogen lamp was used as source, which was collimated using a collimator. The optically polished ADPHB single crystal was placed on a rotary stage between crossed polarizer in the path of polarized light. The thickness of about 0.85 mm ADPHB crystal was used for the measurement. The direction of optical axis of the crystal slab was kept perpendicular to the direction of the collimated light transmitted by the polarizer. Finally, the light was passed through the analyzer and emerging light waves were examined by the microscope. The birefringence was calculated using the relation [32]

$$\Delta n = \frac{k\lambda}{t}, \quad (9)$$

where λ is the wavelength of light, t is the thickness of the crystal and k is the order of interference maximum. The plot drawn between birefringence (Δn) and wavelength (λ) is shown in Figure 9. From this measurement, the dispersion of birefringence existing in the entire visible region observed.

3.4.7. Second Harmonic Generation Test. The second harmonic generation efficiency of ADPHB crystal was measured by using Kurtz-Perry powder SHG technique with potassium dihydrogen phosphate (KDP) crystal as reference material [33]. The fundamental laser beam of 1064 nm wavelength and 8 ns pulse width, with 10 Hz pulse rate, was made to fall normally on the sample cell. The input laser energy incident on the powdered sample was chosen to be 5.65 mJ/pulse. The grown single crystal of ADPHB and reference KDP was ground in a uniform particle size of 125–150 μm, packed in a microcapillary of uniform bore size, and exposed to laser radiations. The fundamental beam was filtered by using an IR filter. A photomultiplier tube was used as detector. The output from the sample was monochromated to collect the intensity of 532 nm component and to eliminate the fundamental radiation. The generation of the second harmonic was confirmed by the emission of green light. The SHG output efficiency for ADPHB and KDP samples was found to be 195 mV and 55 mV, respectively. Thus, it is observed that SHG efficiencies of ADPHB crystal were found to be 3.5 times higher those that of KDP.

4. Conclusion

Nonlinear optical 2-amino-4,6-dimethylpyrimidine p-hydroxybenzoate single crystals were grown by slow evaporation method. The crystalline perfection of the grown crystal was examined by X-ray diffraction studies. The modes of vibration molecule groups present in the grown crystal were confirmed by Fourier transform infrared spectral analysis. The thermogravimetric analyses reveal that the crystal is thermally stable up to 130.4°C without any structural phase transition. The transmission percentage, lower cut-off wavelength, and optical band gap energy (E_g) of the grown crystal were examined by UV-Vis spectral analysis. The linear refractive index, photoluminescence behavior, and birefringence property of the crystal were examined. The surface laser damage threshold value for ADPHB crystal was found to be 4.73 GW/cm^2 and compared with other NLO crystals. The second harmonic generation efficiency of the crystal was found to be 3.5 times higher than that of standard KDP crystal.

Acknowledgment

One of the authors (M. K. Kumar) would like to acknowledge, Council of Scientific and Industrial Research, New Delhi, India, for providing financial support (project no. 03 (1200)/11/EMR-II).

References

[1] T. Kaino, B. Cai, and K. Takayama, “Fabrication of DAST channel optical waveguides,” *Advanced Functional Materials*, vol. 12, no. 9, pp. 599–603, 2002.

[2] P. N. Prasad and D. J. Williams, *Introduction to Nonlinear Optical Effects in Molecules and Polymers*, John Wiley & Sons, New York, NY, USA, 1991.

[3] B. K. Periyasamy, R. S. Jebas, and B. Thailampillai, “Synthesis and spectral studies of 2-aminopyridinium*para*-nitrobenzoate: a novel optoelectronic crystal,” *Materials Letters*, vol. 61, no. 7, pp. 1489–1491, 2007.

[4] R. W. Munn and C. N. Ironside, *Principles and Applications of Nonlinear Optical Materials*, Chapman and Hall, London, UK, 1993.

[5] J. Bardan, R. Hierle, A. Perigaud, J. Zyss, and D. J. Williams, *Nonlinear Optical Properties of Organic Molecules and Polymeric Materials*, vol. 223 of *ACS Series*, American Chemical Society, Washington, 1993.

[6] M. L. Caroline and S. Vasudevan, “Growth and characterization of an organic nonlinear optical material: l-alanine alaninium nitrate,” *Materials Letters*, vol. 62, pp. 2245–2248, 2008.

[7] B. J. Coe J.A. Harris, A. K. Clays, G. Olbrechts, A. Persoons, and J. T. Hupp, “Quadratic nonlinear optical properties of*N*-aryl stilbazolium dyes,” *Advanced Functional Materials*, vol. 12, no. 2, pp. 110–116, 2002.

[8] L. Dalton, "Nonlinear optical polymeric materials: from chromophore design to commercial applications," *Advances in Polymer Science*, vol. 158, pp. 1–86, 2002.

[9] Y. Zhang, J. Martinez-Perdiguero, U. Baumeister et al., "Laterally azo-bridged H-shaped ferroelectric dimesogens for second-order nonlinear optics: ferroelectricity and second harmonic generation," *Journal of the American Chemical Society*, vol. 131, no. 51, pp. 18386–18392, 2009.

[10] D. Frattarelli, M. Schiavo, A. Facchetti, M. A. Ratner, and T. J. Marks, "Self-assembly from the gas-phase: dsign and implementation of small-molecule chromophore precursors with large nonlinear optical responses," *Journal of the American Chemical Society*, vol. 131, no. 35, pp. 12595–12612, 2009.

[11] G. J. Ashwell, "Langmuir-Blodgett films: molecular engineering of non-centrosymmetric structures for second-order nonlinear optical applications," *Journal of Materials Chemistry*, vol. 9, pp. 1991–2003, 1999.

[12] L. R. Dalton, S. J. Benight, L. E. Johnson et al., "Systematic nanoengineering of soft matter organic electro-optic materials," *Chemistry of Materials*, vol. 23, pp. 430–445, 2011.

[13] M. Jazbinsek, O. P. Kwon, C. Bosshard, and P. Gunter, "Organic nonlinear optical crystals and single-crystalline thin films," in *Handbook of Organic Electronics and Photonics*, S. H. Nalwa, Ed., American Scientific Publishers, 2008.

[14] D. Rezzonico, S. J. Kwon, H. Figi, O. P. Kwon, M. Jazbinsek, and P. Günter, "Photochemical stability of nonlinear optical chromophores in polymeric and crystalline materials," *Journal of Chemical Physics*, vol. 128, no. 12, Article ID 124713, 2008.

[15] H. Figi, L. Mutter, C. Hunziker, M. Jazbinšek, P. Günter, and B. J. Coe, "Extremely large nonresonant second-order nonlinear optical response in crystals of the stilbazolium salt DAPSH," *Journal of the Optical Society of America B*, vol. 25, no. 11, pp. 1786–1793, 2008.

[16] M. Jazbinsek, L. Mutter, and P. Günter, "Photonic applications with the organic nonlinear optical crystal DAST," *IEEE Journal on Selected Topics in Quantum Electronics*, vol. 14, no. 5, pp. 1298–1311, 2008.

[17] M. Tonouchi, "Cutting-edge terahertz technology," *Nature Photonics*, vol. 1, no. 2, pp. 97–105, 2007.

[18] K. Balasubramani, P. T. Muthiah, and D. E. Lynch, "2-Amino-4,6-dimethylpyrimidine-4-hydroxybenzoic acid (1/1)," *Acta Crystallographica Section E*, vol. 62, no. 7, pp. o2907–o2909, 2006.

[19] K. A. Nandekar, J. R. Dontulwar, and W. B. Gurnule, "Thermoanalytical studies and kinetics of newly synthesized copolymer derived from P-hydroxybenzoic acid, and semicarbazide," *Rasayan Journal of Chemistry*, vol. 5, no. 3, pp. 261–268, 2012.

[20] R. A. Khatri, S. S. C. Chuang, Y. Soong, and M. Gray, "Thermal and chemical stability of regenerable solid amine sorbent for CO_2 capture," *Energy and Fuels*, vol. 20, no. 4, pp. 1514–1520, 2006.

[21] C. N. R. Rao, *Ultraviolet and Visible Spectroscopy*, Butterworths, London, UK, 3rd edition, 1975.

[22] V. Krishnakumar and R. Nagalakshmi, "Crystal growth and vibrational spectroscopic studies of the semiorganic non-linear optical crystal: bisthiourea zinc chloride," *Spectrochimica Acta A*, vol. 61, no. 3, pp. 499–507, 2005.

[23] R. N. Rai, S. R. Mudunuri, R. S. B. Reddi, V. S. A. Kumar Satuluri, S. Ganeshmoorthy, and P. K. Gupta, "Crystal growth and nonlinear optical studies of *m*-dinitrobenzene doped urea," *Journal of Crystal Growth*, vol. 321, no. 1, pp. 72–77, 2011.

[24] V. Pandey, N. Mehta, S. K. Tripathi, and A. Kumar, "Optical band gap and optical constants in $Se_{85}Te_{15-x}Pb_x$ thin films," *Chalcogenide Letters*, vol. 2, no. 5, pp. 39–44, 2005.

[25] M. Dongol, "Optical absorption and structural properties of as-deposited and thermally annealed As-Te-Ga thin films," *Egyptian Journal of Solids*, vol. 25, no. 1, pp. 33–47, 2002.

[26] R. E. Denton, R. D. Campbell, and S. G. Tomlin, "The determination of the optical constants of thin films from measurements of reflectance and transmittance at normal incidence," *Journal of Physics D*, vol. 5, pp. 852–863, 1972.

[27] A. Ashour, N. El-Kadry, and S. A. Mahmoud, "On the electrical and optical properties of CdS films thermally deposited by a modified source," *Thin Solid Films*, vol. 269, no. 1-2, pp. 117–122, 1995.

[28] G. C. Bhar, A. K. Chaudhary, and P. Kumbhakar, "Study of laser induced damage threshold and effect of inclusions in some nonlinear crystals," *Applied Surface Science*, vol. 161, no. 1, pp. 155–162, 2000.

[29] H. L. Bhat, "Growth and characterization of some novel crystals for nonlinear optical applications," *Bulletin of Materials Science*, vol. 17, no. 7, pp. 1233–1249, 1994.

[30] K. Nagarajan and C. K. Shasidharan Nair, "The use of fractional orders in the determination of birefringence of highly dispersive materials by the channelled spectrum method," *Optics Communications*, vol. 275, pp. 348–353, 2007.

[31] N. Vijayan, K. Nagarajan, A. M. Z. Slawin, C. K. S. Nair, and G. Bhagavannarayana, "Growth of benzimidazole single crystal by sankaranarayanan-ramasamy method and its characterization by high-resolution X-ray diffraction, thermogravimetric/differential thermal analysis, and birefringence studies," *Crystal Growth and Design*, vol. 7, no. 2, pp. 445–448, 2007.

[32] N. Vijayan, G. Bhagavannarayana, K. Nagarajan, and V. Upadhyaya, "Synthesis and growth of nearly perfect single crystal of l-histidine bromide (LHB) and its structural, optical and electrical characterizations," *Materials Chemistry and Physics*, vol. 115, no. 2-3, pp. 656–659, 2009.

[33] S. K. Kurtz and T. T. Perry, "A powder technique for the evaluation of nonlinear optical materials," *Journal of Applied Physics*, vol. 39, no. 8, pp. 3798–3813, 1968.

Stoichiometry Control of ZnO Thin Film by Adjusting Working Gas Ratio during Radio Frequency Magnetron Sputtering

Chaoyang Li,[1] Dapeng Wang,[1] Zeming Li,[1] Xin Li,[1] Toshiyuki Kawaharamura,[1] and Mamoru Furuta[2]

[1] *Institute for Nanotechnology, Kochi University of Technology, 185 Miyanokuchi, Tosayamada-cho, Kami, Kochi 782-8502, Japan*
[2] *Environmental Science and Engineering, Kochi University of Technology, 185 Miyanokuchi, Tosayamada-cho, Kami, Kochi 782-8502, Japan*

Correspondence should be addressed to Chaoyang Li; li.chaoyang@kochi-tech.ac.jp

Academic Editor: Sung-Hoon Kim

ZnO thin films were deposited on quartz glasses by a radio frequency (rf) magnetron sputtering. The mechanism for stoichiometry in the ZnO thin films was investigated by adjusting Ar/O_2 working gas ratio during deposition. The optical emission spectroscopy (OES) in situ measurement revealed the kinetics species variation during rf plasma deposition process. It was found that the intensity of the excited atomic oxygen (O^*) was increased with the oxygen ratio increasing, resulting in enhancing the oxidization effect during ZnO film fabrication. On the contrary, the intensities of atomic zinc emission were gradually decreased, resulting in the zinc ratio in the film were decreased with the oxygen ratio increasing. Therefore, it is possible to control the stoichiometry of ZnO film by simply adjusting the working gas ambient in the rf plasma deposition. The structural and optical properties of ZnO thin films were investigated as well.

1. Introduction

In recent years zinc oxide (ZnO) thin films, a compound semiconductor material with a wide-band gap of (3.37 eV at 300 K), have attracted much attention for applying in different photonic and electronic devices such as optical waveguides, thin film transistor, and gas sensors [1–3], due to its promising electrical, optical, and piezoelectric properties. Among the many fabrication techniques of ZnO film, radio frequency (rf) magnetron sputtering is considered as a convenient method for deposition of homogeneous ZnO thin film on large area because the stability and reproducibility of ZnO films are quite high comparing to other techniques [4]. In our previously work [3, 5], some optimum deposition conditions using rf magnetron sputtering system had been found to prepare high-quality ZnO films for TFTs application.

We already found that the bombardment effect of energetic particles during rf magnetron sputtering was one of the important factors influencing the properties of ZnO thin films [6]. Comparing to a low-pressure deposition [7], we found that the properties of as-deposited ZnO thin films were changed by a high-pressure deposition procedure. However, the structural and optical properties of ZnO films obtained from high-pressure deposition process were still not quite understood. In this research, the effects of different working gas (Ar/O_2) ratios on the fabrication of ZnO films were investigated under a high deposition pressure of 7 Pa. The mechanism for ZnO film stoichiometry was discussed, and the structural and optical properties of ZnO films were evaluated. The experiment result is expected to supply helpful information for controlling the quality of as-deposited ZnO films for future application in the TFT and other optoelectronic devices.

2. Experimental

ZnO films (500 nm in thickness) were deposited onto quartz glasses by rf (13.56 MHz) magnetron sputtering system using a 4-inch ZnO (5 N) target. The target was located on the cathode 60 mm away from the substrate stage. Prior to deposition, the substrate on the anode was preheated to 150°C

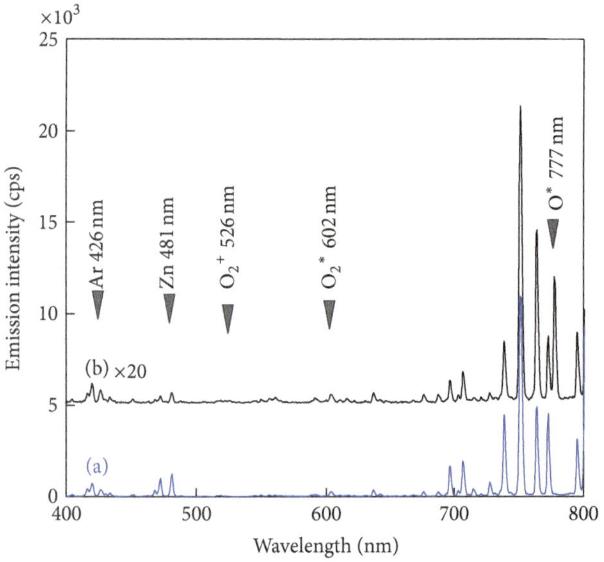

FIGURE 1: (Color online) Typical optical emission spectra obtained from in-situ measurement during ZnO deposition: (a) Ar: 30 sccm and, (b) Ar/O_2 = 10/30 (sccm/sccm) at deposition pressure of 7 Pa.

for 1 hour. The working chamber was evacuated to 5×10^{-5} Pa with a turbo molecular pump. The rf magnetron sputtering deposition process was kept at power of 180 W for the target, under a deposition pressure of 7 Pa. In order to investigate the effects of working gas ratio (Ar/O_2) on the properties of ZnO films, the working gases were set in different Ar/O_2 (10/5, 10/10, and 10/30 sccm/sccm) ratio by adjusting the mass flow controllers (MFCs). Additionally, the pure argon and pure oxygen gases depositions were also carried out for comparison during the deposition processes.

During the rf deposition process, a photonic multichannel spectra analyzer (Hamamatsu C-7473) system was used to in situ monitor the optical emission from excited species in the rf plasma. Before the deposition started, the optical fiber directly touched on the outside of quartz window in the chamber, positioned at the middle of cathode and anode. The crystal structure of the ZnO films was characterized using an X-ray diffraction system (Rigaku ATX-G diffractometer), employing a Cu Kα tube (λ = 0.154178 nm) radiation. The surface morphologies of the films were observed by a field emission scanning electron microscope (FE-SEM) system (JEOL-JSM7400F). The different luminescent characteristics of ZnO films were measured by photoluminescence (PL) measurement, which was performed on an iHR320 Micro-PL/Raman spectroscope (Horiba). An He-Cd laser with a wavelength of 325 nm at a power of 20 mW was used as an excitation light source. All of measurements were carried out at room temperature.

3. Results and Discussion

Figure 1 shows the OES spectra in situ measurement for the energetic species in the rf plasma during the ZnO thin films deposited with the different working gases ambient. The typical emission peaks from different emission particles during the deposition could be detected as argon (426 nm) corresponding to the 5p-4s transition, metal zinc (481 nm) corresponding to the 4p-5s transition [8], excited atomic oxygen (O^*) (777 nm) corresponding to the 5p-5s transition, and molecular ion oxygen (O_2^+) (526 nm) (first negative system) [9], molecular oxygen (O_2^*) (602 nm) corresponding to transitions from $^4\Sigma^-_g$ to $^4\Pi_\mu$ [10]. The OES spectra from the 750 nm Ar I and 428 nm Ar II lines was recorded. When the deposition uses only argon as a working gas, ionic species of oxygen present in the rf plasma come entirely from the ZnO target. It was obviously observed that the intensity of zinc emission was much higher (more than 20 times) in the only argon deposition than that in the Ar/O_2 (10/30 sccm/sccm) deposition.

Figure 2 shows the emission intensities originated from different energetic species: Ar (426 nm), O^* (777 nm), O_2^+ (526 nm), and Zn (481 nm) and emission intensity ratios of Zn/O^* and O^*/Ar as function of oxygen concentration. Comparing the emission intensity from different energetic species peaks, it was found that the excited atomic oxygen O^* in the rf plasma was increased with the oxygen ratio increasing. On the contrary, argon, O_2^+, and metal atomic zinc emission were correspondingly decreased. In Figure 2(b), the dependence of emission intensity ratio of Zn/O^* and O^*/Ar as function of oxygen concentration was compared as well. It was found that the Zn/O^* ratio was decreased with the oxygen increase, which meant that the oxidation effect of metal zinc was suppressed with the oxygen concentration increasing resulting in the zinc ratio in the film stoichiometry decreased with the oxygen ratio increase.

It was reported [11] that the O^* existed in the rf plasma during Ar/O_2 radio frequency sputtering is negative oxygen ion, which contributed to neutralization in plasma on the way to the anode [9] and had major influence on thin film growth. With O^* content increasing, the plasma neutralization will be increased during the film deposition, resulting in deposition rate decrease, which was confirmed by the deposition rate measurement shown in Figure 3 later. In the OES result, the excited atomic oxygen O^* was gradually increased with the oxygen ratio increasing, which contributed to enhancing the oxidation of the ZnO films, so that the crystallinity of ZnO film might be improved with oxygen ratio increasing, which could be confirmed by the XRD measurement result later.

As well known, the argon gas contributed to the bombardment effect to the target, it was found that the argon emission was decreased with the oxygen ratio increasing from OES in situ measurement, which led to the deposition rate of ZnO film being remarkably decreased. In this experiment, the deposition rate was decreased from 13.8, 8.06, 6.66, 5.65 to 5.02 nm/min, respectively, with oxygen ratio increasing, as shown in Figure 3. Meanwhile, O^*/Ar was also increased with the oxygen increasing, which meant that the ZnO oxidization effect was much stronger than argon bombardment effect during the ZnO film deposition.

Figure 4 shows the morphologies of the obtained ZnO thin films deposited with different working gas ambient. It was clearly observed that all of the obtained films existed

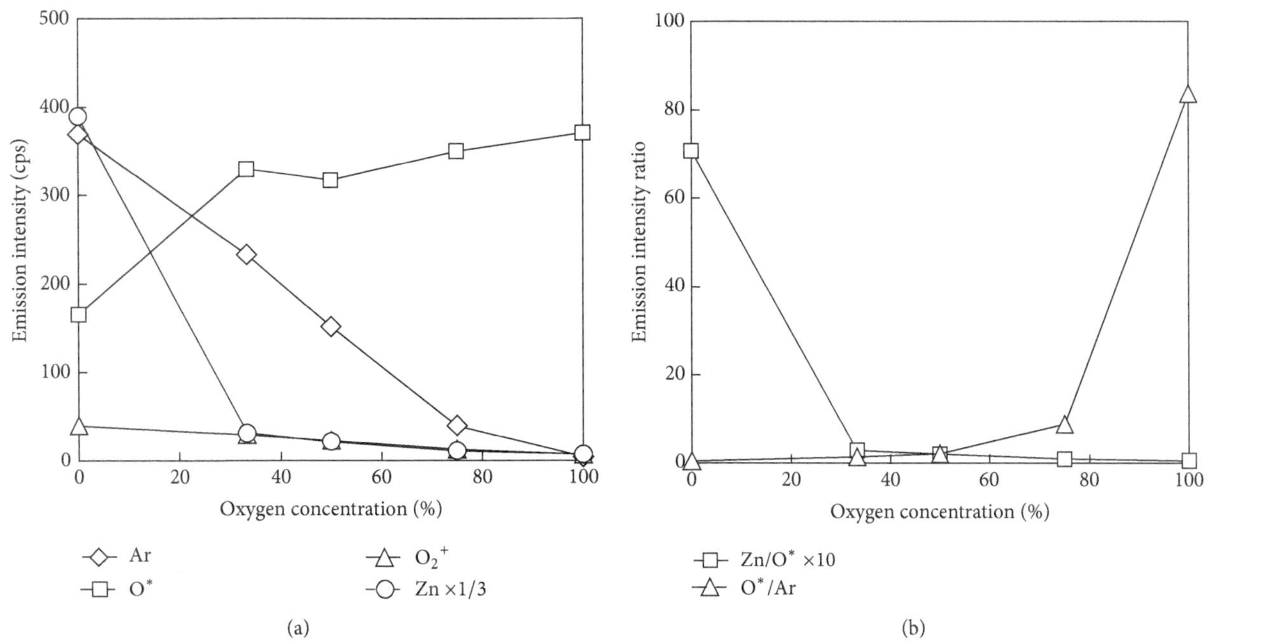

FIGURE 2: (a) Emission intensities originated from Ar (426 nm), O^* (777 nm), O_2^+ (526 nm), and Zn (481 nm) and (b) emission intensity ratios of Zn/O^* and O^*/Ar as function of oxygen concentration.

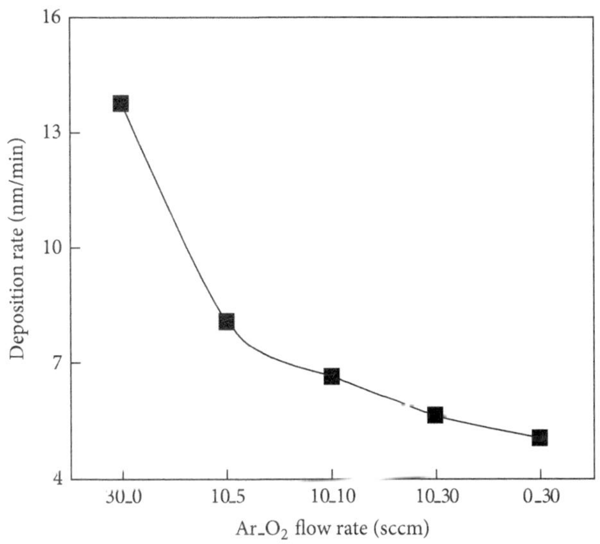

FIGURE 3: The deposition rate comparison for ZnO films deposited with different working gases.

the uniform and flat surface. Furthermore, it was found that the average grain sizes from different as-deposited ZnO films were decreased, respectively, from 170, 95, 90, 70 to 60 nm with the oxygen ratio increasing. According to the result of the deposition rate, it was known that the oxidization effect of the oxygen rich ambient is much strong during deposition, resulting in the formed ZnO crystalline size in c-axis being remarkably increased. In contrast, the average gain size was decreased. Therefore the much flatness surface was obtained from higher oxygen concentration deposition processes.

Figure 5 shows the XRD patterns of the ZnO thin films deposited from different gas ratio during rf magnetron sputtering. It was obvious that the (002) diffraction peak was dominated peak which was oriented highly perpendicular to the plane of the substrate for every sample. It was found that the intensity of c-axis (002) peak was increased with the oxygen ratio increasing, and the full width at half maximum (FWHM) value of (002) diffraction peaks being correspondingly decreased with the oxygen ratio increasing. In addition, by calculation using the Scherrer formula [12],

$$D = \frac{K\lambda}{\beta \cos\theta}, \qquad (1)$$

where K is the shape factor, λ is the X-ray wavelength, β is the line broadening at half the maximum intensity (FWHM) in radians, and θ is the Bragg angle. It was also found that the average crystallite sizes along c-axis in the different as-deposited films increased with oxygen ratio increasing, correspondingly as 21, 35, 36, 38, and 39 nm, as shown in Figure 5(b). Therefore, it was obvious that the crystallinities of the as-deposited ZnO thin films were significantly improved with the oxygen ratio increasing under high-pressure of 7 Pa deposition during the rf magnetron sputtering process.

Figure 6 shows the room temperature PL spectra obtained from as-deposited ZnO films deposited from different working gases (Ar/O_2) ambient. It was observed that the strong UV peak centered at 375 nm (3.31 eV), which is near band edge emission, attributed to excitonic emission due to recombination of photo-generated charge

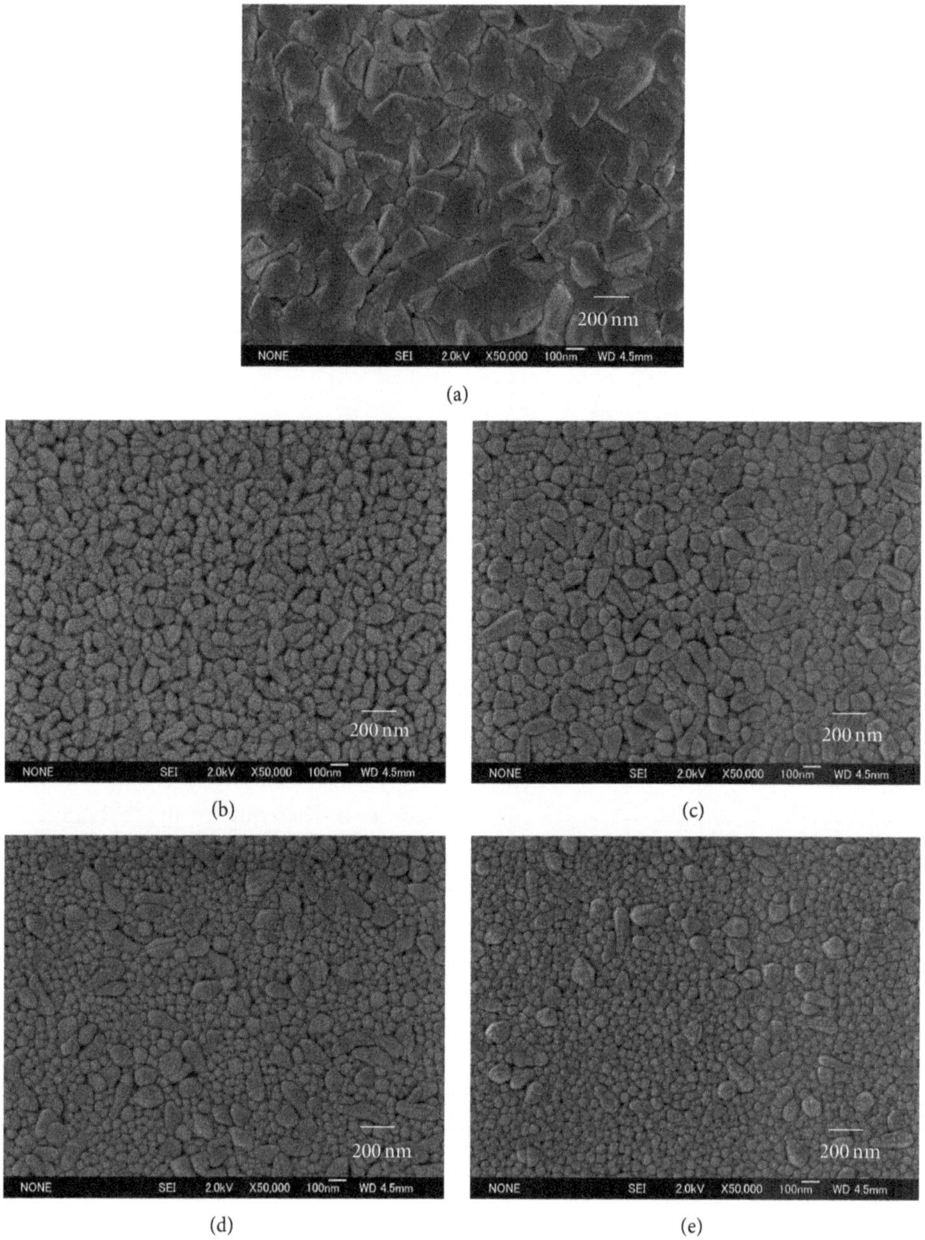

Figure 4: SEM images of as-deposited ZnO films obtained under different gas ratios (a) Ar: 30 (sccm), (b) Ar/O_2: 10/5 (sccm/sccm), (c) Ar/O_2: 10/10 (sccm/sccm), (d) Ar/O_2: 10/30 (sccm/sccm), and (e) O_2: 30 (sccm).

carriers dominating the spectrum; meanwhile suppressed deep level emission peaks around 572~604 nm that covered yellow orange range were also observed from all the samples deposited with the oxygen gas including deposition. It was obviously that the near band edge emission was enhanced with the oxygen ratio decreasing. However, in case of ZnO film deposited in working gas as pure argon (Ar) ambient, the both strong UV emission peaked at 378 nm (3.27 eV) with a shoulder about 424 nm (2.91 eV) and a broad visible emission were observed. According to the calculation for the energy level of various intrinsic defects in ZnO by applying the full-potential linear muffin-tin orbital method [13], the shoulder peak at 2.91 eV indicated that there was much high ratio of interstitial zinc (Zn_i) that existed in the film, which was well consistent with the OES measurement result. It was well accepted that the deep level emission comes from the intrinsic defects in ZnO film including zinc vacancies (V_{Zn}), oxygen vacancy V_O, interstitial zinc (Zn_i), interstitial oxygen O_i, and antisite oxygen O_{Zn}. It was no doubt that there were many defects introduced to the ZnO films by sputtering process. From the PL spectra, it was confirmed that the yellow-orange emission (572–604) might be due to excess oxygen during oxygen including sputtering. The film deposited using only argon gas included much more defects

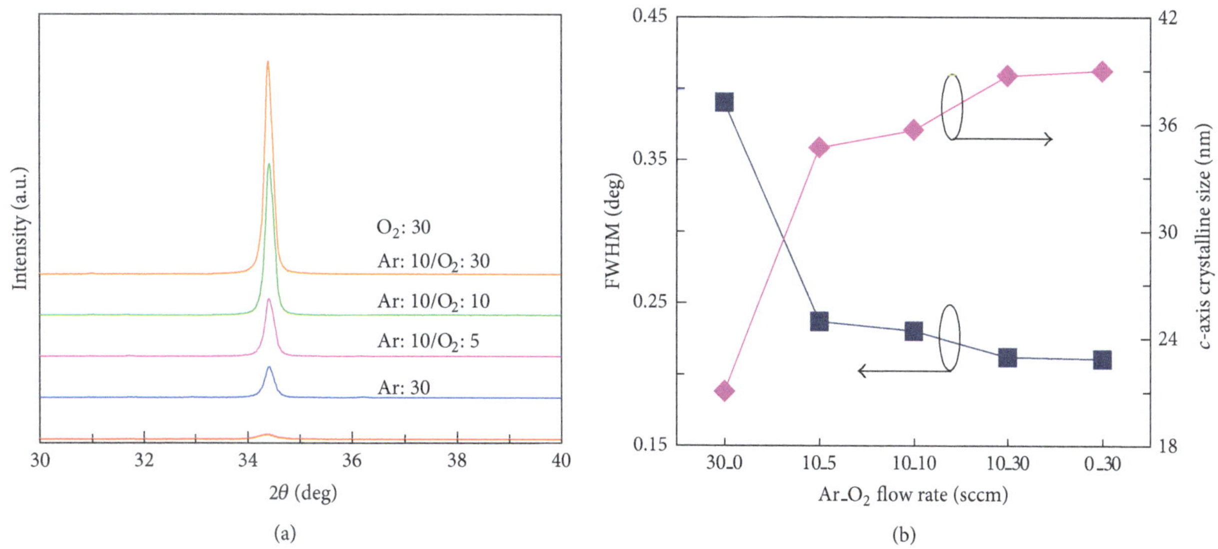

FIGURE 5: (Color online) (a) XRD patterns of ZnO films, (b) comparison of the FWHM values and c-axis crystalline size in the (002) direction for the films deposited with different Ar/O_2 gas ratios (10/5, 10/10, 10/30 sccm/sccm, Ar and O_2: 30 sccm, resp.).

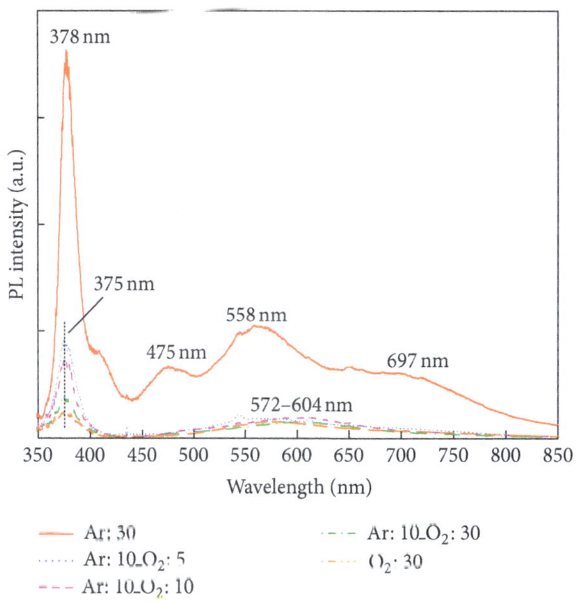

FIGURE 6: (Color online) PL spectra of as-deposited ZnO films prepared under different working gases ambient at deposition pressure of 7 Pa.

due to lack of oxygen during the ZnO film deposition; the defects centers in the film need to be further investigated in the further experiment.

4. Conclusions

The stoichiometry of ZnO films deposited at high-pressure of 7 Pa was significantly dependent on the working gas ambient in rf magnetron sputtering deposition. The intensity of the excited atomic oxygen (O^*) during rf plasma sputtering process was increased with the oxygen ratio increasing, resulting in enhancing the ZnO oxidization during the deposition. It was confirmed by OES measurement that the metal zinc ratio in the ZnO films was decreased with the oxygen ratio increasing. The crystallinities of ZnO films were improved with the oxygen ratio increasing. Therefore, it is possible to control the stoichiometry of ZnO films by simply adjusting the working gas ratio during rf magnetron sputtering process.

Acknowledgment

The authors gratefully acknowledge the support by Grant-in-Aid for Scientific Research (Grant No. 24561060) from the Ministry of Education, Culture, Sports, Science, and Technology, Japan.

References

[1] R. L. Hoffman, B. J. Norris, and J. F. Wager, "ZnO-based transparent thin-film transistors," *Applied Physics Letters*, vol. 82, article 733, 3 pages, 2003.

[2] E. M. C. Fortunato, P. M. C. Barquinha, A. C. B. Pimentel et al., "Wide-bandgap high-mobility ZnO thin-film transistors produced at room temperature," *Applied Physics Letters*, vol. 85, article 2541, no. 13, 3 pages, 2004.

[3] M. Furuta, T. Hiramatsu, T. Matsuda, C. Li, H. Furuta, and T. Hirao, "Oxygen bombardment effects on average crystallite size of sputter-deposited ZnO films," *Journal of Non-Crystalline Solids*, vol. 354, no. 17, pp. 1926–1931, 2008.

[4] T. Shimomura, D. Kim, and M. Nakayama, "Optical properties of high-quality ZnO thin films grown by a sputtering method," *Journal of Luminescence*, vol. 112, no. 1, pp. 191–195, 2005.

[5] C. Li, M. Furuta, T. Matsuda, T. Hiramatsu, H. Furuta, and T. Hirao, "RF power and thermal annealing effect on the properties of zinc oxide films prepared by radio frequency

magnetron sputtering," *Research Letters in Materials Science*, vol. 2007, Article ID 26459, 5 pages, 2007.

[6] M. Furuta, T. Hiramatsu, T. Matsuda, H. Furuta, and T. Hirao, "Effect of energetic particle bombardment on microstructure of zinc oxide films deposited by RF magnetron sputtering," *Japanese Journal of Applied Physics part 1*, vol. 46, no. 7A, pp. 4038–4041, 2007.

[7] T. Hiramatsu, M. Furuta, H. Furuta, T. Matsuda, and T. Hirao, "Influence of thermal annealing on microstructures of zinc oxide films deposited by RF magnetron sputtering," *Japanese Journal of Applied Physics Part 1*, vol. 46, no. 6 A, pp. 3319–3323, 2007.

[8] "Numerical data and functional relationships in science and technology," in *Landolt-Börnstein*, I Band 1. Teil, S. 87, 88, 200, Springer, New York, NY, USA.

[9] T. Nagata, A. Ashida, N. Fujimura, and T. Ito, "The effects of Xe on an rf plasma and growth of ZnO films by rf sputtering," *Journal of Applied Physics*, vol. 95, no. 8, pp. 3923–3927, 2004.

[10] P. W. B. Pearse and A. G. Gaydon, *The Identification of Molecular Spectra*, Chapman and Hall, New York, NY, USA, 4th edition, 1984.

[11] K. Tominaga, S. Iwamura, I. Fujita, Y. Shintani, and O. Tada, "Influence of bombardment by energetic atoms on c-axis orientation of zno films," *Japanese Journal of Applied Physics Part 1*, vol. V21, no. 7, pp. 999–1002, 1982.

[12] M. Birkholz, *Thin Film Analysis by X-Ray Scattering*, Wiley-VCH, Weinheim, Germany.

[13] B. Lin, Z. Fu, and Y. Jia, "Green luminescent center in undoped zinc oxide films deposited on silicon substrates," *Applied Physics Letters*, vol. 79, article 943, no. 7, 2001.

Permissions

The contributors of this book come from diverse backgrounds, making this book a truly international effort. This book will bring forth new frontiers with its revolutionizing research information and detailed analysis of the nascent developments around the world.

We would like to thank all the contributing authors for lending their expertise to make the book truly unique. They have played a crucial role in the development of this book. Without their invaluable contributions this book wouldn't have been possible. They have made vital efforts to compile up to date information on the varied aspects of this subject to make this book a valuable addition to the collection of many professionals and students.

This book was conceptualized with the vision of imparting up-to-date information and advanced data in this field. To ensure the same, a matchless editorial board was set up. Every individual on the board went through rigorous rounds of assessment to prove their worth. After which they invested a large part of their time researching and compiling the most relevant data for our readers. Conferences and sessions were held from time to time between the editorial board and the contributing authors to present the data in the most comprehensible form. The editorial team has worked tirelessly to provide valuable and valid information to help people across the globe.

Every chapter published in this book has been scrutinized by our experts. Their significance has been extensively debated. The topics covered herein carry significant findings which will fuel the growth of the discipline. They may even be implemented as practical applications or may be referred to as a beginning point for another development. Chapters in this book were first published by Hindawi Publishing Corporation; hereby published with permission under the Creative Commons Attribution License or equivalent.

The editorial board has been involved in producing this book since its inception. They have spent rigorous hours researching and exploring the diverse topics which have resulted in the successful publishing of this book. They have passed on their knowledge of decades through this book. To expedite this challenging task, the publisher supported the team at every step. A small team of assistant editors was also appointed to further simplify the editing procedure and attain best results for the readers.

Our editorial team has been hand-picked from every corner of the world. Their multi-ethnicity adds dynamic inputs to the discussions which result in innovative outcomes. These outcomes are then further discussed with the researchers and contributors who give their valuable feedback and opinion regarding the same. The feedback is then collaborated with the researches and they are edited in a comprehensive manner to aid the understanding of the subject.

Apart from the editorial board, the designing team has also invested a significant amount of their time in understanding the subject and creating the most relevant covers. They scrutinized every image to scout for the most suitable representation of the subject and create an appropriate cover for the book.

The publishing team has been involved in this book since its early stages. They were actively engaged in every process, be it collecting the data, connecting with the contributors or procuring relevant information. The team has been an ardent support to the editorial, designing and production team. Their endless efforts to recruit the best for this project, has resulted in the accomplishment of this book. They are a veteran in the field of academics and their pool of knowledge is as vast as their experience in printing. Their expertise and guidance has proved useful at every step. Their uncompromising quality standards have made this book an exceptional effort. Their encouragement from time to time has been an inspiration for everyone.

The publisher and the editorial board hope that this book will prove to be a valuable piece of knowledge for researchers, students, practitioners and scholars across the globe.

List of Contributors

Philip Hülsmann, Markus Heck and Michael Köhl
Fraunhofer ISE, HeidenhofstraBe 2, 79110 Freiburg im Breisgau, Germany

A. Kanni Raj
Faculty of Mechanical Engineering, The Indian Engineering College, Tirunelveli 627116, Tamil Nadu, India

R. Krishnan, C. David, P. K. Ajikumar, R. Nithya, S. Tripura Sundari, S. Dash, B. K. Panigrahi, M. Kamruddin and A. K. Tyagi
Materials Science Group, Indira Gandhi Centre for Atomic Research, Kalpakkam, Tamil Nadu 603102, India

Vikram Jayaram
Department of Materials Engineering, Indian Institute of Science, Bangalore, Karnataka 560012, India

Baldev Raj
President, PSG Institutions, Coimbatore, Tamil Nadu 641004, India

Ajit Behera
Department of Metallurgical & Materials Engineering, Indian Institute of Technology, Kharagpur, West Bengal 721302, India

S. C. Mishra
Department of Metallurgical & Materials Engineering, National Institute of Technology, Rourkela, Odisha 769008, India

Asit Behera
Department of Mechanical Engineering, National Institute of Technology, Rourkela, Odisha 769008, India

J. P. Dhal
Department of Chemistry, National Institute of Technology, Rourkela, Odisha 769008, India

Woo-Hwan Jung
Division of Electrical Engineering, Howon University, 64, Howondae 3 gil, Impi, Gunsan 573-930, Republic of Korea

I. A. Kaurova
Moscow State Open University, 3 Entuziastov Street, Aleksandrov 601655, Russia

G. M. Kuzmicheva and E. N. Domoroshchina
Lomonosov Moscow University of Fine Chemical Technology, 86 Vernadskogo Prospekt, Moscow 119571, Russia

V. B. Rybakov
Lomonosov State University, Vorobyovy Gory, Moscow 119992, Russia

A. Cousson
Laboratoire Leon Brillouin, Cea/Saclay, 91191 Gif-sur-Yvette Cedex, France

O. Zaharko
Laboratory for Neutron Scattering, ETZ Zurich & Paul Scherrer Institute (PSI), CH 5232 Villigen, Switzerland

Aleksandra W. Cyganiuk, Anna Kuci Nska and Jerzy P. Aukaszewicz
Faculty of Chemistry, Nicholas Copernicus University, Ulica Gagarina 11, 87-100 Torun, Poland

Roman Klimkiewicz
Institute of Low Temperature and Structure Research PAN, Ulica Okolna 2, 50-422Wrocław, Poland

Andrzej Olejniczak
Faculty of Chemistry, Nicholas Copernicus University, Ulica Gagarina 11, 87-100 Torun, Poland
Flerov Laboratory of Nuclear Reactions, Joint Institute for Nuclear Research, Dubna 141980, Russia

B. Sivasankari
Department of Physics, Kalasalingam University, Krishnankoil 626 190, India

P. Selvarajan
Department of Physics, Aditanar College of Arts and Science, Tiruchendur 628 216, India

Kun Yang, Min Liu, Kesong Zhou and Changguang Deng
Department of New Materials, Guangdong General Research Institute of Industrial Technology (Guangzhou Research Institute of Non-Ferrous Metals), No. 363 Changxing Road, Tianhe District, Guangzhou 510651, China

Andrzej Wolkenberg
Institute of Electron Technology, Aleja Lotnikow 32/46, 02-668 Warsaw, Poland
Institute of Materials Engineering, Czestochowa University of Technology, Aleja Armii Krajowej 19, 42 200 Czestochowa, Poland

Rajen Patel
Armament Research, Development and Engineering Center, Picatinny, NJ 07806, USA

Frank J. Owens
Department of Physics, Hunter College and Graduate Center, City University of New York, 696 Park Avenue, New York, NY 10065, USA

V. Sangeetha
Department of Physics, D. G. Vaishnav College, Arumbakkam, Chennai 106, India

N. Kanagathara
Department of Physics, Vel Tech Multi Tech Dr. Rangarajan Dr. Sakunthala Engineering College, Avadi, Chennai 62, India

R. Sumathi
Department of Physics, Vel Tech Polytechnic College, Avadi, Chennai 62, India

N. Sivakumar and G. Anbalagan
Department of Physics, Presidency College, Chennai 5, India

Florencia Edith Wiria and Bee Yen Tay
Singapore Institute of Manufacturing Technology, 71 Nanyang Drive, Singapore 638075

Elaheh Ghassemieh
Department of Mechanical Engineering, The University of Sheffield, Sir Frederick Mappin Building, Sheffield S1 3JD, UK

Gabriela Craciun and Elena Manaila
National Institute for Lasers, Plasma and Radiation Physics, Accelerators Laboratory, 409 Atomistilor Street, 077125 Magurele, Romania

Maria-Daniela Stelescu
National Research and Development Institute for Textile, 93 Ion Minulescu Street, 031215 Bucharest, Romania

Jyoti Prakash Dhal
Department of Chemistry, National Institute of Technology, Rourkela, Odisha 769008, India

S. C. Mishra
Department of Metallurgical & Materials Engineering, National Institute of Technology, Rourkela, Odisha 769008, India

S. Chenna Krishna, Narendra Kumar Gangwar, Abhay K. Jha and Bhanu Pant
Materials Processing and Characterization Group, Vikram Sarabhai Space Centre, Trivandrum 695 022, India

K. K. Bamzai, Nidhi Kachroo, Vishal Singh and Seema Verma
Crystal Growth and Material Research Laboratory, Department of Physics & Electronics, University of Jammu, Jammu 180006, India

Md Enamul Hoque, Tan Jie Ye and Leng Chuan Yong
University of Nottingham Malaysia Campus, 43500 Semenyih, Selangor, Malaysia

Khairul Zaman Mohd Dahlan
Malaysian Nuclear Agency, 43000 Kajang, Selangor, Malaysia

A. Frattini and A. Di Loreto
Area Fısica, Departamento de Quımica-Fısica, FCByF, UNR, Suipacha 531, 2000 Rosario, Argentina
Laboratorio de Materiales Ceramicos, FCEIyA, IFIR, UNR, Pellegrini 250, 2000 Rosario, Argentina

O. De Sanctis
Laboratorio de Materiales Ceramicos, FCEIyA, IFIR, UNR, Pellegrini 250, 2000 Rosario, Argentina

Fábio E. F. Silva, Maria Carolina B. Di-Medeiros, Karla A. Batista and Kátia F. Fernandes
Laboratorio de Quımica de Proteınas, Departamento de Bioquımica e Biologia Molecular, ICB 2, Universidade Federal de Goias, Caxia Postal 131, 74001-970 Goiania, GO, Brazil

J. Kayathri, N. Rani Meiyammai, S. Rani, K. P. Bhuvana, K. Palanivelu and S. K. Nayak
Central Institute of Plastics Engineering & Technology, Chennai 600 032, India

Gastón P. Barreto, Graciela Morales and Ma. Luisa López Quintanilla
Centro de Investigacion en Quımica Aplicada, Boulevard Enrique Reyna 140, 25253 Saltillo, COAH, Mexico

Makanjuola Oki
Institute of Petroleum Studies, University of Port Harcourt, PMB 2, Choba, Port Harcourt 500102, Rivers State, Nigeria
Department of Chemical Sciences, Oduduwa University, Ipetumodu, Ile-Ife 220371, State of Osun, Nigeria

Kayode Oki
Technical Department, Greenfield-Oaks Limited, London SE25 4HL, UK

John Otaigbe and Samuel Otikor
Department of Pure and Industrial Chemistry, University of Port Harcourt, PMB 5323, Port Harcourt 500102, Rivers State, Nigeria

Ali Akbar Ashkarran
Department of Physics, Faculty of Basic Sciences, University of Mazandaran, P.O. Box 47416-1467, Babolsar, Iran

Sima Eshghi and Mohammad Reza Nourani
Department of Material Engineering, Science and Research Branch, Islamic Azad University, Tehran, Iran

Douglas C. Hofmann
Jet Propulsion Laboratory, California Institute of Technology, MS 18-105, 4800 Oak Grove Drive, Pasadena, CA 91109, USA

S. Sudhahar, M. Krishna Kumar, A. Silambarasan and R. Mohan Kumar
Department of Physics, Presidency College, Chennai 600 005, India

R. Muralidharan
Department of Science and Humanities, Veltech High Tech Dr. R. Rangarajan Dr. Sakunthala Engineering College, Chennai 600 062, India

Chaoyang Li, Dapeng Wang, Zeming Li, Xin Li and Toshiyuki Kawaharamura
Institute for Nanotechnology, Kochi University of Technology, 185 Miyanokuchi, Tosayamada-cho, Kami, Kochi 782-8502, Japan

Mamoru Furuta
Environmental Science and Engineering, Kochi University of Technology, 185 Miyanokuchi, Tosayamada-cho, Kami, Kochi 782-8502, Japan